D0803532

Structural Effects on Equilibria in Organic Chemistry

Structural Effects on Equilibria in Organic Chemistry

Jack Hine

The Ohio State University

ROBERT E. KRIEGER PUBLISHING COMPANY
HUNTINGTON, NEW YORK
1981

Original Edition 1975
Reprint Edition 1981, with corrections

Printed and Published by
ROBERT E. KRIEGER PUBLISHING COMPANY, INC.
645 NEW YORK AVENUE
HUNTINGTON, NEW YORK 11743

Printed in the United States of America

Library of Congress Cataloging in Publication Data

Hine, Jack Sylvester, 1923-
Structural Effects on Equilibria in Organic Chemistry.

Reprint of the edition published by Wiley, New York.
Includes bibliographical references and index.
1. Chemical equilibrium. 2. Chemical structure.
3. Chemistry, Physical organic. I. Title.
[QD503.H56 1981] 547.1'392 80-11714
ISBN 0-89874-144-0

Preface

This book deals with the effect of structure, including the structure of the solvent as well as that of the reactants and products, on equilibria in the reactions of organic compounds. An understanding of both equilibria and reaction rates is essential for a chemist, but the study of equilibria is simpler and therefore more reasonably emphasized first. Although reaction rates depend on the characteristics of transition states, which are intrinsically incapable of being observed directly, the study of equilibria deals only with true chemical species, many of which can be observed directly in great detail. In many reactions equilibrium is established faster than the chemist can mix the reactants or separate the products, and in many other reactions equilibrium is easily established under relatively mild conditions. In such cases all we may need to know about the reaction rate is the qualitative fact that it is fast enough for equilibrium to be established.

Although many of the generalizations discussed in this book are of a qualitative nature, the best evidence for them comes from quantitative data. Therefore the equilibria dealt with specifically are very largely those for which equilibrium constants have been measured quantitatively. Two decades ago there were very few series of related reactions, except for acid-base processes in solution, for which equilibrium constants had been determined. However, the introduction of modern spectrometric and chromatographic techniques into organic chemistry has resulted in the appearance of reliable equilibrium data on a variety of types of reactions at a steadily increasing rate. In surveying this large and growing field we find areas, such as conformational analysis and acid-base reactions in solution, in which a particularly large amount of work has been done, but in which the salient features may be illustrated by use of a relatively small fraction of the existing data. In other fields, which have been less intensively mined, a large fraction of the available data must be used to illustrate the significant trends.

This book was written for students at the graduate and advanced undergraduate level. Only a good knowledge of the standard first-year courses in organic and physical chemistry is assumed. I believe that the extended tables of experimental data, substituent constants, structural contributions to thermodynamic properties, etc., will make the book of continued use to such students even after they have gone through the text once in a course. It should also be of use to professional chemists, especially those whose previous education in the area is out of date.

Earlier versions of the present manuscript, including the problems at the ends of the chapters, were used as a text in a graduate course in organic chemistry that was taught a number of times at The Ohio State University during the period 1965–1973. I should like to thank the many students in this course for their comments, and Mrs. Nancy W. Flachskam, Dr. Robert L. Flachskam, Jr., and Dr. Stephen S. Ulrey for their assistance. I am also grateful to the National Science Foundation for support of research in the area covered by this book. Above all, I want to acknowledge my wife's help at every step of the way.

JACK HINE

Columbus, Ohio

May 1974

Contents

Structural Effects on Equilibria in Organic Chemistry

1

Some Thermodynamic Considerations

1-1. SYMMETRY CORRECTIONS[1,2a]

It seems intuitively obvious that biphenyl-3,3′-dicarboxylic acid would be twice as strong as biphenyl-3-carboxylic acid, if there were no interaction between the two carboxy groups, simply because it has twice as many carboxy groups. In considering the manner in which structural changes in organic compounds are accompanied by changes in the rate and equilibrium constants for the compounds, we often need to make corrections for such "statistical effects" in order to learn whether any other effects are operative. In the example described, as in many other cases, the nature of the statistical or symmetry correction that should be made is obvious. In many other cases, however, it is less obvious, and hence a brief treatment of such corrections in terms of statistical mechanics is in order.

The equilibrium constant for the reaction

$$\mathrm{A} + \mathrm{B} \overset{K}{\rightleftharpoons} \mathrm{C} + \mathrm{D} \tag{1-1}$$

will obey the equation

$$K = \frac{Q_{\mathrm{C}} Q_{\mathrm{D}}}{Q_{\mathrm{A}} Q_{\mathrm{B}}} e^{-\Delta E_0^\circ / RT} \tag{1-2}$$

where ΔE_0° is the change in energy for the reaction at absolute zero, R is the gas constant, T is the absolute temperature, and the Q's are partition functions, if the compounds A, B, C, and D are incapable of optical activity. Equation 1-2 will also be applicable if A, B, C, and D include optically active compounds as long as each is defined to include only one type of molecule, for example, one optical isomer. However, if C is defined as the racemic form of a compound, namely, as a mixture of equal amounts of two different kinds of molecules, then at equilibrium there will be twice as much racemic material present as there will be of either optical isomer, and thus K will be twice as large as it would be if C were defined as being a given optical isomer. When applied to optically active compounds, Eq. 1-2 may be written in the form

$$K = 2^{p-q} \frac{Q_C Q_D}{Q_A Q_B} e^{-\Delta E_0^\circ/RT} \tag{1-3}$$

where p is the number of racemic products and q the number of racemic reactants.

In most cases the partition functions for a given species may be expressed to a good approximation as the product of its translational, overall rotational, internal rotational, vibrational, electronic, and nuclear spin components.

$$Q = Q_t Q_r Q_{ir} Q_v Q_e Q_{ns}$$

Each of these partition functions may be expressed in terms of various properties of the molecule in question, such as its mass, moments of inertia, vibrational frequencies, etc., but at present we need be concerned only with the rotational partition functions. The overall rotational partition function contains in its denominator the external symmetry number of the molecule, which we denote σ, and the internal rotational partition function analogously contains the internal symmetry number n. (Other symbols are sometimes used for these two symmetry numbers and sometimes their product is referred to as the symmetry number of the molecule.) The external symmetry number is the number of indistinguishable positions that can be reached by rigid rotation of the molecule as a whole. For example, σ is 1 for hydrogen chloride, 2 for molecular bromine, 4 for ethylene, and 6 for ethane. The internal symmetry number is the number of indistinguishable positions that can be reached by rotation around single bonds without rotation of the molecule as a whole. Thus n is 1 for water, 3 for ethane or methanol, 9 for propane, and 81 for neopentane. If these symmetry numbers are separated from the rest of the partition function, we may write

$$Q = \frac{Q'}{\sigma n} \tag{1-4}$$

which may be combined with Eq. 1-3 to give

$$K = \frac{\sigma_A n_A \sigma_B n_B 2^{p-q}}{\sigma_C n_C \sigma_D n_D} \frac{Q_C' Q_D'}{Q_A' Q_B'} e^{-\Delta E_0{}^\circ/RT} \tag{1-5}$$

Since most of the ways in which we shall explain structural effects on rates and equilibria (e.g., resonance effects, electrostatic interactions, and steric hindrance, that is, "chemical" effects) have nothing to do with symmetry numbers and optical activity, it is convenient to have a symbol for the "symmetry-corrected" equilibrium constant. This "chemical" equilibrium constant[1] is denoted K^{chem}.

$$K^{\text{chem}} = \frac{Q_C' Q_D'}{Q_A' Q_B'} e^{-\Delta E_0{}^\circ/RT} \tag{1-6}$$

Then if the symmetry term is called K_σ, where

$$K_\sigma = \frac{\sigma_A n_A \sigma_B n_B 2^{p-q}}{\sigma_C n_C \sigma_D n_D} \tag{1-7}$$

we may write

$$K = K_\sigma K^{\text{chem}} \tag{1-8}$$

Since the symmetry correction is temperature-independent, it is a correction to the entropy of reaction. By analogy to Eqs. 1-6 and 1-7 we may define, for reaction 1-1

$$\Delta S^\circ = \Delta S^\circ{}_{\text{chem}} + \Delta S_\sigma \tag{1-9}$$

$$\Delta S_\sigma = R \ln K_\sigma \tag{1-10}$$

or for a given compound

$$S^\circ = S^\circ{}_{\text{chem}} + S_\sigma \tag{1-11}$$

where

$$S_\sigma = -R \ln(\sigma n) \tag{1-12}$$

unless the compound is a racemate, in which case

$$S_\sigma = -R \ln \frac{\sigma n}{2} \tag{1-13}$$

In principle it should be possible to calculate $\Delta E_0{}^\circ$ and the various partition functions from quantum mechanics and statistical mechanics. In practice, however, the difficulties of making such calculations rigorously are so great that the job has never been done for any organic reaction. For this reason it is useful to consider various empirical and approximate relationships that have been observed between molecular structure and thermodynamic properties, some of which are discussed in the next section.

1-2. ADDITIVITY RULES FOR CORRELATING* THERMODYNAMIC PROPERTIES[3-5]

1-2a. Atomic Contributions to Thermodynamic Properties. Many of the properties of chemical compounds are commonly calculated by adding contributions from constituent parts of the molecules of the compound in question. A rule of additivity of atomic contributions is customarily used in calculating molecular weights. The atomic contributions in this case are the atomic weights. Although the rule is believed to be slightly imperfect (because of the relationship between the energy content of a molecule and its mass), it certainly serves its purpose adequately. If some thermodynamic property, such as the molar enthalpy content, of a chemical compound were in general equal to the sum of the atomic contributions of all the atoms in the molecule, then the change in that property for any chemical reaction would be zero, because the number of atoms of each of the various types is the same for the reactants as for the products. It is therefore not surprising that the role of additivity of atomic contributions works very poorly for the correlation of molar enthalpy and free energy contents.[4] For the correlation of molar entropies and heat capacities the rule is better but still not very good.[4] For the compounds that can be made from n different elements, a strict rule of additivity of atomic contributions would require only n different atomic contributions for each property correlated. Such a rule may be made to give better agreement with experimental data by various modifications that increase the number of disposable parameters used. For example, for a given element a different atomic contribution may be used for each state of hybridization. Other methods of correlation that also use larger numbers of parameters are discussed in the following sections.

1-2b. Electronegativity Corrections to an Atomic Additivity Rule. In an attempt to correlate enthalpies of atomization of various compounds by the rule of additivity of atomic contributions, it seems reasonable to give the atomic contribution for hydrogen the value 56.6 kcal/mole, half the enthalpy of atomization of molecular hydrogen. In fact, monovalent elements would generally be assigned atomic contributions equal to half the enthalpy of atomization of the diatomic molecule. The value for chlorine thus obtained, 28.6 kcal/mole, upon combination with the value for hydrogen, gives a calculated value of 85.2 kcal/mole for the enthalpy of atomization of hydrogen chloride. This is 16.9 kcal/mole smaller than the experimental value. That is, hydrogen chloride is 16.9 kcal/mole more stable than

*Unless otherwise indicated, all correlations in this section are of symmetry-corrected data; that is, the $\Delta G°$, $\Delta S°$, and K values correlated are $\Delta G°_{chem}$, $\Delta S°_{chem}$, and K^{chem} values.

would be expected from the rule of additivity of atomic contributions and the stabilities of molecular hydrogen and chlorine. Pauling noticed that compounds with heteronuclear bonds commonly have such extra stability. He attributed this stability to an interaction between the two atoms of the bond that is dependent on the difference in their electronegativities.[6] In fact, the electronegativities of various elements on Pauling's scale (Table 1-1) were chosen so as to give the best fit to the following equation

$$BE_{AB} = \frac{BE_{AA} + BE_{BB}}{2} + 23\,(X_A - X_B)^2 \tag{1-14}$$

where the *BE*'s are the bond energies of the single-bonded molecules AB, AA, and BB, and the *X*'s are the electronegativities of A and B. The constant 23 is required if the bond energies are expressed in kcal/mole but should be deleted if they are expressed in electron-volts. Thus two parameters are required for each element, a *BE* and an *X*. More parameters would be needed to correlate the energies of double and triple bonds.

Although the electronegativity corrections certainly produce a marked improvement in the atomic additivity rule, there are still major imperfections in Eq. 1-14. Pearson has pointed to a number of large deviations, most of which deal with fluorides and oxides, especially metal fluorides and oxides.[7]

1-2c. Bond Contributions to Thermodynamic Properties. Thermodynamic properties may be correlated more reliably by a rule of additivity of bond contributions than by an atomic contribution scheme, but more parameters are required. For all the compounds made from *n* different elements using only single bonds, $n(n-1)/2$ different bond contributions would be required. Furthermore, all the contributions for bonds between monovalent

Table 1-1. Pauling's Electronegativities for Various Elements[10]

Element	Electronegativity	Element	Electronegativity
Fluorine	4.0	Phosphorus	2.1
Oxygen	3.5	Hydrogen	2.1
Chlorine	3.0	Boron	2.0
Nitrogen	3.0	Arsenic	2.0
Bromine	2.8	Silicon	1.8
Iodine	2.5	Aluminum	1.5
Sulfur	2.5	Magnesium	1.2
Carbon	2.5	Lithium	1.0
Selenium	2.4	Sodium	0.9

elements would be trivial. Thus the contribution of the hydrogen-chlorine bond to any thermodynamic property may simply be set equal to the value of that property for hydrogen chloride, the only compound that has a hydrogen-chlorine bond. There are several ways in which multiple bonds may be treated. In the system devised by Benson and Buss,[4] certain groups, known as units, are treated like atoms in the sense that there are contributions for the bonds leading to them but no contributions for bonds within the unit. This point may be illustrated by calculating the molar entropy of gaseous vinyl chloride from the bond contributions of Benson and Buss (as modified by Benson[2b]) listed in Table 1-2. In this compound $-\overset{|}{C}=\overset{|}{C}-$ is treated as a tetravalent unit. There are three C_d—H bond contributions (13.8 eu each) and one C_d—Cl contribution (21.2), from which a molar entropy of 62.6 eu may be calculated for $S°_{chem}$. Since vinyl chloride has internal and external symmetry numbers of 1, this is also the calculated value for $S°$. The experimental value is 63.0 eu. The average deviation of the calculated values for acyclic compounds for which data were available from the experimental values of $C_p°$ and $S°$ was about 1 eu, which corresponds to about 0.3 kcal/mole at 25° C. The average deviation in the correlation of $\Delta H_f°$ values was about 2 kcal/mole for the compounds treated by Benson and Buss, but it may exceed 30 kcal/mole for other compounds, such as tetramethyl orthocarbonate.[8] Relatively large deviations may also be found for cyclic compounds (to which the correlation was not intended to be applicable), for compounds with very electronegative groups such as nitro and fluoro, and for new types of compounds that were not represented in the compounds from which the parameters were obtained. Like atomic contributions, bond contributions correlate $S°$ and $C_p°$ values much more reliably than they correlate $\Delta H_f°$ values.

It should be noted that since the entropy contributions in Table 1-2 are based on a standard state of 1 atm pressure, if any equilibrium constants that have concentration units are calculated from these contributions the concentration units will be atmospheres. For example, the equilibrium constant for addition of methane to ethylene to give propane would have the dimensions atm^{-1}.

The bond contribution scheme of Benson and Buss is equivalent to one in which there are bond contributions for carbon-carbon double bonds and carbon-oxygen double bonds and different contributions for single bonds to carbon, depending on whether the carbon was sp^3-hybridized, part of a carbon-carbon double bond, or part of a carbonyl group. Thus the entropy of vinyl chloride could be expressed

$$S° (CH_2{=}CHCl) = C_{C=C} + 3C_{C_d-H} + C_{C_d-Cl}$$

Table 1-2. Bond Contributions to Thermodynamic Properties in the Gas Phase at 25°C and 1 atm.[2b,4]

Bond[a]	$\Delta H_f^{\circ\,b}$	$S^{\circ}_{chem}{}^{c}$	C_p°	Bond	$\Delta H_f^{\circ\,b}$	$S^{\circ}_{chem}{}^{c}$	C_p°
C—H	−3.83	12.90	1.74	N—H	−0.8	17.7	2.3
C—D	−4.73	13.60	2.06	C_d—C^d	6.7	−14.3	2.6
C—C	2.73	−16.40	1.98	C_d—H^d	3.2	13.8	2.6
C—F		16.90	3.34	C_d—F^d		18.6	4.6
C—Cl	−7.4	19.70	4.64	C_d—Cl^d	−0.7	21.2	5.7
C—Br	2.2	22.65	5.14	C_d—Br^d	9.7	24.1	6.3
C—I	14.1	24.65	5.54	C_d—I^d	21.7	26.1	6.7
C—O	−12.0	−4.0	2.7	CO—H^e	−13.9	26.8	4.2
C—N	9.3	−12.8	2.1	CO—C^e	−14.4	−0.6	3.7
C—S	6.7	−1.5	3.4	CO—O^e	−50.5	9.8	2.2
O—H	−27.0	24.0	2.7	CO—F^e		31.6	5.7
O—D	−27.9	24.8	3.1	CO—Cl^e	−27.0	35.2	7.2
O—O	21.5	9.1	4.9	C_B—H^f	3.25	11.7	3.0
O—Cl	9.1	32.5	5.5	C_B—C^f	7.25	−17.4	4.5
O—NO_2[g]	−3.0	43.1		S—H	6.7	27.0	3.2
O—NO[g]	9.0	35.5		S—S		11.6	5.4

[a] C by itself refers to sp^3-hybridized carbon.
[b] Contribution to the enthalpy of formation (kcal/mole) from the elements in their standard states.
[c] Based on the ideal gas at 1 atm pressure as the standard state.
[d] C_d refers to the tetravalent $-\overset{|}{C}=\overset{|}{C}-$ unit. That is, each C_d—X bond contribution may be considered to contain one-fourth of the carbon-carbon double-bond contribution.
[e] CO refers to the divalent carbonyl unit. Each CO-X bond contribution contains half the carbon-oxygen double-bond contribution.
[f] C_B represents an aromatic carbon atom. Each C_B-X contribution contains one contribution from the bond between two carbon atoms in an aromatic ring.
[g] The nitro and nitroso groups are treated as univalent units. For example, the O—NO bond contribution contains the nitro-oxygen double-bond contribution also.

where the C's are contributions for various bonds. However, since allenes, ketenes, etc., were not included in the correlation, the bond contributions for a carbon-carbon double bond will always be accompanied by four contributions of the type C_{C_d-X}. Hence $C_{C=C}$ might just as well be divided equally among the four contributions.

$$S^{\circ}\,(CH_2{=}CHCl) = 3\,(\tfrac{1}{4}\,C_{C=C} + C_{C_d-H}) + (\tfrac{1}{4}\,C_{C=C} + C_{C_d-Cl})$$

The parameters ($\frac{1}{4}\ C_{C=C} + C_{C_d-H}$) and ($\frac{1}{4}\ C_{C=C} + C_{C_d-Cl}$) in the preceding equation are simply the bond contributions for the C_d—H and C_d—Cl bonds, respectively, in the Benson-Buss scheme.

1-2d. Interaction Corrections to Bond Contribution Schemes. Just as the Pauling electronegativity equation (1-14) improved on the rule of additivity of atomic contributions by correcting for interactions between adjacent atoms, we may improve on the rule of additivity of bond contributions by correcting for interactions between adjacent bonds. Schemes that allow only for pairwise interactions between adjacent bonds, like the electronegativity correction, increase the required number of disposable parameters by no more than a factor of two. Since by adjacent bonds we mean bonds to the same atom, adjacent bond interactions may be referred to as interactions of given pairs of atoms across a given atom. Thus the interaction of the carbon-fluorine and carbon-chlorine bonds in chlorofluoromethane may be referred to as the interaction of chlorine and fluorine across carbon, and denoted Γ_{FCCl}. It is possible in such a correlation scheme to set certain interactions arbitrarily equal to zero without affecting the reliability of the correlation (just as carbon-carbon and carbon-oxygen double-bond contributions may be considered to be arbitrarily set equal to zero in the Benson-Buss bond contribution scheme). In the correlation devised by Bernstein,[9] interactions between like atoms (e.g., Γ_{ClCCl}) are defined as zero. It seems more convenient, however, to define the interaction as zero in cases in which one or both of the interacting atoms is hydrogen, since the required data are more commonly available and since all interactions may then be thought of as being relative to interaction with the common atom hydrogen. This has been done in the scheme devised by Allen, which also contains terms to allow for trios of nonhydrogen atoms attached to a given atom.[10] Such terms are now usually denoted Δ with the three interacting atoms subscripted. If there is no superscript the interaction is across carbon. Thus Δ_{COO} is the interaction of a carbon and two oxygen atoms attached to the same carbon atom, and Δ_{CCC}^{N} is the interaction of three carbon atoms attached to the same nitrogen atom. Various interactions and bond contributions,[3a,11,12] denoted B(X—Y), are listed in Table 1-3. The subscripts *d* and *t* refer to double bonded and triply bonded atoms, respectively. The scheme may be applied to a relatively simple molecule, 1,1,2-trichloroethane, as follows.

$$\Delta H_f^\circ\ (ClCH_2CHCl_2) = 3B(\text{C—H}) + 3B(\text{C—Cl}) + B(\text{C—C}) + \Gamma_{ClCCl} + 3\ \Gamma_{CCCl} + \Delta_{ClCCl}$$

$$= -36.60\ \text{kcal/mole}$$

Table 1-3. Contributions for an Interaction Correlation of Enthalpies of Formation in the Gas Phase at 25°C[a]

		Bond Contributions			
B(C—H)	−4.47	B(C—I)	16.84	B(S—H)	−2.24
B(C—C)	6.61	B(C—NO_2)[b]	−4.42	B(S—S)	7.01
B(C—N)	15.30	B(C—NO_3)[b]	−15.67	B(C—SO)[b]	−3.49
B(C—O)	−5.64	B(O—H)	−28.90	B(C—SO_2)[b]	−31.31
B(C—F)	−42.84	B(O—O)	25.30	B(C=C)	30.36
B(C—S)	10.28	B(N—H)	−3.68	B(C≡C)	63.30
B(C—Cl)	−7.11	B(N—N)	37.49	B(C=O)	−17.04
B(C—Br)	4.34			B(C≡N)	36.78
		Pairwise Interactions			
Γ_{CCC}	−2.58	Γ_{CCN}	−4.10	$\Gamma_{CC(SO)}$[b]	−5.74
$\Gamma_{CC_dC_d}$[c]	−5.35	Γ_{CNC}	−4.30	$\Gamma_{CC(SO_2)}$[b]	0.67
$\Gamma_{CC_tC_t}$	−7.89	Γ_{CNN}	−5.80	Γ_{FCF}	−13.64
Γ_{CCO}	−5.66	Γ_{NCN}	∼−5[d]	Γ_{CCF}	−4.14
Γ_{COC}	−5.90	$\Gamma_{CC_tN_t}$	−12.00	Γ_{ClCCl}	0.23
Γ_{OCO}	−13.09	$\Gamma_{CC(NO_2)}$[b]	−4.42	Γ_{CCCl}	−3.20
Γ_{COO}	−8.63	$\Gamma_{CC(NO_3)}$[b]	−5.27	Γ_{CCBr}	−4.30
$\Gamma_{CC_dO_d}$	−11.45	Γ_{CCS}	−3.30	Γ_{BrCBr}	−1.45[e]
$\Gamma_{OC_dO_d}$	−34.50	Γ_{CSC}	−2.47	Γ_{CCI}	−3.15
Γ_{OCN}	∼−10[d]	Γ_{CSS}	−3.35	Γ_{ICI}	2.25[e]
		Trio Interactions			
Δ_{CCC}	0.55	Δ_{CCN}	1.00	Δ_{FFF}	7.42
Δ_{CCC_d}	1.25	Δ_{CCC}^{N}	1.63	Δ_{CFF}	3.46
Δ_{CCO}	1.43	Δ_{CCN}^{N}	2.30	Δ_{CCF}	3.58
Δ_{COO}	3.67	$\Delta_{CC(NO_2)}$[b]	0.72	Δ_{ClClCl}	0.40
Δ_{CCO_d}	4.20	$\Delta_{CC(NO_3)}$[b]	1.27	Δ_{ClCCl}	0.90
Δ_{COO_d}	7.96[f]	Δ_{CCS}	1.20	Δ_{CCCl}	0.50
Δ_{OOO}	6.6[g]	$\Delta_{CC(SO_2)}$[b]	0.67	Δ_{CCBr}	1.45
				Δ_{CCI}	0.80

[a] From the elements in their standard states. Values (kcal/mole) are from Ref. 3a unless otherwise noted.

[b] The NO_2 (nitro, not nitrite) and NO_3 groups are treated as univalent atoms, and SO (sulfoxide) and SO_2 (sulfone) are treated as divalent atoms.

[c] Represents the interaction in the system $-\overset{|}{\underset{|}{C}}-\overset{|}{C}=\overset{|}{C}-$, not that in $-\overset{|}{\underset{|}{C}}-\overset{\|}{C}-\overset{|}{\underset{|}{C}}=$.

[d] Estimated from the solution-phase data of Ref. 11.

[e] A. W. Klueppel, unpublished calculations.

[f] From the sum of $\Gamma_{OC_dO_d}$ and Γ_{COO_d} obtained in Ref. 12.

[g] Average of four values from Ref. 8.

The compound is reported to be about 1.5 kcal/mole less stable,[3] but the calculated value is good enough for some purposes.

Enthalpies of formation calculated from the parameters in Table 1-3 may be markedly in error when there are significant resonance interactions, steric repulsions, ring strains, or interactions between polar bonds not leading to a common atom.

1-2e. Group Contributions to Thermodynamic Properties. Further improvement in the empirical correlation of thermodynamic properties, at the expense of a further increase in the number of parameters used, may be achieved by using a rule of additivity of group contributions. The group contributions that will be described here are essentially those of Benson and co-workers,[2,5] whose symbolism will be modified slightly, however. The scheme may be thought of as letting the contribution of a given atom depend on the nature of all the other atoms to which it is attached. Then the contributions of all the univalent atoms are included with that of the polyvalent atom(s) to which they are attached. Hence the symbol $[CH_3(O)]$* represents the contribution of an oxygen-bound methyl group and [OH(C)] that of a carbon-bound hydroxy group. Thus

$$\Delta H_f^\circ\ (CH_3OH) = [CH_3(O)] + [OH(C)]$$

It happens that the method of defining parameters leaves some free to be assigned arbitrarily (just as in certain treatments some parameters may be arbitrarily defined as zero). It was therefore possible to give $[CH_3(O)]$, $[CH_3(N)]$, etc., the same value as that for $[CH_3(C)]$. In Table 1-4 are given enough group contributions to illustrate the use of a group contribution scheme; many more may be found in the tables of Benson and co-workers.[2,5,13] The following calculations exemplify their use.

$$S^\circ\ (CH_3CH_2CH_2Br) = S^\circ_{\text{chem}} + S_\sigma$$

$$S^\circ_{\text{chem}} = [CH_3(C)] + [CH_2(C)_2] + [CH_2Br(C)]$$

$$= 30.41 + 9.42 + 40.8 = 80.63$$

$$S_\sigma = -R\ \ln\ (\sigma n) = -4.576 \log 1\cdot 3 = -2.18$$

$$S^\circ(CH_3CH_2CH_2Br) = 78.45 \text{ eu (observed 79.2 eu)}$$

$$\Delta H_f^\circ\ (\textit{cis}\text{-2-hexene}) = [CH_3(C_d)] + 2[C_dH(C)] + [CH_2(C_d)(C)] + [CH_2(C)_2] + [CH_3(C)] + \textit{cis}\text{-contribution}$$

*Benson and co-workers use the symbol $[C\text{-}(O)(H)_3]$ for this contribution, but their symbol may suggest that the contribution of the parenthesized hydrogen atoms is not included.

$$= -10.08 + 2(8.59) - 4.76 - 4.95 - 10.08 + 1.0$$

$$= -11.69 \text{ kcal/mole (observed } -12.51 \text{ kcal/mole)}$$

$$\Delta H_f^\circ\,(p\text{-}H_2NC_6H_4CH_2CH_2SO_2CH_3) = [NH_2(C_B)] + 4[C_BH] + [CH_2(C)(C_B)] + [CH_2(C)(SO_2)] + [SO_2(C)_2] + [CH_3(SO_2)]$$

$$= 4.8 + 4(3.3) - 4.86 - 7.68 - 69.74 - 10.08 \text{ kcal/mole}$$

$$= -74.36 \text{ kcal/mole (no observed value available)}$$

It might be noted that missing parameters may often be approximated well by available parameters. For example, in the calculation of ΔH_f° (*cis*-2-hexene), if $[CH_2(C_d)(C)]$ had not been available $[CH_2(C)_2]$ could have been used instead without greatly changing the result.

Table 1-4 Group Contributions to Thermodynamic Properties in the Gas Phase at 25° C[a]

Group[b]	ΔH_f°	S°_{chem}[c]	C_p°	Group[b]	ΔH_f°	S°_{chem}[c]	C_p°
Hydrocarbons and Organic Oxygen Compounds							
$CH_3(X)$[e]	−10.08	30.41	6.19	$CH_2(C)(O)$	−8.1	10.3	4.99
$CH_2(C)_2$	−4.95	9.42	5.50	$CH(C)_2(O)$	−7.2	−11.0	4.80
$CH(C)_3$	−1.90	−12.07	4.54	$C(C)_3(O)$	−6.6	−33.56	4.33
$C(C)_4$	0.50	−35.10	4.37	$CH_2(C_B)(O)$	−8.1	9.7	
C_dH_2	6.26	27.61	5.10	$CH(C)_2(CO)$	−1.7		
$C_dH(C)$	8.59	7.97	4.16	$CH_2(C)(CO)$	−5.2	9.6	6.2
$C_d(C)_2$	10.34	−12.70	4.10	$C_B(O)$	−0.9	10.2	3.9
$CH_2(C_d)(C)$	−4.76	9.80	5.12	$C_dH(O)$	8.6	8.0	4.16
$C_dH(C_d)$	6.78	6.38	4.46	$C_d(C)(CO)$	7.5		
$C_d(C)(C_d)$	8.88	−14.6	4.40	$C_dH(CO)$	5.0		
$CH_2(C_d)_2$	−4.29	10.2	4.7	$CHO(C)$	−29.1	34.93	7.03
$CH_2(C)(C_t)$	−4.73	10.30	4.95	$CO(C)_2$	−31.4	15.01	5.59
$CH_2(C)(C_B)$	−4.86	9.34	5.84	$CO(C)(O)$	−35.1	14.78	5.97
$CH(C)_2(C_d)$	−1.48	−11.69	4.16	$OH(C)$	−37.9	29.07	4.33
C_tH	26.93	24.7	5.27	$O(C)_2$	−23.2	8.68	3.4
$C_t(C)$	27.55	6.35	3.13	$OH(C_B)$	−37.9	29.1	4.3
C_BH	3.30	11.53	3.24	$OH(O)$	−16.27	27.85	5.17
$C_B(C)$	5.51	−7.69	2.67	$O(C)(O)$	−4.5	9.4	3.7
$C_t(C_d)$	29.20	6.43	2.57	$OH(CO)$	−58.1	24.52	3.81
$C_B(C_d)$	5.68	−7.80	3.59	$O(C_d)_2$	−33.0	8.1	
$C_B(C_B)$	4.96	−8.64	3.33	$O(C)(CO)$	−43.1	8.39	

(*Continued*)

Table 1-4—Continued

Group[b]	ΔH_f°	S°_{chem}[c]	C_p°	Group[b]	ΔH_f°	S°_{chem}[c]	C_p°
			Organic Nitrogen and Sulfur Compounds				
$CH_3(X)$[e]	−10.08	30.41	6.19	$CH_2(C)(S)$	−5.65	9.88	5.38
$CH_2(C)(N)$	−6.6	9.8	5.25	$CH(C)_2(S)$	−2.64	−11.32	4.85
$CH(C)_2(N)$	−5.2	−11.7	4.67	$C(C)_3(S)$	−0.55	−34.41	4.57
$C(C)_3(N)$	−3.2	−34.1	4.35	$C_B(S)$	−1.8	10.20	3.90
$C_B(N)$	−0.5	−9.69	3.95	$C_dH(S)$	8.56	8.0	4.16
$CH_2CN(C)$	22.5	40.20	11.10	$C_d(C)(S)$	10.93	−12.41	3.50
$CHCN(C)_2$	25.8	19.80	11.00	$CH_2(C)(SO)$	−7.72		
C_dHCN	37.4	36.58	9.80	$CO(C)(S)$	−31.56	15.43	5.59
C_BCN	35.8	20.50	9.8	$CS(N)_2$	−31.56	15.43	5.59
C_tCN	63.8	35.40	10.30	$CH_2(C)(SO_2)$	−7.68		
$CH_2NO_2(C)$	−15.1	48.4		$SH(C)$	4.62	32.73	5.86
$CHO(N)$	−29.6	34.93	7.03	$S(C)_2$	11.51	13.15	4.99
$CO(C)(N)$	−32.8	16.2	5.37	$S(C)(S)$	7.05	12.37	5.23
$NH_2(C)$	4.8	29.71	5.72	$SH(C_B)$	11.96	12.66	5.12
$NH_2(C_B)$	4.8	29.71	5.72	$S(C)(C_d)$	9.97		
$NH(C)_2$	15.4	8.94	4.20	$S(C_d)_2$	−4.54	16.48	4.79
$N(C)_3$	24.4	−13.46	3.48	$S(S)_2$	3.01	13.4	4.7
$NH_2(N)$	11.4	29.13	6.10	$SH(CO)$	−1.41	31.20	7.63
$NH(C)(N)$	20.9	9.61	4.82	$SO(C)_2$	−14.41	18.10	8.88
$N(C)_2(N)$	29.2	−13.8		$SO_2(C)_2$	−69.74	20.90	11.52
$NH_2(CO)$	−14.9	24.69	4.07	$SO_2(C_B)_2$	−68.58		
$NH(C)(CO)$	−4.4	3.9		$SCN(C)$	37.18	41.06	9.51
$ONO(C)$	−5.9	41.9	9.10	$NH_2(CS)$	12.78	29.19	6.07
$ONO_2(C)$	−19.4	48.50					
			Organic Halogen Compounds				
$CF_3(C)$	−158.4	42.5	12.7	$CCl(C)_3$	−12.8	−5.4	9.3
$CHF_2(C)$	−102.3	39.1	9.9	C_dCl_2	−1.8	42.1	11.4
$CH_2F(C)$	−51.8	35.4	8.1	C_dHCl	2.1	35.4	7.9
$CF_2(C)_2$	−97.0	17.8	9.9	C_BCl	−3.8	18.9	7.4
$CHF(C)_2$	−48.4	14.0		$CHCl(C)(O)$	−20.2	19.5	9.0
$CF_2Cl(C)$	−106.3	40.5	13.7	$CH_2Br(C)$	−5.4	40.8	9.1
C_dF_2	−77.5	37.3	9.7	$CBr(C)_3$	−0.4	−2.0	9.3
C_dHF	−37.6	32.8	6.8	C_dHBr	12.7	38.3	8.1
C_BF	−42.8	16.1	6.3	C_BBr	10.7	21.6	7.8
$CF_3(C_B)$	−162.7	42.8	12.5	$CH_2Br(C_B)$	−6.9		
$NF_2(C)$	−7.8			$CH_2I(C)$	7.95	42.5	9.2
$CCl_3(C)$	−20.7	50.4	16.3	$CHI(C)_2$	10.7	22.2	8.7
$CHCl_2(C)$	−18.9	43.7	12.1	$CI(C)_3$	13.0	0.0	9.7
$CH_2Cl(C)$	−15.6	37.8	8.9	C_dHI	24.5	40.5	8.8
$CHCl(C)_2$	−12.8	17.6	9.0	C_BI	24.0	23.7	8.0

(Continued)

Table 1-4—Continued

Structural Feature	*Structural Contributions* ΔH_f	S°_{chem} [c]	C_p°
Cyclopropane ring	27.6	32.1	−3.05
Cyclopropene ring	53.7	33.6	
Cyclobutane ring	26.2	29.8	−4.61
Cyclobutene ring	29.8	29.0	−2.53
Cyclopentane ring	6.3	27.3	−7.50
Cyclopentene ring	5.9	25.8	−5.98
Cyclohexane ring	0.0	18.8	−6.40
Cyclohexene ring	1.4	21.5	−4.28
Cycloheptane ring	6.4	15.9	
Cyclooctane ring	9.9	16.5	
Bicyclo[1.1.0]butane ring	68.4	69.2	
Spiropentane ring	63.5	67.6	
Ethylene oxide ring	26.9	30.5	−2.0
Ethylene imine ring	27.7	31.6	
Pyrrolidine ring	6.8	26.7	−6.17
Ethylene sulfide ring	19.72	29.47	−2.85
gauche-R—C—C—R [f]	0.80		
cis-R—C=C—R [f]	1.00	0.0	−1.34
ortho-R_2 [f]	0.57	−1.61	1.12

[a] Taken from Refs. 2 and 5 with minor modification.

[b] C refers to sp^3-hybridized carbon except in CO, CS, and CN, which are carbonyl, thiocarbonyl, and cyano groups, respectively, C_d is one of the carbon atoms in a carbon-carbon double bond, and C_t one in a carbon-carbon triple bond. C_B is a carbon atom in a benzene ring. O is a single-bonded oxygen except in CO, SO, and SO_2; N is single-bonded nitrogen except in CN, NO_2, and ONO; and S is single-bonded sulfur except in SO, SO_2, and CS. In some cases adjacent atoms that must automatically be present are not shown explicitly. For example, C_BH could be written more fully as $C_BH(C_B)_2$ and C_dH_2 more fully as $C_dH_2(C_d)$.

[c] Based on the ideal gas at 1 atm pressure as the standard state.

[d] Most of the entries in this column are from Ref. 13.

[e] Where X is any polyvalent atom or group.

[f] Where the R groups are nontertiary alkyl groups.

The values of $\Delta H_f°$ calculated from the group contributions of Benson and co-workers deviate from the literature values by about 0.5 kcal/mole, on the average; the average deviation for $S°$ and $C_p°$ is about 0.3 cal/degree-mole. Some deviations up to six times as large are observed for heavily substituted compounds. It seems plausible that both steric and polar interactions between substituents contribute to these larger deviations.

The statement that the Allen scheme is equivalent to the group contribution scheme[3b] is incorrect. To correlate a property of the five alkanes of the type Me_nCH_{4-n}, the Allen scheme uses four parameters and the group contribution scheme would require five. If a set of terms for interactions among quartets of atoms attached to a common atoms were added, the Allen scheme would become equivalent to the group contribution scheme.

1-3. QUANTUM MECHANICAL CALCULATION OF THERMODYNAMIC PROPERTIES

The rigorous quantum mechanical calculation of thermodynamic properties presents such mathematical difficulties that it has not yet been done, even for the simplest organic molecule. There are many empirical and approximate methods of calculating thermodynamic properties (or differences in thermodynamic properties) that are more or less related to quantum mechanics.[14–16] Such methods will probably become increasingly important as they are refined and as computer technology continues to develop. Adequate treatment of such methods is a project of book-length proportions. We limit ourselves here to consideration of the promise shown by one of the less approximate of the methods currently being used.

Synder and Basch discussed the calculation of enthalpies of reaction from self-consistent field (SCF) energies of closed-shell molecules.[17] They made calculations by the Hartree-Fock-Roothan SCF method of the energies of 26 molecules containing up to 24 electrons. As a first step in a Hartree-Fock SCF calculation, the wave function for each electron in the molecule is estimated. Then the effect on one electron of the time-average charge distribution of all the other electrons (and the nuclei) is used to calculate an improved wave function for this electron. The process is repeated for a second electron, a third one, etc., until improved wave functions for all the electrons have been obtained. With this set of once-improved wave functions, a second improvement can be made in the wave function for the first electron, the second, etc. This method of successive approximations is stopped when the changes in the wave functions become relatively small. if enough disposable parameters (called basis functions) are used to describe the various wave functions and enough calculations made, it is possible to come as close as desired to what is known as the Hartree-Fock energy. The Hartree-Fock energy underestimates the stability of a molecule (with respect to nuclei and electrons, the usual reference point in quantum chemical calculations) because it allows only for the *time-average* electrostatic interactions between electrons, and because it does not allow for relativistic effects. Actually, the movement of any electron is correlated with that of any other electron in the vicinity. When two electrons are in orbitals around the same nucleus and one of the electrons is near the nucleus, the second electron is somewhat less likely to be near the nucleus than it would otherwise be. Since this correlation of the movement of electrons automatically decreases electrostatic repulsions, it stabilizes the molecule.

The electronic energy of a molecule may be expressed as the sum of the Hartree-Fock energy, the electron correlation energy, and the relativistic energy.

$$E_{el} = E_{HF} + E_{corr} + E_{rel} \qquad (1\text{-}15)$$

The total energy of a molecule is subdivided into electronic, vibrational, rotational, and translational components.

$$E_T{}^\circ = E_{el} + E_{vib} + E_{rot} + E_{trans} \qquad (1\text{-}16)$$

The energy change ΔE for any reaction is equal to the difference in energy between reactants and products.

$$\Delta E = \Delta E_{el} + \Delta E_{vib} + \Delta E_{rot} + \Delta E_{trans} \qquad (1\text{-}17)$$

For the molecules under consideration, the vibrational, rotational, and translational energies could be calculated reliably by the methods of statistical mechanics. This reduces the problem of calculating ΔE (from which ΔH may be calculated readily) to one of calculating ΔE_{el}. If E_{calc} is defined as the electronic energy calculated from the Hartree-Fock-Roothan SCF calculation used, then it follows from Eq. 1-15 that

$$\Delta E_{el} = \Delta E_{calc} + (E_{HF} - E_{cacl})^{prod} + (E_{HF} - E_{calc})^{react} + \Delta E_{corr} + \Delta E_{rel} \qquad (1\text{-}18)$$

On the basis of evidence that the relativistic energy of an atom is essentially independent of the nature of the molecule in which it may be situated[18] ΔE_{rel} was assumed to be negligible (compared with the other terms in the equation, at least). If the correlation energies of the products and reactants are essentially equal ($\Delta E_{corr} \sim 0$), and if the calculations miss the Hartree-Fock energy for the products by about the same amount that they do for the reactants, then ΔE_{calc} will be a good approximation for ΔE_{el}. It may also be seen that ΔE_{calc} will be a good approximation for ΔE_{el} if ΔE_{corr} is significant in size but essentially cancelled by the terms $(E_{HF} - E_{calc})^{prod} - (E_{HF} - E_{calc})^{react}$, but it seems less likely that this would be generally true. To learn how good an approximation ΔE_{calc} is for ΔE_{el}, Snyder and Basch calculated the enthalpies of a number of reactions of the 26 compounds under consideration via Eq. 1-17, using ΔE_{calc} in place of ΔE_{el}.[17] For 16 reactions involving hydrogen that included almost all the possible reactions of one molecule of hydrogen with one molecule of another of the 26 compounds (H_2, CH_4, NH_3, H_2O, HF, C_2H_6, N_2H_4, H_2O_2, F_2, C_2H_4, CH_2O, C_2H_2, HCN, N_2, CO, $HCONH_2$, HCO_2H, HCOF, CO_2, cyclopropane, ethylene oxide, ethylenimine, diaziridine, cyclopropene, diazirine, and diazomethane) to give products that were among the 26 compounds, the average deviation between the calculated and experimental values of $\Delta H°$ at 25° C was 8.3 kcal/mole. Since electron correlation energies for the molecules under consideration are on the order of hundreds of kilocalories per mole, it follows that the correlation energies of the products were nearly equal to those of the reactants. Evidence was described that the deviations of the calculated from the experimental enthalpies of reaction are due partly to changes in electron correlation energies and partly to differences between E_{calc} and E_{HF} (which could be decreased greatly by the expenditure of enough additional computation time). Subsequent calculations on other molecules by Matcha[19] suggest that the cancellation of relativistic energies may have been much less complete than was assumed. There is also evidence that much poorer results would have been obtained if molecules containing atoms with unfilled outer electronic shells (a category that covers all species with unpaired electrons) had been included.

Ab initio SCF calculations at a uniform level of approximation have also been carried out on the complete set of acyclic molecules that contain one, two, or three first-row atoms (C, N, O, F) and any number of hydrogen atoms and that can be written as classical valence structures without formal charges or incomplete outer electronic shells.[20] In addition, 14 trisubstituted methanes of the type CHXYZ where X, Y, and Z are Me, NH_2, OH, and F were treated.[21] The level of the approximation used for this group of more than 100 compounds was considerably poorer than that used by Snyder and Basch for their smaller set of compounds,[17] but the conclusions drawn were rather similar. In almost every case the isomer or conformer known to be the most stable experimentally was the one calculated to be most stable. In addition, interesting results were obtained for compounds that have not yet been isolated in the laboratory.

Although the quantum mechanical calculation of enthalpies of reaction described yields poorer agreement with experimental data and is vastly more time consuming than the more empirical methods described in preceding sections, it might well be a more reliable method for a compound for which no reference data for a related compound exist from which the parameters of the type used in the preceding sections could be estimated. The development of improved methods for the quantum mechanical calculation or empirical correlation of correlation energies (especially interatomic correlation energies) should greatly improve the applicability of SCF methods to the calculation of thermodynamic properties.

1.4. SOLVENT EFFECTS ON THERMODYNAMIC PROPERTIES

The correlations of thermodynamic properties considered so far refer to the gas phase and hence to the properties of isolated molecules. The practicing chemist usually deals with reactions in solution where the interaction of the various molecules with surrounding molecules is also of importance. To go from thermodynamic properties in the gas phase to the same properties in a given solvent, there are two alternative approaches. The less direct approach considers the condensation of the given compound in the vapor phase to give the liquid or solid compound and then considers the process of dissolving the compound in the solvent in question. For this purpose heats of vaporization, vapor pressures, heats of solution, solubilities, and activity coefficients are needed. Wadsö discussed relationships between molecular structure and heats of vaporization and boiling points.[22] A more direct approach would be to measure the heats of solution of the gaseous compound in the solvent in question and the partial pressures of the compounds over their solutions in the solvent, but a much smaller number of measurements of this type have been made. More measurements have been made on the solvent water than any other, but even here there are many more free energies of transfer between the aqueous and gaseous phases (partial pressures of solutes over aqueous solutions) available than there are heats of solution. Butler and Ramchandani noted that such free energies of transfer tend to be additive functions of the various groups in the solute molecule.[23]

It is convenient to consider the transfer of a compound between the gas phase and solution in terms of the activity coefficient in the gas phase relative to dilute solution. The activity coefficient γ, as shown in Eq. 1-19, is equal to the concentration of the compound (c) in solution divided by its concentration in a gas phase that is in equilibrium with the solution.

$$\gamma = \frac{[c]_s}{[c]_g} \tag{1-19}$$

Mookerjee has correlated values of log γ for various compounds in aqueous solution at 25° C in terms of a bond contribution scheme and also in terms of a group contribution scheme.[24] The values of the parameters obtained, expressing the concentrations in Eq. 1-19 in moles per liter, are listed in Tables 1-5 and 1-6. The bond contributions are from a set that correlated 245 experimental values of log γ with a standard deviation of 0.42. Large deviations are noted with polyhydroxy compounds, which were not used in determining the parameter values listed, because monohydroxy alcohols are not ordinarily hydrogen bonded in the gas phase whereas the polyhydroxy compounds ordinarily are.

A number of rather large deviations were noted that could be attributed to polar substituent effects. For example, 1,2-diethoxyethane had a smaller value of log γ (i.e., a smaller affinity for water) than would be expected from data on alkanes and monoethers. The strongest interaction of an ether with water is probably via hydrogen bonding between the basic ether oxygen atom and the hydrogen atoms of water. In 1,2-diethoxyethane each oxygen atom should act as an electron-withdrawing substituent and decrease the basicity of the other oxygen atom. This would lower log γ.

Table 1-5. Bond Contributions to log γ in the Vapor Phase Referred to Dilute Aqueous Solution at 25°C[a]

Bond	Contribution	Bond	Contribution
C—H	−0.11	C_t—H[c]	0.00
C—C	0.04	C_t—C[c]	0.64
C—Cl	0.30	CO—H	1.19
C—Br	0.87	CO—C	1.78
C—I	1.03	CO—O	0.28
C—O	1.00	C_B—H	0.11
C—NH_2[b]	3.97	C_B—C	0.44
C—CN[b]	3.28	C_B—Cl	0.19
C—NO_2	3.10	C_B—Br	0.54
C—S	1.11	C_B—NO_2	2.16
O—H	3.21	C_B—O	−0.41
C_d−H	−0.15	C_B—S	0.86
C_d—C	0.15	S—H	0.23

[a] Data from Ref. 24. See footnotes *a*, *b*, *d*, *e*, *f*, and *g* of Table 1-2.

[b] The NH_2 and CN groups are treated as univalent units.

[c] C_t refers to the divalent —C≡C— unit.

Table 1-6. Group Contributions to log γ in the Vapor Phase Referred to Dilute Aqueous Solution at 25°C[a]

Group	Contribution	Group	Contribution
$CH_3(X)$	−0.62	$O(C)_2$	2.93
$CH_2(C)_2$	−0.15	$O(CO)(C)$	0.53
$CH(C)_3$	0.24	$CH_2(C)(N)$	−0.08
$C(C)_4$	0.71	$NH_2(C)$	4.15
C_dH_2	−0.41	$NH(C)_2$	4.37
$C_dH(C)$	0.22	$N(C)_3$	4.14
$CH_2(C_d)(C)$	−0.23	$CH_2NO_2(C)$	3.27
$C_dH(C_d)$	0.18	$CH_2CN(C)$	3.43
$C_d(C)(C_d)$	0.86	C_BNO_2	2.18
$CH_2(C)(C_B)$	−0.19	$CH_2(C)(S)$	−0.02
C_BH	0.11	$SH(C)$	1.56
$C_B(C)$	0.70	$S(C)_2$	2.35
$CH_2(C)(O)$	−0.13	$CCl_3(C)$	0.80
$CH(C)_2(O)$	0.12	$CHCl_2(C)$	1.33
$C(C)_3(O)$	0.78	$CH_2Cl(C)$	1.05
$CH_2(C)(CO)$	0.15	$CHCl(C)_2$	1.46
$CHO(X)$	3.23	C_dHCl	0.05
$CO(C)_2$	4.03	C_BCl	0.21
$CO(C)(O)$	4.09	$CH_2Br(C)$	1.10
$OH(C)$	4.45	$CHBr(C)_2$	1.59
$OH(CO)$	1.44	C_BBr	0.49
		$CH_2I(C)$	1.14

[a] Data from Ref. 24. See footnote *b* of Table 1-4.

To minimize such polar effects the group contributions were based on a restricted set of compounds. All the atoms in a molecule that were not carbon or hydrogen atoms had to appear in the same group or the compound was not included in the set. Thus 1,1-dichloroethane was included because both chlorine atoms are in the group [$CHCl_2(C)$], so that allowance for interaction between the chlorine atoms is assured. The two chlorine atoms in 1,2-dichloroethane are in two different [$CH_2Cl(C)$] groups, however, and this compound was not included in the restricted set. The resulting group contributions correlated 212 log γ values with a standard deviation of 0.11. Considerably larger deviations would be expected if these contributions were applied to compounds containing polar interactions of the type that were omitted in obtaining the parameter values.

The negative contributions are from bonds or groups that tend to make compounds hydrophobic and the large positive contributions from those that tend to make compounds hydrophilic. However, these data refer to equilibria between aqueous solution and the gas phase, whereas the terms hydrophobic and hydrophilic are most commonly used in connection with equilibria between solution in water and solution in some nonaqueous solvent. For this reason the contributions give an imperfect picture of what is usually meant by hydrophobic or hydrophilic character.

The range of experimental values of log γ used in the correlations is large, going from -2.12 for *n*-octane to 7.77 for *p*-nitrophenol.[24] It is much smaller, however, than the range of enthalpies and free energies of formation of the compounds involved.

Pierotti, Deal, and Derr described an approach that may be used for the estimation of the activity coefficient of any compound in any solvent if data on chemically related compounds and solvents are available.[25] Hansch and co-workers have collected a large number of data for an additivity scheme for correlating distribution coefficients between water and octanol (or some similar organic solvent).[26,27]

1-5. POTENTIAL ENERGY AND KINETIC ENERGY EFFECTS OF SUBSTITUENTS[28–32]

Many of the reactions on which substituent effect studies have been carried out may be written as follows:

$$\text{Y—N—A} + \text{C} \xrightleftharpoons{K_Y} \text{Y—N—B} + \text{D} \tag{1-20}$$

where Y is the substituent, A is the reaction center in the reactant condition, B is the reaction center in the product condition, N is a bivalent radical that joins the substituent to the reaction center, C is a co-reactant, and D is a co-product. Thus in the ionization of *p*-chlorobenzoic acid in water, Y is chlorine, N is the *p*-phenylene group, A is the carboxy group, and B is the carboxylate anion group. We could let D be H^+ and C be nonexistent, or D could be H_3O^+ and C be H_2O; since C and D will later cancel, it makes no difference. The equilibrium constant for reaction 1-20 is then compared with that for another reaction of the same type but with a different substituent Z.

$$\text{Z—N—A} + \text{C} \xrightleftharpoons{K_Z} \text{Z—N—B} + \text{D}$$

In this comparison we examine the quotient

$$\frac{K_Y}{K_Z} = \frac{[\text{YNB}][\text{ZNA}]}{[\text{YNA}][\text{ZNB}]} \tag{1-21}$$

which is the equilibrium constant for the reaction

$$\text{Y—N—A} + \text{Z—N—B} \rightleftharpoons \text{Y—N—B} + \text{Z—N—A} \qquad (1\text{-}22)$$

Organic chemists have suggested a wide variety of ways in which the change of substituent from Y to Z can affect the equilibrium constant for reaction 1-20, that is, ways in which the difference between Y and Z can affect the magnitude of the equilibrium constant for reaction 1-22. Many of these "effects," for example, resonance stabilization and electrostatic interactions, must change the potential energy contents of the various molecules involved. Another category of effects, which is probably not so frequently considered by organic chemists, depends on the masses of the molecules, the distribution of atomic masses within the molecules, etc. These effects, for example, interference with internal rotations, change the kinetic energies of the molecules involved. At absolute zero the kinetic energy change for reaction 1-22 is simply equal to the difference in zero-point vibrational energies of the products and reactants. (Actually, what we are referring to simply as kinetic energy is really equal to the total amount of vibrational energy; in the strict physical definition of the term, kinetic energy is transformed to potential energy and back again as the vibrations take place.) Thus the zero-point vibrational energy (E_z) of a given compound with j fundamental modes of vibration may be expressed

$$E_z = \sum_{i=1}^{j} \tfrac{1}{2}h\nu_i \qquad (1\text{-}23)$$

where h is in Planck's constant and ν_i is the frequency of the ith vibrational mode. For reaction 1-22

$$\Delta E_z = (E_z)_{\text{YNB}} + (E_z)_{\text{ZNA}} - (E_z)_{\text{YNA}} - (E_z)_{\text{ZNB}} \qquad (1\text{-}24)$$

An analogous expression may be written for $\Delta E_p°$, the change in potential energy at 0° K, and for the overall energy change.

$$\Delta E_0° = \Delta E_p° + \Delta E_z \qquad (1\text{-}25)$$

In order to confront rationalizations of structural effects on equilibria more squarely with experimental facts, it would be desirable to know the extent to which a given substituent effect appears in the $\Delta E_p°$ term and the extent to which it appears in the ΔE_z term of Eq. 1-25. To learn this would require a thorough analysis of the vibrational spectra (infrared and Raman) of the four compounds involved and enough of the isotopically substituted derivatives to permit the application of Eq. 1-23. Then a calorimetric study of the reaction would have to be made at some temperature at which equilibrium could be established in a reasonable time, followed by heat capacity measurements from this temperature down to near 0° K in order to learn $\Delta E_0°$. As an alternative to at least some of the calorimetric work, partition function calculations could be carried out, but this is often impractical. Such studies would be quite difficult and time consuming and are almost never made. The available data ordinarily consist of the equilibrium constant for reaction 1-22 defined by Eq. 1-21, from which the free energy change $\Delta G°$ for the reaction may be calculated,

$$\Delta G° = -RT_r \ln K \tag{1-26}$$

(T_r is the absolute temperature at which K for the reaction was measured), and in many cases the enthalpy $\Delta H°$ and entropy of reaction $\Delta S°$.

$$\Delta G° = \Delta H° - T_r \Delta S° \tag{1-27}$$

(When the only experimental data are equilibrium constants at various temperatures, the uncertainty in $\Delta H°$ and $T_r \Delta S°$ is much greater than that in $\Delta G°$.) These terms may be related to ΔE_p and ΔE_z via the equation

$$\Delta H° = \Delta E_0° + \int_0^{T_r} \Delta C_p dT \tag{1-28}$$

where ΔC_p is the difference between the heat capacities of products and reactants (a function of temperature). Combining Eqs. 1-25, 1-27, and 1-28,

$$\Delta G° = \Delta E_p° + \Delta E_z + \int_0^{T_r} \Delta C_p dT - T_r \Delta S° \tag{1-29}$$

or, since

$$\Delta S° = \int_0^{T_r} \frac{\Delta C_p}{T} dT \tag{1-30}$$

$$\Delta G° = \Delta E_p° + \Delta E_z + \int_0^{T_r} \Delta C_p dT - T_r \int_0^{T_r} \frac{\Delta C_p}{T} dT \tag{1-31}$$

Equations 1-29 and 1-31 are applicable to any equilibrium process, not just the ones that can be written in the form of Eq. 1-22. There is no simple experimental way of evaluating ΔE_z, as would be necessary in order to learn whether a given $\Delta G°$ was due to a potential energy effect (i.e., to the $\Delta E_p°$ term) or to a kinetic energy effect (i.e., to the last three terms of the equations). However, it may be seen that with a sufficient knowledge of ΔC_p as a function of temperature, we could tell whether a given ΔG° was due to $\Delta E_0°$ or not. Unfortunately, there are no very simple useful absolute limitations on the way in which ΔC_p can vary with temperature. However, many of the factors that would tend to make ΔC_p positive (or negative) at one temperature would also tend to make it positive (or negative at any other temperature. If ΔC_p does not change algebraic sign between 0 and T_r, the terms $\int_0^{T_r} \Delta C_p dT$ and $T_r \int_0^{T_r} (\Delta C_p/T)\, dT$ will have the same algebraic sign, and the latter will have the larger absolute magnitude. When this is the case it follows that

$$|\Delta G^\circ - \Delta E_0{}^\circ| < |T_r \Delta S^\circ|$$

which in turn means that if ΔS° is small, ΔG° is a good reflection of $\Delta E_0{}^\circ$.

If we narrow our attention slightly to equilibria in which there is no change in the number of molecules, we may note that a fairly likely reason why ΔC_p would be positive (negative) at room temperature is that there are more (fewer) low-frequency vibrations in the products than in the reactants. In such a case the sum of the last two terms of Eq. 1-29 or 1-31 will be negative (positive). If this is true and if the total number of fundamental vibrational modes is the same for the products as for the reactants, then there will be more (fewer) high-frequency vibrations in the reactants than in the products, tending to make ΔE_z negative. Inasmuch as

$$\Delta H^\circ = \Delta E_p{}^\circ + \Delta E_z + \int_0^{T_r} \Delta C_p dT \qquad (1\text{-}32)$$

the last two terms in eq. 1-32 will be of opposite algebraic sign and will therefore tend to cancel. This will tend to make ΔH° a good measure of $\Delta E_p{}^\circ$ and in the most favorable possible case, where the last two terms of Eq. 1-32 are of the same absolute magnitude, ΔH° would be equal to $\Delta E_p{}^\circ$.

There are cases in which ΔH° is a better measure of $\Delta E_p{}^\circ$ than ΔG° is and also cases in which the reverse is true. It seems that ΔH° is more commonly the better measure of $\Delta E_p{}^\circ$ for gas phase reactions, at least, but ΔG° is often a more useful measure because it is known so much more reliably.

If we narrow our attention further to equilibria that may be represented by Eq. 1-22, we see that there are now added reasons why ΔS° and ΔE_z should not be very large. The equilibrium cannot involve the net formation or cleavage of a ring, and the number of fundamental vibrational modes in the products is probably the same as (and cannot differ by more than one from) the number in the reactants. From such generalizations as those of Benson and Buss we may state probable limitations on the absolute values of ΔS° and ΔC_p (which are functions largely of the nature of the group N in Eq. 1-22). Before consideration of the numerical values of such limitations in various cases, let us discuss the matter in general terms. Let us consider a case in which we have empirical reasons to believe that the absolute value of ΔS° is quite unlikely to be larger than the limit L_s and the absolute value of ΔC_p is quite unlikely to be larger than L_c at the temperature at which the equilibrium constant has been determined. Inasmuch as ΔS° and ΔC_p both approach zero as the temperature approaches 0° K, it is reasonable to assume that these limits are also applicable at all temperatures below the temperature at which K was determined. (In fact, if ΔC_p did not approach zero at 0° K, ΔS° would be infinite.) In order to learn the consequences of some simple, but not implausible, relationship between ΔC_p and temperature, let us assume that ΔC_p is proportional to the absolute temperature and becomes L_c at T_r. When this is the case, it may be shown that L_c must be equal to or smaller in absolute magnitude than L_s, and that

$$\Delta H^\circ = \Delta E_p{}^\circ + \Delta E_z + \tfrac{1}{2} T_r L_c$$

and

$$\Delta G^\circ = \Delta E_p{}^\circ + \Delta E_z - \tfrac{1}{2} T_r L_c$$

That is, ΔH° and ΔG° will be equally good measures of $\Delta E_0{}^\circ$. Either could be a better measure of $\Delta E_p{}^\circ$, depending on the magnitude of L_c. On the other hand, consider a more complicated case in which ΔC_p varies according to a sine function, changing sign once at $T_r/2$; specifically, if

$$\Delta C_p = \frac{TL_c}{T_r} \sin \frac{2\pi T}{T_r}$$

then

$$\int_0^{T_r} \Delta C_p dT = -\frac{T_r L_c}{2\pi}$$

and

$$T_r \Delta S^\circ = T_r \int_0^{T_r} \frac{\Delta C_p}{T} dT = 0$$

so that

$$\Delta H^\circ = \Delta E_p{}^\circ + \Delta E_z - \frac{T_r L_c}{2\pi} = \Delta G^\circ$$

and again ΔH° and ΔG° are equally good measures of $\Delta E_0{}^\circ$. Or if

$$\Delta C_p = L_c \sin \frac{2\pi T}{T_r}$$

then

$$\int^{T_r} \Delta C_p dT = 0$$

and

$$T_r \Delta S^\circ = T_r \int_0^{T_r} \frac{\Delta C_p}{T} dT = 5.94\, T_r L_c$$

(The latter would be possible, of course, only if $5.94L_c \leq L_s$.) Therefore

$$\Delta H^\circ = \Delta E_p{}^\circ + \Delta E_z$$

and

$$\Delta G^\circ = \Delta E_p{}^\circ + \Delta E_z - 5.94\, T_r L_c$$

with ΔH° being a perfect measure of $\Delta E_0{}^\circ$. ΔG° would not be, but it would be a better measure of $\Delta E_p{}^\circ$ in cases where L_c has the same sign as ΔE_z and is not too large.

The three functions chosen for ΔC_p in the preceding paragraph cover a major part of the plausible range of possibilities in terms of variations in the relative and absolute magnitudes of $\int_0^{T_r} \Delta C_p dT$ and $T_r \Delta S^\circ$ that may result.

The question of what we may learn about potential energy changes from data on enthalpies and/or free energies of reaction may be discussed in terms of stuctural correlations of entropies, heat capacities, and zero-point vibrational energies. Correlations of these three properties in terms of atomic contributions for acyclic compounds with atoms in normal valence states have been found to agree with experimental data in the gas phase at 25° C with standard deviations around 2.0 cal/degree-mole,[4] 1.5 cal/degree-mole,[4] and 1.2 kcal/mole,[33] respectively. According to these correlations in terms of atomic contributions ΔS°, ΔC_p, and ΔE_z should all be zero for any chemical reaction. According to the standard deviations for the correlations just given, however, standard deviations of 4.0 cal/degree-mole, 3.0 cal/degree-mole, and 2.4 kcal/mole from a value of zero would be expected for processes in which two molecules of reactants yield two molecules of products without ring formation or cleavage, if there is no reason for any better approximation than a correlation in terms of atomic contributions to be applicable. (In most chemical reactions there are a number of the same kinds of bonds, or even groups, in the products as in the reactants; this offers a reason why ΔS°, ΔC_p, and ΔE_z in such reactions should be nearer zero than might be expected from the correlation in terms of atomic contributions.) If we assume that the deviation for ΔC_p decreases at least as rapidly as the absolute temperature, the third term on the right-hand side of Eqs. 1-29, 1-31, and 1-32 will differ from zero with a standard deviation of about 0.4 kcal/mole at 25° C. Combining this with the standard deviations of the other terms, using the usual methods of calculating the propagation of uncertainties (which assumes that the various deviations are independent of each other, which is not really true), leads to the conclusion that the standard deviation of ΔG° from ΔE_p will be about 3 kcal/mole and that of ΔH° will be almost as large. The same sort of calculation for a reaction in which there are no changes in the types of bonds present gives values of around 1 kcal/mole. For a correlation in terms of group contributions or bond contributions plus α-interaction contributions the deviations drop to around 0.3 kcal/mole.

From the preceding paragraph it follows that for a gas phase reaction for which ΔG° is known experimentally and ΔE_z, ΔC_p, and ΔS° may be calculated from bond and α-interaction contributions, the absolute value of ΔE_p that is obtained from Eq. 1-29 will have to be significantly larger than 0.3 kcal/mole (corresponding to an increase of 60%, for example, in an equilibrium constant at 25° C) before we can in general feel very sure that a potential energy effect has been observed. In special cases, of course, where not only all α interactions but all β interactions in the products are the same as those in the reactants and where no significant changes in resonance interactions and freedom of internal rotations have occurred, significance may be attributed to smaller calculated values of ΔE_p. Data on nonpolar reactants and products in nonpolar solvents may probably be treated rather similarly to gas phase data, but where hydrogen bonding and ion solvation become important the correlations of thermodynamic properties in terms of structural parameters become less reliable, and larger changes in rate and equilibrium constants are required to provide good evidence that potential energy effects are important.

PROBLEMS

1. Give both internal and external symmetry numbers for each of the following species.

   a. Ethylene
   b. Propyne
   c. *p*-Chlorotoluene
   d. Dimethyl ether
   e. Methylene chloride
   f. Phosphorus pentachloride
   g. Hydrogen cyanide
   h. *t*-Butyl chloride

2. Calculate K_σ for each of the following reactions.

   a. $CH_3CH_2I + HI \rightleftharpoons CH_3CH_3 + I_2$
   b. $CH_3OH + CH_3SH \rightleftharpoons (CH_3)_2S + H_2O$

3. Calculate K^{chem} in terms of K for each of the following reactions (e.g., $K^{chem} = 6K$).

   a. *cis*-$CH_3CH{=}CHCH_3 + H_2O \rightleftharpoons CH_3CH_2CH(OH)CH_3$
   b. $(CH_3)_2C{=}CH_2 + H_2O \rightleftharpoons (CH_3)_3COH$
   c. $CH_3CH_2CH_2CH_3 \rightleftharpoons (CH_3)_3CH$
   d. $CH_3Br + C_6H_6 \rightleftharpoons C_6H_5CH_3 + HBr$

4. Calculate the enthalpy of reaction of methanol with methanethiol to give

dimethyl sulfide and water in the gas phase at 25°C from Benson's bond contributions.

5. Calculate the equilibrium constant for the isomerization of *n*-propylamine to trimethylamine in the gas phase at 25° C from Benson's group contributions.

6. If K is the equilibrium constant for the isomerization of dimethyl ether to ethanol, use the constants in Table 1-6 to calculate the ratio of K in the gas phase to K in aqueous solution at 25° C.

REFERENCES

1. S. W. Benson, *J. Amer. Chem. Soc.*, **80,** 5151 (1958).
2. S. W. Benson, *Thermochemical Kinetics*, Wiley, New York, 1968. (a) Sec. 2.11. (b) Sec. 2.4.
3. J. D. Cox and G. Pilcher, *Thermochemistry of Organic and Organometallic Compounds*, Academic Press, New York, 1970. (a) pp. 593–594. (b) p. 540.
4. S. W. Benson and J. H. Buss, *J. Chem. Phys.*, **29,** 546 (1958).
5. S. W. Benson, F. R. Cruickshank, D. M. Golden, G. R. Haugen, H. E. O'Neal, A. S. Rodgers, R. Shaw, and R. Walsh, *Chem. Rev.*, **69,** 279 (1969).
6. L. Pauling, *The Nature of the Chemical Bond*, 3rd ed., Cornell University Press, Ithaca, N.Y., 1960, pp. 88–105.
7. R. G. Pearson, in *Survey of Progress in Chemistry*, A. F. Scott, Ed., Vol. 5, Academic Press, New York, 1969, Chap. 1.
8. J. Hine and A. W. Klueppel, *J. Amer. Chem. Soc.*, **96,** 2924 (1974).
9. H. J. Bernstein, *J. Phys. Chem.*, **69,** 1550 (1965).
10. T. L. Allen, *J. Chem. Phys.*, **31,** 1039 (1959).
11. F. E. Rogers and R. J. Rapiejko, *J. Phys. Chem.*, **78,** 599 (1974).
12. M. Månsson, *J. Chem. Thermodyn.*, **1,** 141 (1969).
13. H. K. Eigenmann, D. M. Golden, and S. W. Benson, *J. Phys. Chem.*, **77,** 1687 (1973).
14. A. Streitwieser, Jr., *Molecular Orbital Theory for Organic Chemists*, Wiley, New York, 1961.
15. L. Salem, *The Molecular Orbital Theory of Conjugated Systems*, Benjamin, New York, 1966.
16. M. J. S. Dewar, *The Molecular Orbital Theory of Organic Chemistry*, McGraw-Hill, New York, 1969.
17. L. C. Snyder and H. Basch, *J. Amer. Chem. Soc.*, **91,** 2189 (1969).
18. E. Clementi and A. D. McLean, *Phys. Rev.*, **133**(2A), 419 (1963).
19. R. L. Matcha, *J. Amer. Chem. Soc.*, **95,** 7505 (1973).
20. L. Radom, W. J. Hehre, and J. A. Pople, *J. Amer. Chem. Soc.*, **93,** 289 (1971).

21. W. A. Lathan, L. Radom, W. J. Hehre, and J. A. Pople, *J. Amer. Chem. Soc.*, **95,** 699 (1973).
22. I. Wadsö, *Acta Chem. Scand.*, **20,** 544 (1966); **22,** 2438 (1968).
23. J. A. V. Butler and C. N. Ramchandani, *J. Chem. Soc.*, 952 (1935).
24. P. K. Mookerjee, unpublished observations, The Ohio State University, 1974.
25. G. J. Pierotti, C. H. Deal, and E. L. Derr, *Ind. Eng. Chem.*, **51,** 95 (1959).
26. A. Leo, C. Hansch, and D. Elkins, *Chem. Rev.*, **71,** 525 (1971).
27. C. Hansch, A. Leo, S. H. Unger, K. H. Kim, D. Nikaitani, and E. J. Lien, *J. Med. Chem.*, **16,** 1207 (1973).
28. Cf. L. P. Hammett, *Physical Organic Chemistry*, McGraw-Hill, New York, 1940, Chap. III.
29. M. G. Evans and M. Polanyi, *Trans. Faraday Soc.*, **32,** 1333 (1936).
30. R. W. Taft, Jr., in *Steric Effects in Organic Chemistry*, M. S. Newman, Ed., Wiley, New York, 1956, Chap. 13.
31. R. P. Bell, *The Proton in Chemistry*, 2nd ed., Cornell University Press, Ithaca, N.Y., 1973, pp. 79–82.
32. C. D. Ritchie and W. F. Sager, *Progr. Phys. Org. Chem.*, **2,** 323 (1964).
33. A. W. Klueppel, Ph.D. Dissertation, The Ohio State University, 1971.

2

Interactions Within and Between Molecules

2-1. INTERACTIONS OF PERMANENT POLES

2-1a. The Magnitude of Electrostatic Interactions. According to Coulomb's law, the force f acting between two point charges may be expressed

$$f = -\frac{e_1 e_2}{Dr^2}$$

where e_1 and e_2 are the magnitudes of the charges, r is the distance between them, and D is the dielectric constant of the surrounding medium. Integration of this expression gives E, the energy of interaction of the two charges (relative to the energy at infinite separation).

$$E = \frac{e_1 e_2}{Dr} \tag{2-1}$$

Electrically neutral bodies may also interact electrostatically with charges and with each other if the balanced positive and negative charges in the neutral bodies are arranged asymmetrically. When the center of positive charge of a neutral body does not coincide with its center of negative charge the body is a dipole, and it has a dipole moment equal to the magnitude of either charge multiplied by the distance between the two centers of charge. The energy of interaction between a charge and a dipole may be calculated from Eq. 2-1. Consider a dipole consisting of the two equal and opposite point charges e_2 and $-e_2$ separated by the distance l interacting with a charge e_1 located at the distance r from the center of the dipole. The angle θ shown is the angle between the line joining the charge

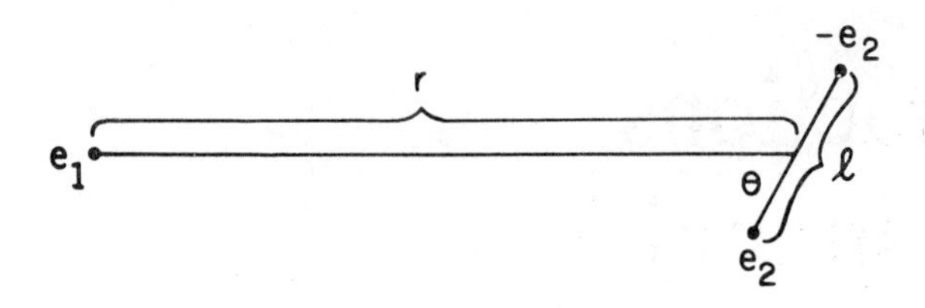

and the middle of the dipole and the line joining the middle of the dipole and that end of the dipole having a charge of the same algebraic sign as the interacting charge. This definition of θ permits us to write equations for the energy of interaction that give the proper algebraic sign. It is probably simpler, however, to let θ be the smaller of the two angles, that is, the acute angle, and to assign the algebraic sign of the energy of interaction by application of common sense. For example, if the positive end of a dipole is nearer an interacting positive charge than the negative end is, the interaction will increase the potential energy of the system and E will be positive. In terms of e_1, e_2, r, l, and θ, the equation for the energy of interaction is rather complicated. However, when r is large compared to l, this complicated relationship may be very well approximated by Eq. 2–2. Since

$$E = \frac{e_1 e_2 l \cos\theta}{Dr^2} \tag{2-2}$$

the dipole moment μ of the dipole is equal to the product of the charge e_2 and the separation l, Eq. 2-2 may be further simplified to give

$$E = \frac{e_1\, \mu \cos\theta}{Dr^2} \tag{2-3}$$

For two dipoles separated by the distance r, the energy of the interaction

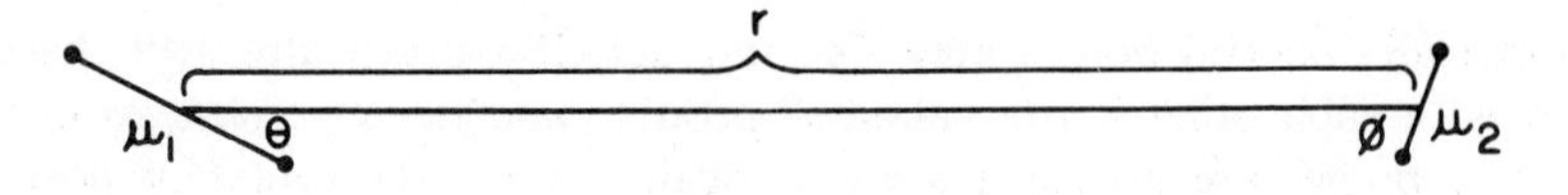

may also be expressed fairly simply if r is large compared to the lengths of the dipoles.

$$E = \frac{2\mu_1\mu_2 \cos\theta \cos\phi}{Dr^3} \tag{2-4}$$

Equations 2-3 and 2-4 are the relationships ordinarily used in calculating energies of interaction between a charge and a dipole or between two dipoles. In a number of cases in which such calculations have been carried out, especially for intramolecular interactions, the conditions assumed in the derivation of the equations (that the length of the dipole is small compared to the distance between the dipole and the charge or dipole with which it is interacting, and that the dipole may be treated satisfactorily as two point charges separated by a given distance) have not held. Complications caused by r not being large compared to l are fairly easily handled simply by applying Eq. 2-1 to each interaction between any two point charges. To allow for the fact that molecular dipoles are not ordinarily just point charges separated by a given distance is much more complicated and will not be attempted here. An equally important complication in applying any of these equations to inter- and intramolecular interactions is that we ordinarily do not know what value should be used for D, the dielectric constant of the medium in which the interaction is taking place. Further attention is given to this point in the next section.

Even molecules without dipoles may interact electrostatically with charges, dipoles, and with similar molecules. Carbon dioxide is a linear molecule. Its oxygen atoms are partially negative, but its partially positive carbon atom is halfway between these oxygen atoms so that the center of positive charge and the center of negative charge coincide. Although such a molecule is not a dipole, it is a quadrupole. It is qualitatively obvious that two carbon dioxide molecules located with respect to each other as

$$\overset{-\delta}{O}=\overset{+2\delta}{C}=\overset{-\delta}{O}$$

$$\underset{-\delta}{O}=\underset{+2\delta}{C}=\underset{-\delta}{O}$$

shown will attract each other and hence interact in a stabilizing manner with each other. Just as a molecule that is not a dipole may be a quadrupole, one that is not a quadrupole may be an octapole, etc. We have seen that the energy of interaction of a charge with a charge (Eq. 2-1) is proportional to r^{-1} and that the energy of interaction of a charge with a dipole (eq. 2-3) is proportional to r^{-2}. It is therefore not surprising that the energy of interaction of a charge with a quadrupole (whose dimensions are small compared to the distance from the charge) is proportional to r^{-3}. Similarly, the energy of dipole-dipole interaction (Eq. 2-4) is proportional to r^{-3}, that of dipole-quadrupole interaction to r^{-4}, etc. It therefore follows that as we go from charges to dipoles to quadrupoles to octapoles, etc., there is a strong decrease in the effective range of action. Only charges and, to a lesser extent, dipoles are capable of acting over a fairly long distance.

It is obvious both intuitively and from the equations that have been given that dipoles (and also quadrupoles, etc.) may interact in a stabilizing manner with charges and other dipoles if the two interacting species are suitably oriented with respect to each other. Orientations that lead to stabilization, and hence attraction, are shown below.

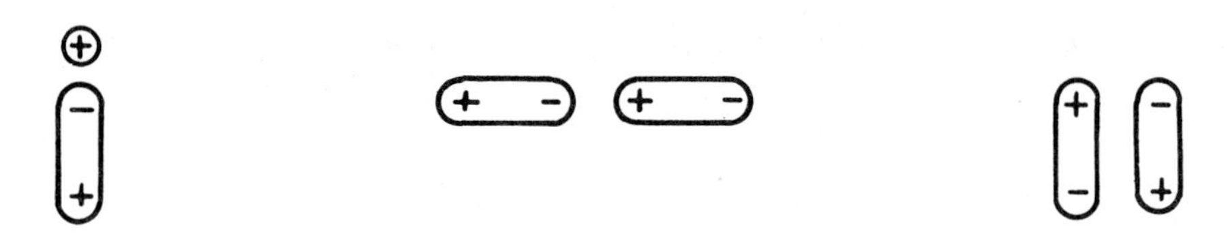

When ions and dipolar molecules are present in a solution they tend to attract and stabilize each other by assuming such relative configurations. This tendency must work against the tendency of thermal motion to randomize all relative configurations. When charges and dipoles are present in the same molecule they may not be free to change their relative orientations significantly. Hence their interaction may be either stabilizing or destabilizing. In many cases substituent effects on equilibrium constants and reaction rates seem at least partly explicable in terms of such interactions. Such substituent effects are known as *field effects.* To illustrate the techniques and problems involved in trying to assess quantitatively the magnitude of field effects, the next section is devoted to consideration of field effects on Brønsted acidity.

2-1b. Field Effects on Acidity. As an example of the effect of an electrically charged substituent on acidity, we might consider the first and second ionization constants of adipic acid in water at 25° C.[1]

$$K_1 = \frac{[H^+][HO_2C(CH_2)_4CO_2^-]}{[HO_2C(CH_2)_4CO_2H]} = 3.70 \times 10^{-5}\ M$$

$$K_2 = \frac{[H^+][^-O_2C(CH_2)_4CO_2^-]}{[HO_2C(CH_2)_4CO_2^-]} = 3.86 \times 10^{-6}\ M$$

The fact that the first ionization constant is 9.6 times as large as the second is in part the result of statistical effects. If symmetry corrections are made as described previously (Section 1-1), it is found that

$$K_1^{\text{chem}} = 3.70 \times 10^{-5}\ M$$

$$K_2^{\text{chem}} = 1.54 \times 10^{-5}\ M$$

The 2.4-fold difference in ionization constants remaining corresponds to a 520 cal/mole difference in free energies of reaction. It seems reasonable to attribute at least part of this effect to electrostatic interactions. If we may view a carboxy group as a carboxylate anion group with a proton adjacent to it, we may see that there are four types of electrostatic interactions in adipic acid. These are labeled *a*, *b*, *c*, and *d* in the following formulas:

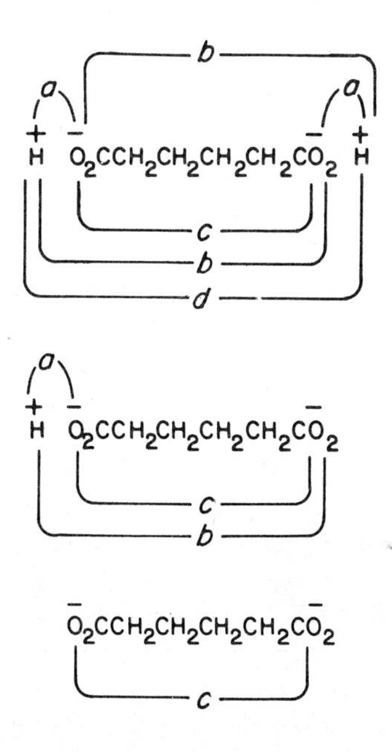

From comparison of these formulas it may be seen that dissociation of adipic acid is accompanied by the loss of one type *a* interaction, one type *b* interaction, and one type *d* interaction. The dissociation of the hydrogen adipate monoanion is accompanied by loss of one type *a* and one type *b* interaction. Thus the difference between these two dissociations, in an electrostatic sense, is that the former is accompanied by the loss of a type *d* interaction that does not occur in the latter. For this reason the electrostatic contribution to the difference in free energies of dissociation will be equated to the energy of interaction of the two acidic protons of adipic acid. If we assume that the carbon skeleton of adipic acid has the fully extended structure determined by X-ray crystallographic analysis for the anionic part of hexamethylenediammonium adipate,[2] and that the carboxy groups have the same structure as those in formic acid,[3] there are still many structures possible, depending on which oxygen atoms the protons are bonded to and on the dihedral angle between the plane of the carboxy group and that of the carbon skeleton. The distance between the average possible positions of the two protons is about 9.14 Å. According to Eq. 2-1 the energy of interaction between two protons separated by this distance is

$$E = \frac{(4.8 \times 10^{-10}\ \text{esu})(4.8 \times 10^{-10}\ \text{esu})}{D(9.14 \times 10^{-8}\ \text{cm})}$$

$$E = \frac{2.52 \times 10^{-12}}{D}\ \text{ergs/molecule}$$

$$E = \frac{36{,}260}{D}\ \text{cal/mole}$$

Taking D as the dielectric constant of water (the solvent), which is 79 at 25° C, an energy effect of 460 cal/mole may be calculated. This value is somewhat lower than, but perhaps within the experimental uncertainty of, the observed value.

Bjerrum appears to have been the first investigator to use Eq. 2-1 in calculating the magnitude of field effects on ionization constants.[4] However, he used the experimentally observed difference in free energies of ionization as E and treated r as the unknown. The value for r obtained by Bjerrum's method in comparing the first and second ionization constants of adipic acid is 8.2 Å. This is shorter than the distance (9.14 Å) for the fully extended molecule and is a more reasonable distance for use in Eq. 2-1. The adipate carbon skeleton must exist in a number of conformations, all but one of which have end-to-end distances of less than 9.14 Å. Thus the Bjerrum model gives a satisfactory rationalization of the relative magnitudes of the first and second ionization constants of adipic acid. The same may be said for other fairly long chain dicarboxylic acids, for which plausible values of r may be calculated. However, with short-chain dicarboxylic acids the values calculated for r become implausibly small (e.g., 0.69 Å for diethylmalonic acid). The application of the Bjerrum method to the effect of dipolar substituents on ionization constants, when made for dipolar substituents fairly near the acidic proton (only in such cases is there an effect large enough to be measured reliably), also leads to unreasonably small values for r. Reasonable values for r may be obtained in some cases by using much smaller dielectric constants, such as those characteristic of nonpolar organic compounds. In a compound like chloroacetic acid, the volume between and around the carbon-chlorine dipole and the acidic proton is occupied largely by the chloroacetic acid molecule rather than by the solvent. Therefore it would seem more sensible to use the dielectric constant of the organic molecule than that of the solvent. Dielectric constants of organic compounds vary widely, of course. Essentially nonpolar compounds like alkanes, carbon tetrachloride, and benzene have dielectric constants near 2.0. Polar organic compounds may have much larger dielectric constants, at least largely because of the ability of polar molecules to change their orientation in an applied electric field. Since that part of a chloroacetic acid molecule between the acidic proton and the carbon-chlorine dipole is not free to reorient itself significantly with respect to the fields of the dipole and the proton, it seems most logical to assume that the interior of the molecule has an effective dielectric constant near 2.0. Reasonable values for r may thus be obtained sometimes by using a dielectric constant of 2.0 and sometimes by using one of 79 (in aqueous solution). In other cases, however, it seems that an intermediate dielectric constant is needed. A major step toward telling just what value to use for the dielectric constant was made by Kirkwood and Westheimer.[5,6] They treated the organic acid as a cavity of low dielectric constant in a medium of high dielectric constant. In the case of an interaction between the ends of a long slender molecule (Fig. 2-1a) most of the lines of force pass through the solvent, whereas if the interacting charges and/or dipoles are located close to each other and deep within a bulky molecule (Fig. 2-1b) most of the

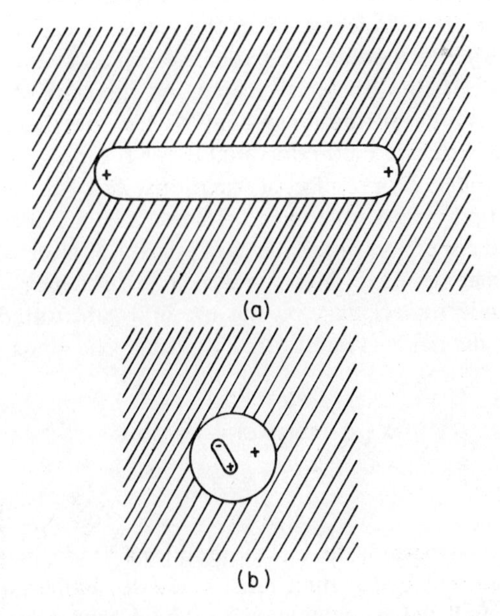

Figure 2-1. Charges and dipoles in cavities of low dielectric constant surrounded by media of high dielectric constant.

lines of force pass only through the molecule. Because of mathematical difficulties their treatment was limited to cavities that are spheres[5] and to those that are ellipsoids with the charges and/or centers of the dipoles on the axis passing through the foci[6] (and even in these cases the mathematics is not simple). For any such cavity with a given location of internal charges and dipoles it is possible to calculate an effective dielectric constant D_E, which can be used in equations like 2-1 and 2-3. Edward, Farrell, and Job give tables of effective dielectric constants.[7] Using an internal dielectric constant of 2.0, plausible values of r were calculated for a variety of acids with electrically charged and dipolar substituents.[5–8] Thus with a cavity like that in Fig. 2-1*a*, D_E has a value about equal to that of the solvent, and with a cavity like that in Fig. 2-1*b* D_E has a value near 2.0.

There are certainly deficiencies in the Kirkwood-Westheimer model, as the authors have pointed out. Not all molecules may be represented plausibly as spheres or even as ellipsoids, much less as ellipsoids whose internal charges are all on the bifocal axis. Even a long chain molecule in its extended conformation is represented more plausibly by a shape like that shown in Fig. 2-1*a* than by an ellipsoid. It is not at all clear that the small number of water molecules surrounding the organic acid will act as a continuous medium with a dielectric constant equal to the bulk dielectric constant of water. The choice of 2.0 as the dielectric constant of the cavity is rather arbitrary. With some substituents, such as cyano, it is not clear where the center of the dipole is located. At the time of Kirkwood and Westheimer's work conformational analysis had not been invented, and so there was no plausible method available for the *a priori* calculation of r for flexible molecules. By use of conformational analysis and other modern techniques it might be possible to make some useful improvements in the Kirkwood-Westheimer treatment. Several studies of electrostatic effects on equilibrium constants have been carried out in which conformational analysis was made unnecessary by the use of 4-substituted bicyclo[2.2.2]-octane-1-carboxylic acids (**1**) and other rigid bicyclic acids.

$$X-C\begin{matrix} /CH_2-CH_2\backslash \\ -CH_2-CH_2- \\ \backslash CH_2-CH_2/ \end{matrix}C-CO_2H$$

1

Although these compounds fit the assumptions made in the Kirkwood-Westheimer treatment better than most acids do, Roberts and Moreland found that the ΔpK values calculated for the 4-bromo and 4-cyano substituents, 0.40 and 0.30, respectively, were smaller and of different relative magnitude from the experimental values of 0.67 and 0.85, respectively.[9]

The original treatment and modifications of it[10,11] have been used in calculations on other 4-substituted bicyclo[2.2.2]octane-1-carboxylic acids with no better than fair agreement with the experimental data.[12,13] Nevertheless, it seems generally agreed that the results show that the field effect is an important factor in determining the effects of polar substituents on acidity. In fact, it seems possible that if the difficulty in applying the "cavity" treatment could be overcome, substituent effects fairly near the observed ones could be calculated in saturated systems.

2-2. POLARIZATION EFFECTS

2-2a. Induced Poles. A species with a permanent pole, that is, an ion or a molecule with a dipole, quadrupole, etc., is capable of inducing poles into nearby nonpolar molecules. For example, when a lithium cation is adjacent to an argon molecule,

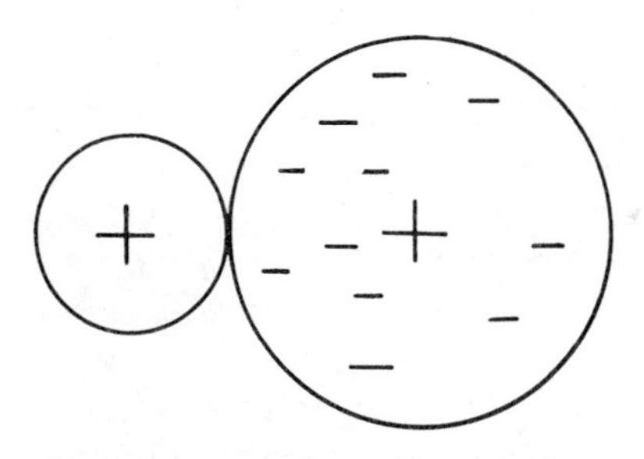

the electrons in the argon atom will, on the average, spend more than half their time on the side of the argon atom nearer the lithium cation. Hence the center of negative charge of the argon atom will not coincide with the center of positive charge. The lithium cation will have induced into the argon molecule a dipole whose negative end is closer to the lithium cation than its positive end is. The net interaction will thus be an attractive and stabilizing one. When r, the distance between the nucleus of a single charged cation and the nucleus of a spherical molecule with which it is interacting, is considerably larger than the radius of the cation or that of the molecule, the energy of interaction is given approximately by the equation

$$E = -\frac{e^2 d^3}{8 D r^4} \tag{2-5}$$

where e is the charge on an electron and d is the diameter of the spherical molecule. Although this equation overestimates the energy of interaction of a small ion *in contact with* a molecule, it is correct in showing that this energy increases quite sharply with decreasing internuclear distance. Interactions due to induced dipoles often contribute significantly to the total energy of interaction of polar species with both polar and nonpolar molecules.

2-2b. Dipole Moments and the Inductive Effect. The dipole moments of molecules are often treated as being equal to the vector sum of the *bond dipoles* of the various bonds in the molecules. Such a treatment fits in nicely with the qualitative fact that molecules like methane, carbon tetrachloride, and *p*-dichlorobenzene have no dipole moments, whereas molecules like methylene chloride and *m*-dichlorobenzene do. The vector sum treatment could be made to agree quantitatively with all known dipole moments if the bond moments were treated as variables whose exact values depended on the nature of the particular molecule in which the bonds were located. However, such a treatment would be trivial and would have no predictive value. Instead, it is customary to treat bond moments as being constant for a given type of bond. If it is assumed that the moment of a bond between an sp^3-hybridized carbon atom and a hydrogen atom is always the same, it follows that any saturated hydrocarbon whose bond angles are all 109°28′ will have a dipole moment of zero. Although alkanes do not fit the bond angle requirements exactly, no alkane has yet been found to have a dipole moment large enough to detect experimentally. In this and many other cases the concept of vector additivity of constant bond moments agrees well with experimental observations. There are cases of significant deviations from experimental observations, however, and such cases have probably taught chemists as much as the cases in which the agreement was nearly perfect.

If bond moments are constant and all bond angles are 109°28′, all alkyl chlorides should have the same dipole moment. Experimentally the following dipole moments have been observed in the gas phase: methyl chloride, 1.86; ethyl chloride, 2.00; *n*-propyl chloride, 2.04; isopropyl chloride, 2.15; *n*-butyl chloride, 2.11; *t*-butyl chloride, 2.13 *D*.[14] The tendency for the dipole moment to increase with increasing size of the alkyl group has been explained in terms of an idea suggested by Lewis.[15] Because of the large nuclear charge of chlorine, the bonding electrons in a carbon-chlorine bond spend more time near the chlorine nucleus than near the carbon nucleus. This is the reason why the chlorine atom in an alkyl chloride bears a partial negative charge and the carbon atom bears a partial positive charge. Because of this partial positive charge, the carbon atom to which chlorine is attached attracts electrons more strongly than does a carbon atom without a strongly electron-withdrawing substituent. Thus the chlorine atom causes the displacement of the bonding electrons as shown by the arrows in the formula below (relative to their average position in the absence of the chlorine atom).

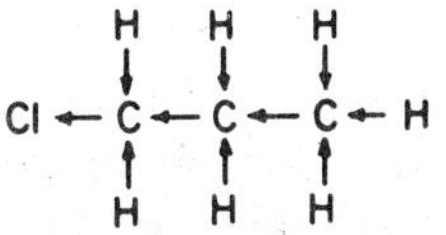

```
      H    H    H
      ↓    ↓    ↓
Cl ← C ← C ← C ← H
      ↑    ↑    ↑
      H    H    H
```

The chlorine atom induces a partial positive charge on the carbon atom to which it is attached. This charge induces a smaller positive charge (positive relative to the charge that would be present in the absence of the chlorine, at least) on the next carbon atom, which induces a still smaller positive charge on the next carbon atom, etc. This tendency of an electron-withdrawing substituent to pass its effect down a chain of atoms, known as the *inductive effect*, has been used to rationalize the effect of electron-withdrawing substituents on rate and equilibrium constants. In β-chloroethylamine, for example,

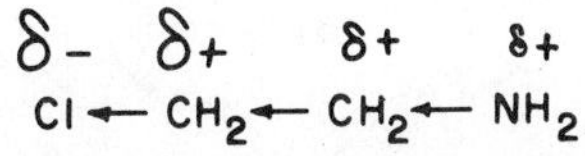

$$\overset{\delta-}{Cl} \leftarrow \overset{\delta+}{CH_2} \leftarrow \overset{\delta+}{CH_2} \leftarrow \overset{\delta+}{NH_2}$$

the inductive effect of the chloro substituents makes the nitrogen atom more positive than it would be otherwise. The nitrogen atom therefore attracts its unshared electron pairs more strongly and does not share them as readily with a proton. Hence the chloro substituent decreases the basicity of the amino group. This experimentally observed effect of the chloro substituent may also be explained qualitatively by the field effect. Since the positive end of the carbon-chlorine dipole is nearer the amino nitrogen atom than the negative end is (in the trans conformation, at least), the dipole interacts in a destabilizing manner with a positive charge on nitrogen. In many specific cases it is clear that the inductive effect and the field effect of a substituent will act in the same direction. Because of difficulties in sorting out the two effects, they are often lumped together. We refer to the combined effect as the polar effect.

There is evidence that in most reactions the polar effect of substituents is largely a field effect rather than an inductive effect. There are many observations that show that the magnitude of substituent effects on rate and equilibrium constants usually decreases by between two- and threefold for every additional methylene group (up to the rather small number of methylene groups beyond which substituent effects cannot be measured reliably) between the substituent and the reaction center. The manner in which dipole moments of aliphatic compounds change with the size of the alkyl groups shows that the inductive effect falls off much more sharply than this. Waters estimated that along a saturated hydrocarbon chain the induced charge effect decreases by about 40-fold in being passed from one atom to the next.[16] According to Dewar and Grisdale, the observed dipole moments of alkyl halides show that the charges induced on atoms in a saturated chain fall off by at least fivefold on going one atom farther down the chain; quantum chemists are said usually to use an eightfold "fall-off factor" in their work.[17]

As the following pK values of the conjugate acids show, when the γ-methylene group of n-butylamine is replaced by an oxygen atom the pK_a decreases by 1.17, but when the γ-methylene group of piperidine is replaced by an oxygen atom the pK_a decreases by 2.79.[18]

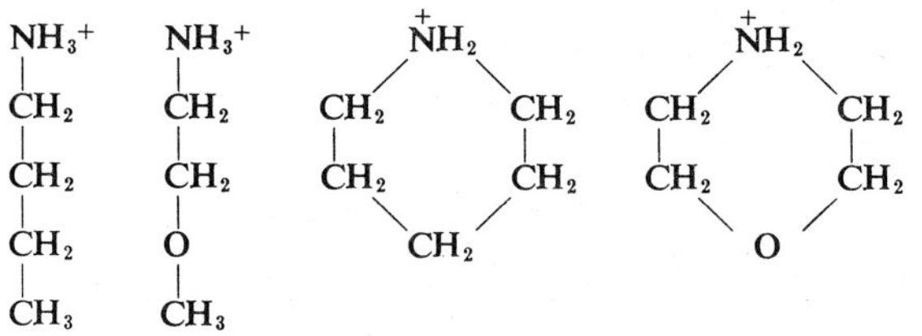

	NH_3^+	NH_3^+	$\overset{+}{N}H_2$	$\overset{+}{N}H_2$
	CH_2	CH_2	CH_2 CH_2	CH_2 CH_2
	CH_2	CH_2	CH_2 CH_2	CH_2 CH_2
	CH_2	O	CH_2	O
	CH_3	CH_3		
pK_a	10.78	9.61	11.12	8.33
Temperature	20° C	20° C	25°C	25° C

These observations, and many other analogous ones, might be thought to show that the oxygen atom is lowering the basicity largely by a field effect since it is nearer the positively charged nitrogen atom in the morpholinium ion than it is in the β-methoxyethylammonium ion. However, the data may also be rationalized in terms of the inductive effect; there are two carbon chains through which the inductive effect of the oxygen atom may be transmitted to the nitrogen atom in the morpholinium ion but only one in the β-methoxyethylammonium ion. Baker, Parish, and Stock compared the acidities of some 4-substituted bicyclo[2.2.2]octane-1-carboxylic acids (**2**) and some 4-substituted cubanecarboxylic acids (**3**), a case in which the

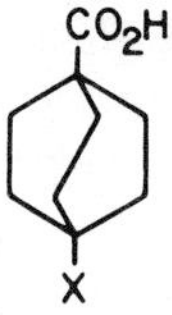

CO_2H X **2** CO_2H X **3**

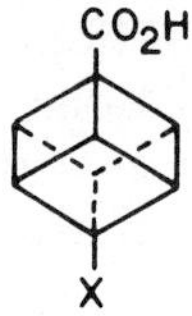

preceding ambiguity is absent.[12] The distances between the substituent and the carboxy group are identical, within 0.1 Å, in the two cases, as are the orientations of the C—X bonds with respect to the carboxy groups. On the other hand there are three three-bond pathways for transmitting the inductive effect of the substituent to the carboxy group in the bicyclooctanecarboxylic acids but six (overlapping) three-bond pathways in the cubanecarboxylic acids. Thus to a first approximation, at least, if only the field effect is operating substituent effects should be of the same magnitude in each of the two series of acids, but if only the inductive effect is operating substituent effects should be twice as large with the cubanecarboxylic acids as with the bicyclooctanecarboxylic acids. The only X groups for which pK_a values were determined in both series were H, CO_2H, and CO_2^-. In Fig. 2-2, the pK_a values for the two sets of acids are plotted against each other. The points fall within the experimental uncertainty of the line of slope 1.0 shown in the figure. Thus the field effect appears to be dominant.

A major part of the difficulty in subdividing polar effects into their inductive- and field-effect components arises from the fact that the inductive and field effects would ordinarily be expected to operate in the same direction. For this reason, special cases in which the inductive and field effects would be expected to act in opposite directions are of interest. In

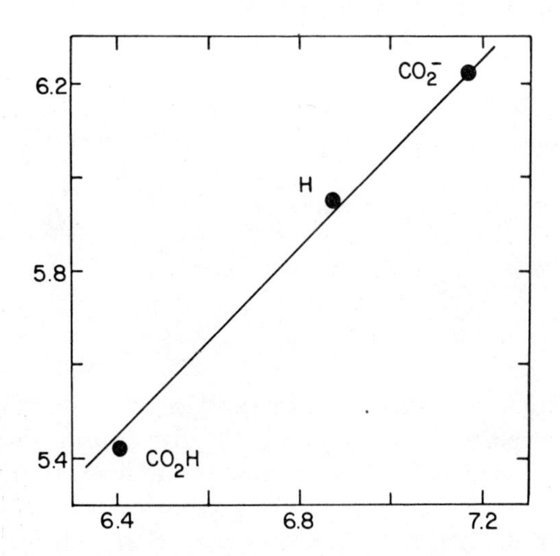

Figure 2-2. Plot of statistically corrected pK_a values of 4-X-cubanecarboxylic acids (ordinate) versus 4-X-bicyclo[2.2.2]octane-1-carboxylic acids (abscissa) in 50% (weight) ethanol-water at 25°C.

the following compound, for example, the carbon-chlorine bond dipole has its negative end aimed almost directly at the positively charged nitrogen atom (all ring fusions being trans).

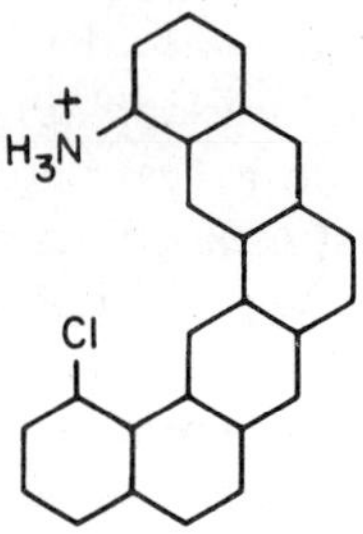

Hence the field effect of the chloro substituent should decrease the acidity of the ammonium ion, but the inductive effect should increase the acidity. Unfortunately, it is much easier to write formulas for such compounds, in which the inductive and field effects are opposed, than it is to synthesize and study the compounds.

Data on arylpropiolic acids were reported simultaneously by Newman and Merrill[19] and by Roberts and Carboni[20] and subsequently by Solomon and Filler.[21] The chloro, fluoro, nitro, and trifluoromethyl substituents clearly should have electron-withdrawing field effects when located in the meta and para positions. The corresponding ortho-substituted compounds would be expected to be species in which the field effect has not been so much reversed as it has been minimized. In *o*-chlorophenylpropiolic acid (**4**), the angle θ used in the applications of Eq. 2-3 may be calculated to be about 100° if one assumes "normal" bond angles and bond lengths. The

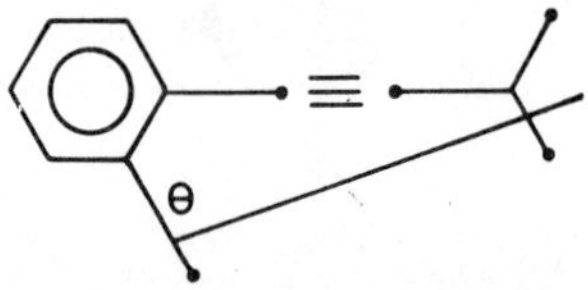

4

observed ionization constants, listed in Table 2-1, show that fluoro, chloro, nitro, and trifluoromethyl substituents in the ortho position all increase the acidity. With the nitro and trifluoromethyl substituents, this may plausibly be attributed to an electron-withdrawing resonance effect. However, with the *o*-halo substituents, which should have an electro-donating resonance effect and a very small and perhaps electron-donating field effect, it seems reasonable to accept the conclusion of Roberts and Carboni that an electron-withdrawing inductive effect is operating. These workers pointed out

Table 2-1. Substituted Phenylpropiolic Acids: Ionization Constants and Rate Constants for Reactions with Diphenyldiazomethane

Substituent	pK_a at 25° C[a]		k_2 in Dioxane at 30° C ($M^{-1}sec^{-1}$)
	35 wt% Dioxane-Water	50 vol% Ethanol-Water	
H	3.24[b]	3.58[c]	0.055[c]
	3.27[d]	3.60[d]	0.053[d]
o-Cl	3.08[b]	3.51[c]	0.094[c]
m-Cl	3.00[b]	3.43[c]	0.106[c]
p-Cl	3.07[b]	3.47[c]	0.088[c]
o-O_2N	2.83[b]	3.39[c]	0.175[c]
m-O_2N	2.73[b]	3.28[c]	0.233[c]
p-O_2N	2.57[b]	3.26[c]	0.315[c]
o-F	3.17[d]	3.48[d]	0.095[d]
m-F	3.10[d]	3.50[d]	0.093[d]
p-F	3.24[d]	3.52[d]	0.078[d]
o-CF_3	3.10[d]	3.49[d]	0.108[d]
m-CF_3	3.08[d]	3.36[d]	0.135[d]
p-CF_3	3.02[d]	3.33[d]	0.130[d]
o-MeO	3.37[b]		
m-MeO	3.21[b]		
p-MeO	3.44[b]		

[a] The pK_a values listed in Ref. 20 are "apparent" pK_a values determined by use of a pH meter calibrated for aqueous solutions. The pK_a was taken as the pH at half neutralization, a procedure that gives pK_a values that are too small by an amount that increases with increasing strength of the acid being titrated and increasing dilution of this acid. The pK_a values from Ref. 21 are also said to be apparent values, but is not clear whether they were set equal to pH's at half neutralization or not. The assumption that pK_a is equal to pH at half neutralization was not made in Ref. 19. These differences in techniques may explain the fact that the best straight line through a plot of the pK_a values in 50% ethanol from Ref. 20 versus the values in 35% dioxane from Ref. 19 has a slope of about 1.9, while a plot of the pK values in 50% ethanol versus those in 35% dioxane, both from Ref. 21, has a slope near 1.0.

[b] Data from Ref. 19.

[c] Data from Ref. 20.

[d] Data from Ref. 21.

that an alternative explanation of the effect of *o*-halo substituents on the ionization constants of arylpropiolic acids is that the lines of force from the negative (halogen) end of the carbon-halogen dipole pass through the high-dielectric solvent to a much greater extent than do those from the positive end. To test this alternative, the rates of reaction of the acids with diphenyldiazomethane,

$$RCO_2H + Ph_2CN_2 \rightarrow RCO_2CHPh_2 + N_2$$

a reaction whose rate is known to increase with increasing strength of the carboxylic acid used, were measured in dioxane (dielectric constant 2). The rate constants for the ortho-substituted compounds (Table 2-1) were found to be at least as large as would be expected (from a plot of log k_2 versus pK_a, for example) from the ionization constants of the acids in solvents of much higher dielectric constant. Although the substituents appear to have inductive effects, field effects seem required to explain the fact that the chloro, nitro, and trifluoromethyl substituents increase the acidity more from the para position, where the inductive effect should be weaker, than they do from the ortho position.

It might also be suggested that inductive effects are more important in the case of arylpropiolic acids than in many other cases because they are conducted more efficiently through π-electron systems than through saturated carbon chains. Baker, Parish, and Stock obtained evidence that π-electron systems do not necessarily transmit polar effects more efficiently than saturated systems.[12] They found that substituent effects were transmitted to the reaction center no more than 10% more efficiently in 4-substituted bicyclo[2.2.2]-2-octene-1-carboxylic acids (**5**) than in the corresponding saturated compounds (**2**), and, if anything, less efficiently in 4-substituted

CO_2H CO_2H

X   X

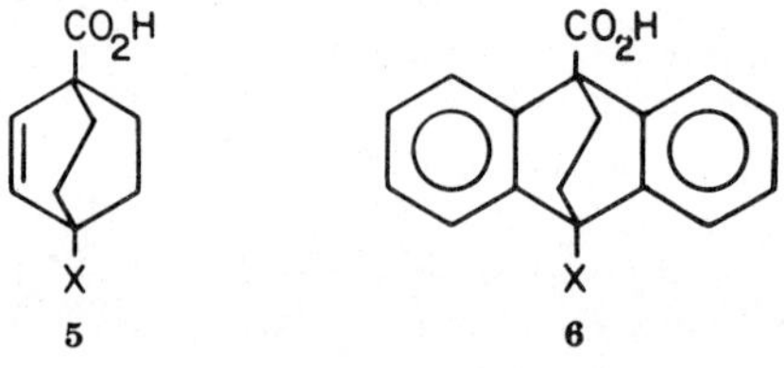

5   **6**

dibenzobicyclo[2.2.2]-2,5-octadiene-1-carboxylic acids (**6**) than in the corresponding bicyclooctanecarboxylic acids (**2**). Nevertheless, there are other observations (cf. Section 6-2) that have been interpreted as showing that polar effects are transmitted significantly more efficiently through unsaturated and aromatic than through saturated systems.

The *reversed* field effect has been reported in several instances. Golden and Stock, for example, determined the acidities of 8-substituted 9,10-

ethanoanthracene-1-carboxylic acids (7) in 50% aqueous ethanol at 25°C.[22] The chloro derivative (pK 6.24), the methoxy derivative (pK 6.15), and

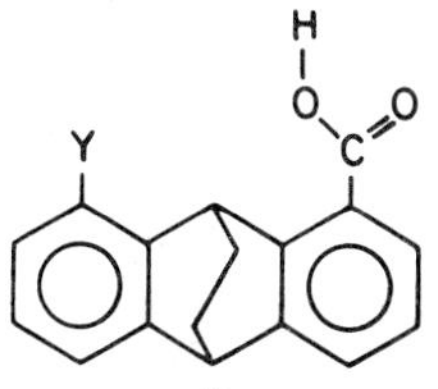

7

the carbomethoxy derivative (pK 6.16) are all weaker acids than the unsubstituted compound (pK 5.99). In the case of the chloro compound, at least, it is clear that the negative end of the C—Y dipole must be nearer the acidic proton than the positive end is. It seems unlikely that the results are being influenced greatly by hydrogen bonding between the carboxy group and the Y substitutents. Such hydrogen bonding would be more plausible in the case where Y is CO_2^-, and yet the ratio K_1/K_2 (33) for the dicarboxylic acid (the compound in which Y is CO_2H) is reasonable for a dicarboxylic acid with its carboxy groups separated by about the distance in question. Unfortunately, none of these reversed dipole effects are very large (perhaps because the negative end of the dipole is never aimed very directly at the acidic proton). For a number of 8-substituents (fluorine, cyano, methyl, hydroxy, and carboxy) the substituent effect is less than 0.1 pK unit.

2-2c. London Forces. The fact that such nonpolar molecules as those of the noble gases attract each other and aggregate to form liquids and solids when the temperature is low enough shows that possession of a permanent pole by one or both of the interacting species is not a prerequisite for intermolecular attraction. The nature of this attraction was explained by London, who pointed out that although a noble gas molecule is nonpolar on a time-average basis, at any given instant it will be only an extremely unlikely coincidence if the electrons are located so that the center of negative charge is at the nucleus. In other words, at most instants any molecule has a finite dipole moment, even though these instantaneous dipole moments may average zero over periods of time that are long relative to the movement of the electrons around the nucleus. When two molecules are near each other their electrons doe not move independently. If, at a given instant, most of the electrons of one molecule are on the side nearer the second molecule, then the electrons of the second molecule will probably be largely on the side away from the first molecule. That is, the motion of the electrons in the two molecules will tend to be correlated;

if the first molecule is an instantaneous dipole whose negative end is nearer the second molecule, then the second molecule will probably be a dipole whose positive end is nearer the first molecule. Thus the rapidly fluctuating dipoles of the two molecules will attract each other a higher percentage of the time than they will repel each other. London showed that the energy of interaction between two noble gas molecules will vary approximately with the inverse sixth power of the internuclear distance. The forces of attraction are known as London forces and as dispersion forces. The energies of interaction, which are always negative, are known as dispersion energies or electron-correlation energies and have already been mentioned in Section 1-3.

2-3. STERIC REPULSION, VAN DER WAALS RADII, AND BOND LENGTHS

All the intermolecular (but not necessarily interionic) forces that have been discussed so far in this chapter have been forces that are attractive and stabilizing or can be, with the proper relative orientation of the interacting molecules. From the fact that matter has finite volume, however, it is obvious that there are repulsive forces that become strong enough to overwhelm these attractive forces when the molecules get close enough together. The repulsion is principally that expected from the Pauli exclusion principle when two electronic orbitals interpenetrate. Several empirical equations have been used for the potential energy of interaction.[23,24] The ordinary reference point for the energy is infinite separation of the molecules. For interactions between monoatomic molecules, r is defined as the internuclear distance and r_m as the value of r at the distance of minimum potential energy. In the Lennard-Jones equation the energy is expressed

$$E = \epsilon\left[\frac{6}{n-6}\left(\frac{r_m}{r}\right)^n - \frac{n}{n-6}\left(\frac{r_m}{r}\right)^6\right] \tag{2-6}$$

where ϵ is the depth of the energy minimum and n is usually taken as 12, although values from 8 to 20 have been used. The sixth power term represents the stabilizing interactions, largely due to London forces, and the nth power term represents the destabilizing interactions. The Kihara function expresses the energy

$$E = \epsilon\left(\frac{r_m - r_0}{r - r_0}\right)^{12} - 2\epsilon\left(\frac{r_m - r_0}{r - r_0}\right)^6 \tag{2-7}$$

It may be thought of as treating the atoms as having incompressible cores, the sum of whose radii is r_0. The Morse function is an exponential

$$E = \epsilon e^{-2a[(r/r_m)-1]} - 2\epsilon e^{-a[(r/r_m)-1]} \tag{2-8}$$

where a is usually set equal to 6. Much of the available data on intermolecular potential functions has been summarized critically.[23,24]

Often one is interested in the energies of interaction only over a certain

Table 2-2. van der Waals Radii and Covalent Bond Radii (Å)

Atom	van der Waals Radius[a]	Bond Radius[b]
H—	1.20	0.30
F—	1.47	0.64
Cl—	1.75	0.99
Br—	1.85	1.14
I—	1.98	1.33
O—	1.52	0.66
O=	1.50	0.55
S—	1.80	1.04
Se—	1.90	1.17
Te—	2.06	1.37
N—	1.55	0.70
N=	1.55	0.60
N≡	1.60	0.55
P—	1.80	1.10
As—	1.85	1.21
C—	1.70[c]	0.77
C=	1.77[d]	0.665
C≡	1.78	0.60
Si—	2.1[e]	1.17

[a] Ref. 26.

[b] L. Pauling, *Nature of the Chemical Bond*, Cornell University Press, Ithaca, N.Y., 1945, p. 164.

[c] Radius of a methyl or methylene group.

[d] Average half-width of aromatic rings.

[e] Radius of an —SiH_3 group.

range of r values. The equation of Jordan and Amdur,[25] for example,

$$E = 4520e^{-4.21r(\text{Å})} \text{ kcal/mole}$$

fits the experimental data on the energy of interaction of helium atoms within 1.8% from an internuclear distance of 0.61 Å (where the energy is 367 kcal/mole) to 1.1 Å (where it is 45 kcal/mole) and within 20% up to 2.3 Å (where it is only 0.23 kcal/mole).

The distance at which attractive and repulsive forces between atoms in different molecules balance is known as the van der Waals distance. For a given pair of atoms it depends on the nature of the other atoms to which they are attached, on the relative orientations of the atoms with respect to their covalent bonds, etc. Nevertheless, to a useful extent van der Waals distances may be expressed as the sums of parameters called van der Waals radii for the two atoms in question. A number of van der Waals radii likely to be of interest in organic chemistry are listed in Table 2-2. They are from a compilation by Bondi,[26] whose paper should be consulted for a discussion of their significance and limitations. Also included are covalent bond radii, which are also needed in most calculations relating to steric effects. Since the effective bond radii of atoms vary somewhat with the difference in electronegativities between the bonded atoms, the state of hybridization of the atoms, etc., lengths of typical single (Table 2-3) and multiple (Table 2-4) bonds of various kinds are also listed.

Table 2-3 Typical Single Bond Lengths (Å)[a]

	sp^3-C	sp^2-C	sp-C	N	O	F	Si	P	S	Cl	Br	I
H	1.09	1.08	1.06	1.01	0.96	0.92	1.48	1.44	1.34	1.27	1.41	1.61
sp^3-C	1.54	1.51	1.46	1.47	1.43	1.37	1.87	1.84	1.82	1.77	1.94	2.16
sp^2-C		1.48	1.43	1.43	1.36	1.33	1.84	1.83	1.76	1.71	1.87	2.09
sp-C			1.38	1.35		1.29	1.82	1.78	1.70	1.64	1.80	1.99
N				1.45	1.41	1.36	1.74	1.67	1.70	1.77		
O					1.48	1.42	1.63	1.57	1.60	1.70		1.89
F						1.42	1.58	1.54	1.56	1.63	1.76	1.80
Si							2.33	2.25	2.15	2.02	2.18	2.44
P								2.21	2.10	2.03	2.14	2.46
S									2.05	2.03	2.26	
Cl										1.99	2.14	2.32
Br											2.28	2.48
I												2.67

[a] Based largely on data from Refs. 2 and 3. In specific compounds the bonds may vary from the values shown here because of differences in oxidation state, other substituents in the molecule, resonance effects, etc.

Table 2-4. Typical Multiple Bond Lengths[a]

Bond	Length (Å)
sp^2-C=C-sp^2	1.34
sp^2-C=C-sp	1.31
sp-C=C-sp	1.28
C≡C	1.20
sp^2-C=O	1.22
sp-C=O	1.16
sp^2-C=N	1.30
sp-C=N	1.21
C≡N	1.16
sp^2-C=S	1.70
sp-C=S	1.56
N=N	1.25
N=O	1.17

[a] See Table 2-3, Footnote *a*.

2-4. RESONANCE INTERACTIONS

Any attempt to prepare a species with an —O^- attached to the carbon atom of a carbonyl group yields instead a species in which the two oxygen atoms are bonded identically and bear identical charges. Employing a symbolism introduced by the English school of physical organic chemists, this can be written as an interaction between the —O^- and the =O groups.

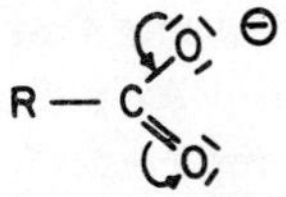

This symbolism starts with a structural formula of a type that gives descriptions of simple species in which resonance is not important and adds curved arrows indicating movement of electron pairs in a given direction. The actual molecule is then thought of as having the electron pairs moved somewhat (in the present case, halfway) in the directions indicated by the arrows.* (Movement of the electron pairs completely as shown by the

*Such arrows are also used, in a somewhat different symbolism, to indicate the direction in which the electron pairs (or in some cases unpaired electrons) move during a given reaction.

arrow would reverse the roles of the two oxygen atoms and indicate a structure just as unsymmetrical as that in which the arrows are not written.) In any case in which curved arrows may be written with the above significance, it is also possible to write two or more valence bond structures and to describe the actual species as a resonance hybrid of the contributing structures.

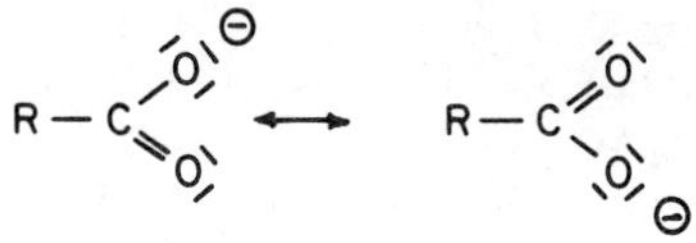

A symbolism that may make it clearer that the actual species has two equivalent oxygen atoms and is not a mixture of tautomers is the following,

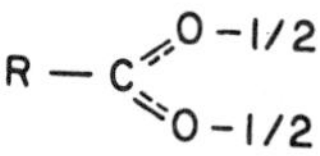

in which the dotted lines are "half-bonds." Each of the three types of symbolism has certain advantages. In many cases a molecular orbital description has advantages over any of the three.

A treatment in terms of the double-headed arrow symbolism is useful in considering the question of when two different valence bond structures represent two different chemical species and when they are merely two of the contributing structures for a resonance hybrid species. This treatment will be applied first to the case of benzene, for which two important valence bond structures may be written. We shall plot the energy to be expected from each of these two contributing structures against the geometric configuration. In our plot (Fig. 2-3), the left end of the abscissa represents the geometric configuration of atoms that would be expected if valence bond structure *A* were a perfect description of the benzene molecule, and the right end of the abscissa represents the geometric configuration of atoms that would be expected if valence bond structure *B* were a perfect description of the benzene molecule.

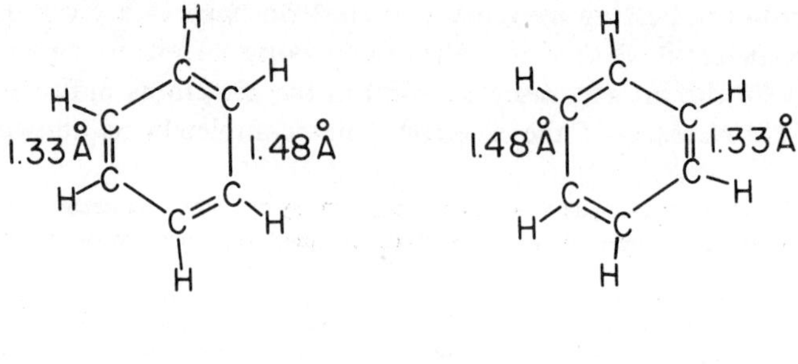

A　　　　　B

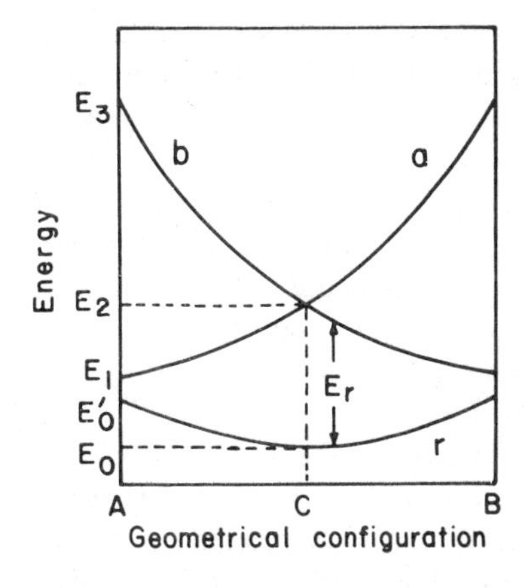

Figure 2-3. Graphical representation of resonance between two Kekulé structures for benzene.

Line *a* represents the potential energy expected of the system if valence bond structure *A* were a perfect description, and line *b* represents the potential energy expected if valence bond structure *B* were a perfect description. Since the left end of the abscissa is defined as corresponding to the optimum geometry for valence bond structure *A*, line *a* must have a minimum at this point; for any other geometric configuration the expected energy is higher. When curve *a* reaches the right end of the abscissa, the potential energy is that to be expected of valence bond structure *A* in the geometric configuration optimum for *B*.

1.48 Å 1.33 Å

Since the double bonds in this structure are 0.15 Å longer than the optimum lengths for such double bonds and the single bonds are 0.15 Å shorter than the optimum length for such single bonds, line *a* should be considerably higher at the right end than at the left end of the abscissa. The difference between E_1, the potential energy corresponding to valence bond structure *A* in geometric configuration *A*, and E_3, the potential energy corresponding to valence bond structure *A* in geometric configuration *B*, will be equal to the energy required to stretch three double bonds by 0.15 Å each and to compress three single bonds by 0.15 Å each. In the harmonic oscillator approximation line *a* should be a segment of a parabola. In a similar manner, curve *b*, the energy to be expected from valence bond structure *B*, may be seen to rise parabolically from a minimum of E_1 at the right side of the plot, to a value of E_3 at the left side. According to the basic concept of valence bond resonance, the actual molecule will be more stable than would be expected from any of the contributing structures. The actual potential energy content will be lower than that of the most stable structure by an amount (the resonance energy, E_r) that depends on how many contributing structures there are, how nearly equal their contributions are, and other factors. In Fig. 2-3 the actual energy is represented by line *r*. Line *r* lies below line *a* or *b*, whichever is lower, by the amount E_r. On going from either the left- or the right-hand side of the plot toward the middle, E_r will increase because contributions of the two contributing structures become more nearly equal. From the fact that benzene exists as a single resonance hybrid rather than a tautomeric species, it follows that the increase in E_r on going toward the middle of the plot is more rapid than the increase in line *a* or *b*; that is, it follows that line *r* has a minimum in the middle of the plot. This seems to be the usual result in cases where the geometric configurations to be expected from the contributing structures differ by only a few tenths of an angstrom unit or less, and where the expected electronic configurations do not differ greatly. (For example, if one valence bond structure is a nonradical and the other is a diradical or triplet, there will always be two species rather than a single hybrid species.) In cases in which the differences in bond distances to be expected from the contributing structures are about 1 Å or more, there are ordinarily two or more species rather than one resonance hybrid. In such cases the increase in E_r that occurs on moving from either side of the figure corresponding to Figure 2-3 cannot be steep enough to counteract the large rise in lines *a* and *b* that must accompany large changes in molecular geometry.

PROBLEMS

1. Assume that electrostatic interactions and statistical effects are the only reasons why the equilibrium constants for reactions of the type

$$2C_6H_5Br \overset{AlBr_3}{\rightleftharpoons} C_6H_6 + C_6H_4Br_2$$

differ from 1.000 and calculate the three equilibrium constants for the formation of *o*-, *m*-, and *p*-dibromobenzene at 35° C. Assume that all bond angles to sp^2 carbon are 120° and that all carbon-carbon bond distances in a benzene ring are 1.40 Å. The dipole moment of bromobenzene is 1.57 *D*.[14] The following dielectric constants[27] may be useful: benzene, 2.28 (20° C), 2.07 (liquid, 129° C); bromobenzene, 5.40 (25° C); *o*-dibromobenzene, 7.35 (20° C); *m*-dibromobenzene, 4.80

(20° C); *p*-dibromobenzene, 2.57 (95° C); chlorobenzene, 5.62 (25° C), 4.21 (130° C). Since *p*-dibromobenzene melts at 88° C, its dielectric constant as a liquid is not available below that temperature. Something about the way in which dielectric constants change with temperature may be seen in the values for benzene and chlorobenzene. Make one set of calculations using Eq. 2-1 and another set using Eq. 2-4 and compare the results. Discuss your choice of a dielectric constant. Experimental values of log K are -2.55, -1.43, and -1.72 for the ortho, meta, and para compounds, respectively.[28]

2. If necessary, learn more about charge-transfer complexes (also called donor-acceptor complexes).[29–32] Note that even saturated organic halides such as carbon tetrachloride and methylene chloride are capable of acting as electron acceptors in the formation of such complexes.[33,34] Then, discuss the possibility that the acid-strengthening effect of ortho substituents in arylpropiolic acids indicated in Table 2-1 is due to donor-acceptor interactions between the triple bond and the ortho substituents.

3. Do you expect K^{chem} for the following reaction[35] to be more or less than 1.0?

$$2CH_3\overset{\overset{O}{\|}}{C}CH_2Br \rightleftharpoons BrCH_2\overset{\overset{O}{\|}}{C}CH_2Br + CH_3\overset{\overset{O}{\|}}{C}CH_3$$

Give your reasons.

REFERENCES

1. G. Kortüm, W. Vogel, and K. Andrussow, *Dissociation Constants of Organic Acids in Aqueous Solution*, Butterworths, London, 1961.
2. *Tables of Interatomic Distances and Configurations in Molecules and Ions*, L. E. Sutton, Ed., Special Publication No. 11, The Chemical Society, London, 1958.
3. *Tables of Interatomic Distances and Configurations in Molecules and Ions, Supplement*, L. E. Sutton, Ed., Special Publication No. 18, The Chemical Society, London, 1965.
4. N. Bjerrum, *Z. Phys. Chem.*, **106,** 219 (1923).
5. J. G. Kirkwood and F. H. Westheimer, *J. Chem. Phys.*, **6,** 506 (1938).
6. F. H. Westheimer and J. G. Kirkwood, *J. Chem. Phys.*, **6,** 513 (1938).
7. J. T. Edward, P. G. Farrell, and J. L. Job, *J. Chem. Phys.*, **57,** 5251 (1972).
8. F. H. Westheimer and M. W. Shookhoff, *J. Amer. Chem. Soc.*, **61,** 555 (1939).
9. J. D. Roberts and W. T. Moreland, Jr., *J. Amer. Chem. Soc.*, **75,** 2167 (1953).
10. C. Tanford and J. G. Kirkwood, *J. Amer, Chem. Soc.*. **79,** 5333 (1957).
11. C. Tanford, *J. Amer. Chem. Soc.*, **79,** 5340. 5348 (1957).
12. F. W. Baker, R. C. Parish, and L. M. Stock, *J. Amer. Chem. Soc.*, **89,** 5677 (1967).

13. C. F. Wilcox and C. Leung, *J. Amer. Chem. Soc.*, **90,** 336 (1968).
14. L. E. Sutton, in *Determination of Organic Structures by Physical Methods*, E. A. Braude and F. C. Nachod, Eds., Academic Press, New York, 1955, Chap. 9.
15. G. N. Lewis, *Valence and the Structure of Atoms and Molecules*, The Chemical Catalog Co., New York, 1923, p. 139.
16. W. A. Waters, *J. Chem. Soc.*, 1551 (1933).
17. M. J. S. Dewar and P. J. Grisdale, *J. Amer. Chem. Soc.*, **84,** 3548 (1962).
18. pK_a values from D. D. Perrin, *Dissociation Constants of Organic Bases in Aqueous Solution*, Butterworths, London, 1965.
19. M. S. Newman and S. H. Merrill, *J. Amer. Chem. Soc.*, **77,** 5552 (1955).
20. J. D. Roberts and R. A. Carboni, *J. Amer. Chem. Soc.*, **77,** 5554 (1955).
21. I. J. Solomon and R. Filler, *J. Amer. Chem. Soc.*, **85,** 3492 (1963).
22. R. Golden and L. M. Stock, *J. Amer. Chem. Soc.*, **88,** 5928 (1966); **94,** 3080 (1972).
23. R. B. Bernstein and J. T. Muckerman, *Advan. Chem. Phys.*, **12,** 389 (1967).
24. G. C. Maitland and E. B. Smith, *Chem. Soc. Rev.*, **2,** 181 (1973).
25. J. E. Jordan and I. Amdur, *J. Chem. Phys.*, **46,** 165 (1967).
26. A. Bondi, *J. Phys. Chem.*, **68,** 441 (1964).
27. A. A. Maryott and E. R. Smith, *Tables of Dielectric Constants of Pure Liquids*, National Bureau of Standards Circular 514, U.S. Government Printing Office, Washington, D.C., 1951.
28. J. Hine and H. E. Harris, *J. Amer. Chem. Soc.*, **85,** 1476 (1963).
29. L. J. Andrews and R. M. Keefer, *Molecular Complexes in Organic Chemistry*, Holden-Day, San Francisco, 1964.
30. E. M. Kosower, *Progr. Phys. Org. Chem.*, **3,** 81 (1965).
31. O. Hassel and C. Rømming, *Quart. Rev.* (London), **16,** 1 (1962).
32. R. S. Mulliken and W. B. Person, *J. Amer. Chem. Soc.*, **91,** 3409 (1969).
33. R. Anderson and J. M. Prausnitz, *J. Chem. Phys.*, **39,** 1225 (1963).
34. R. F. Weimer and J. M. Prausnitz, *J. Chem. Phys.*, **42,** 3643 (1965).
35. C. Rappe, *Ark. Kemi*, **23,** 81 (1964).

3

The Hammett and Taft Equations

3-1. THE HAMMETT EQUATION[1–9]

By the mid-1930s a fairly satisfactory qualitative theory of substituent effects on equilibria and reaction rates in organic chemistry had evolved. Substituents that encouraged the development of negative charges and discouraged the development of positive charges at the reaction center were known as electron-withdrawing substituents, and substituents of the opposite type were known as electron-donating substituents, the reference substituent ordinarily being hydrogen. Empirically it was observed that quite often although not invariably, a substituent found to be an electron withdrawer in one reaction turned out to be an electron withdrawer in another, and electron donors in one reaction tended to be electron donors in others. This is another way of saying that for a large number of reactions, when the rate or equilibrium constants were arranged in the order of increasing (or decreasing) magnitude the rate or equilibrium constant for the unsubstituted, that is, hydrogen-substituted, compound fell at the same position relative to that of the other substituents. Not only was the position of hydrogen invariant in many reaction series, but so was that of most of the other substituents. Regularities in substituent effects were observed most commonly with aromatic compounds, particularly meta- and para-substituted ones.[10] Thus, for example, whatever effect the *p*-nitro substituent had on an equilibrium constant, the *m*-chloro substituent had an effect in the same direction but smaller, the *p*-chloro substituent also had an effect in the same direction but still smaller than that of *m*-chloro, and the *p*-methyl substituent had an effect in the opposite direction.

Hammett examined a number of log-log plots of rate and equilibrium constants of side-chain reactions of meta- and para-substituted aromatic compounds.[11,12] In such plots the log of the rate or equilibrium constant in *one* reaction for a compound with a given substituent is used as the abscissa of a point, and the log of the rate or equilibrium constant in another reaction for a compound with the *same* substituent is used as the ordinate for this point. For a specific example, let us consider the plot of the log of the ionization constants of substituted phenylacetic acids against the log of the ionization constants of substituted benzoic acids. The point for the unsubstituted case, that is, for the hydrogen substituent, has as its abscissa the log of the ionization constant for benzoic acid, at a given temperature in a given solvent, and for its ordinate the log of the ionization constant of phenylacetic acid at a given temperature and in a given solvent (not necessarily the same solvent and temperature as that in which the ionization constant of benzoic acid was determined). The point for the *p*-nitro substituent has for its abscissa the log of the ionization constant of *p*-nitrobenzoic acid under the same conditions as those used for benzoic acid itself and for its ordinate the log of the ionization constant of *p*-nitrophenylacetic acid under the same conditions as those used for phenylacetic acid itself.

The qualitative observation of an invariant order of substituent effects in many reactions of meta- and para-substituted aromatic compounds, which has already been mentioned, does not tell us just what shape our plot of log K' versus log K is going to have, but it does rule out a lot of possibilities for such reactions. That is, the line must be a monotonic function and hence cannot be a parabola or contain a loop, for example.

Actually, as shown in Fig. 3-1 for the pair of reactions just mentioned, Hammett found that such plots for meta- and para-substituted compounds usually give a good approach to a straight line, from which the points for the ortho-substituted compounds often deviate substantially. Thus log K for one reaction of meta- and para-substituted compounds is usually an approximately linear function of log K for another. For the pair of reactions to which Fig. 3-1 refers, this function may be expressed as follows:

$$\log (K_{X_1})^{\mathrm{ArCH_2CO_2H}} = \rho \log (K_{X_1})^{\mathrm{ArCO_2H}} + Y \tag{3-1}$$

where ρ is the slope of the line, Y is the intercept, and X_1 is a meta or para substituent on the aromatic ring. For a second substituent X_2 an equation analogous to 3-1 may be written and when 3-1 is subtracted from it the result is

$$\log\left(\frac{K_{X_2}}{K_{X_1}}\right)^{ArCH_2CO_2H} = \rho \log\left(\frac{K_{X_2}}{K_{X_1}}\right)^{ArCO_2H} \qquad (3\text{-}2)$$

In Eq. 3-2, X_1 may be considered to be a reference substituent whose equilibrium constant is being compared with that of X_2. Hammett used hydrogen as the reference substituent so that the substituted compounds were compared with the "unsubstituted" ones. With the possibility of comparing any two of a large number of reactions with each other, it is also convenient to choose one reaction as a standard with which all other reactions are compared. For this purpose Hammett chose the ionization of benzoic acids in water at 25° C, mainly because of the large number of relatively reliable data that were available. With these assignments the Hammett equation may be written as follows:

$$\log\frac{K_X}{K_H} = \rho \log\left(\frac{K_X}{K_H}\right)^{ArCO_2H} \qquad (3\text{-}3)$$

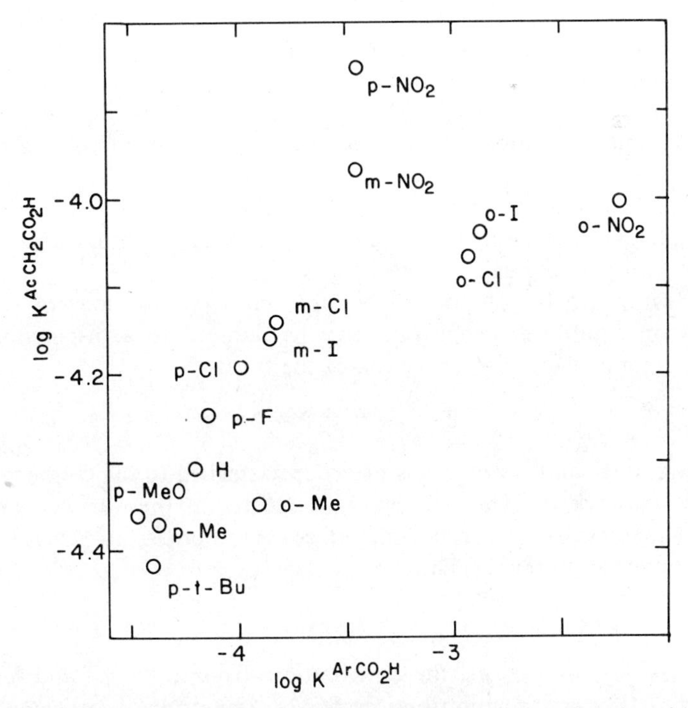

Figure 3-1. Log-log plot of the ionization constants of phenylacetic acids versus those of the corresponding benzoic acid, both in water at 25°C.

where K_X and K_H are the equilibrium constants for the X-substituted (X being a meta or para substituent) and the unsubstituted compounds in any reaction. Since the term log $(K_X/K_H)^{ArCO_2H}$ has a fixed value for a given substituent X, and since this term will be used for any reaction of any compound with this given substituent, it has been abbreviated as σ_X.

$$\sigma_X = \log\left(\frac{K_X}{K_H}\right)^{ArCO_2H} \tag{3-4}$$

This gives the ordinary form of the Hammett equation

$$\log\frac{K_X}{K_H} = \rho\sigma_X \tag{3-5}$$

where σ_X depends only on the nature of the substituent and ρ depends on the nature of the reaction (including the reaction temperature and solvent). A completely analogous equation is used for the correlation of reaction rates.

3-2. SOME GENERAL APPROACHES TO LINEAR FREE ENERGY RELATIONSHIPS

The Hammett equation is a specific example of a linear free energy relationship. This term is applicable because log K, which is proportional to $\Delta G°$ for the reaction, is expressed as a linear function of ρ and σ. There are several lines of reasoning that lead to the expectation that linear free energy relationships should exist in various cases.

In Section 1-5 it was pointed out that most comparisons of structural effects on equilibrium constants may be reduced to a discussion of the equilibrium constant for a reaction of the type

$$\text{Y—N—A} + \text{Z—N—B} \rightleftharpoons \text{Y—N—B} + \text{Z—N—A} \tag{3-6}$$

where A, B, Y, and Z are atoms or groups attached to the common molecular framework N. The free energy contents of the various species involved may be expressed as sums of contributions from structural components (cf. Section 1-2). Thus

$$G^{\circ}_{YNA} = G_Y + G_N + G_A + G_{YN} + G_{NA} + G_{YAN} \tag{3-7}$$

where G_Y, G_N, and G_A are the contributions from the Y, N, and A groups, G_{YN} and G_{NA} are the contributions for interaction between Y and N and N and A (including Y—N and N—A bond contributions), and G_{YAN} is the contribution for the interaction of Y and A across N (in a given medium at a given temperature).[3] We know that for reaction 3-6,

$$\Delta G^\circ = G^\circ_{\mathrm{YNB}} + G^\circ_{\mathrm{ZNA}} - G^\circ_{\mathrm{YNA}} - G^\circ_{\mathrm{ZNB}} \tag{3-8}$$

If the four appropriate expressions of the type of 3-7 are substituted into 3-8 most of the terms cancel, and we are left with

$$\Delta G^\circ = G_{\mathrm{Y}_{\mathrm{B}}\mathrm{N}} + G_{\mathrm{Z}_{\mathrm{A}}\mathrm{N}} - G_{\mathrm{Y}_{\mathrm{A}}\mathrm{N}} - G_{\mathrm{Z}_{\mathrm{B}}\mathrm{N}} \tag{3-9}$$

Consideration of the ways in which the substituents Y and A might interact with each other suggests expressing each interaction as a sum of several contributing terms, for example,

$$G_{\mathrm{YAN}} = G^{\mathrm{bonding}}_{\mathrm{YAN}} + G^{\mathrm{polar}}_{\mathrm{YAN}} + G^{\mathrm{resonance}}_{\mathrm{YAN}} + G^{\mathrm{steric}}_{\mathrm{YAN}} \tag{3-10}$$

The bonding term allows for the formation of an ordinary chemical bond or a hydrogen bond between Y and A, perhaps reversibly. The polar term covers charge-charge, charge-dipole, charge-induced dipole, dipole-dipole, etc., interactions between Y and A. The resonance term covers direct resonance interactions between Y and A, the effect that resonance interactions between Y and N have on resonance interactions between N and A, but not simple resonance interactions between Y and N or N and A, which are covered by the G_{YN} and G_{NA} terms. The steric term allows for steric interactions between Y and A, including van der Waals attractions as well as repulsions and including mutual interferences with internal rotations, steric interference with solvation, etc.

Equation 3-10 may be generally true, but it is of little use in the absence of methods for estimating with reasonable reliability the four terms on the right-hand side of the equation. A particularly useful simplification occurs when the group N is a relatively rigid structure of such a nature that the groups Y and A are not able to touch each other and to undergo direct resonance interactions. Reactions in which Y—N—A— is 1-A-4-Y-bicyclo[2.2.2]octane, or perhaps *m*-$\mathrm{YC_6H_4A}$, should be cases of this type.

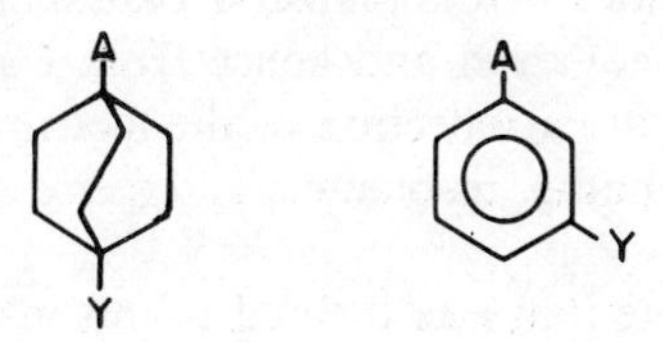

In such cases the bonding, resonance, and steric terms may be negligible and only the polar term significant. Let us assume that the free energy of polar interaction of substituents may be expressed as the product of terms (σ's) called polar substituent constants, characteristic of the substituents, and a proportionality constant (τ) that is a measure of the efficiency with which the influences of the groups are transmitted to each other across the N group in question in a given medium at a given temperature.

$$G_{\mathrm{YAN}}^{\mathrm{polar}} = 2.3RT\,\tau_{\mathrm{N}}\sigma_{\mathrm{Y}}\sigma_{\mathrm{A}} \tag{3-11}$$

The values of σ will be taken as constant for a given substituent, a given N, and a given set of reaction conditions. (The necessity for these restrictions will be considered later.) For a given N group and a given set of reaction conditions, τ_{N} is taken as a constant. The form of Eq. 3-11 shown was chosen because it is the same as the equation for the energy of interaction of two charges, two dipoles, a charge and a dipole, etc. (cf. Section 2-1). The σ constants take the place of the charge or dipole moment of a substituent and τ may be thought of as a function of the distance between substituents, the angle that the dipoles make with each other, and the dielectric constant of the medium through which the substituents interact. The factor $2.3RT$ is separated from the other terms to permit simplification later, of course. When Eqs. 3-9, 3-10, and 3-11 are combined and the bonding, resonance, and steric terms in Eq. 3-10 are neglected, the result is

$$\Delta G^{\circ} = 2.3RT\,\tau_{\mathrm{N}}(\sigma_{\mathrm{Y}}\sigma_{\mathrm{B}} + \sigma_{\mathrm{Z}}\sigma_{\mathrm{A}} - \sigma_{\mathrm{Y}}\sigma_{\mathrm{A}} - \sigma_{\mathrm{Z}}\sigma_{\mathrm{B}}) \tag{3-12}$$

or

$$\log K = \tau_{\mathrm{N}}(\sigma_{\mathrm{A}} - \sigma_{\mathrm{B}})(\sigma_{\mathrm{Y}} - \sigma_{\mathrm{Z}}) \tag{3-13}$$

It is interesting to compare Eq. 3-12 with the relationship that may be obtained by application of Eq. 1-14, Pauling's equation, to reaction 3-6 in the case where N does not exist, that is, when A and C are bonded directly to each other. After a wholesale canceling of terms the result is

$$\Delta E^{\circ} = 23\,(X_{\mathrm{Y}}X_{\mathrm{B}} + X_{\mathrm{Z}}X_{\mathrm{A}} - X_{\mathrm{Y}}X_{\mathrm{A}} - X_{\mathrm{Z}}X_{\mathrm{B}}) \tag{3-14}$$

Thus if ΔS and $\Delta(PV)$ are zero, so that ΔE° and ΔG° are equal, then the assumptions that led to Eq. 3-12 and Pauling's equation give the same result, with the substituent constants having the same type of meaning as Pauling's electronegativities.[13] It is plausible that some of the major deviations from the Pauling equation and hence from Eq. 3-14 (and there are some rather large ones) are a reflection of the fact that when N is so small as to be nonexistent, bonding, resonance, and steric effects are particularly likely to be important.

It is worthwhile to note that τ as defined in this manner will be expected to be a positive number. This follows from the fact that polar interactions involve attraction (or stabilization) between unlike species and repulsion (or destabilization) between like species. In Eq. 3-11 if σ_{A} and σ_{Y} are electrical charges, the product $\sigma_{\mathrm{A}}\sigma_{\mathrm{Y}}$ will be positive if they are like charges. Hence

τ must be positive since an interaction between like charges should destabilize and thus add to the free energy content of a molecule. Empirically it may be seen that in the case where N is nonexistent and Eq. 3-12 may be related to the Pauling equation, τ_N is a positive number.

The equations that have been derived are related not only to the Pauling equation but also to the Hammett equation. In applying the Hammett equation to the ionization constants of *m*-chlorobenzoic acid and *m*-nitrobenzoic acid in water at 25° C we may write

$$\log \frac{K_{m\text{-Cl}}}{K_{\text{H}}} = \rho\sigma_{m\text{-Cl}} \tag{3-15}$$

and

$$\log \frac{K_{m\text{-NO}_2}}{K_{\text{H}}} = \rho\sigma_{m\text{-NO}_2} \tag{3-16}$$

where ρ is the reaction constant for the ionization of benzoic acids, $K_{m\text{-Cl}}$ is the ionization constant of *m*-chlorobenzoic acid, $K_{m\text{-NO}_2}$ is that of *m*-nitrobenzoic acid, and K_{H} that of benzoic acid itself (all in water at 25° C), and $\sigma_{m\text{-Cl}}$ and $\sigma_{m\text{-NO}_2}$ are the substituent constants for the *m*-chloro and *m*-nitro substituents, respectively. If $K_{m\text{-Cl}}$ is divided by K_{H}, the term $[H^+]$ cancels and the quotient is simply the equilibrium constant for the reaction

$$m\text{-ClC}_6\text{H}_4\text{CO}_2\text{H} + m\text{-HC}_6\text{H}_4\text{CO}_2^- \rightleftharpoons m\text{-ClC}_6\text{H}_4\text{CO}_2^- + m\text{-HC}_6\text{H}_4\text{CO}_2\text{H} \tag{3-17}$$

Similarly, the quotient $K_{m\text{-NO}_2}/K_{\text{H}}$ is the equilibrium constant for the reaction

$$m\text{-O}_2\text{NC}_6\text{H}_4\text{CO}_2\text{H} + m\text{-HC}_6\text{H}_4\text{CO}_2^- \rightleftharpoons m\text{-O}_2\text{NC}_6\text{H}_4\text{CO}_2^- + m\text{-HC}_6\text{H}_4\text{CO}_2\text{H} \tag{3-18}$$

Reactions 3-17 and 3-18 may be seen to be of the form of reaction 3-6, with N being the *m*-phenylene group, A the carboxy group, B the carboxylate an ion group, etc. If Eq. 3-13 is then applied to reactions 3-17 and 3-18 we get

$$\log \frac{K_{m\text{-Cl}}}{K_{\text{H}}} = \tau_m(\sigma_{m\text{-CO}_2\text{H}} - \sigma_{m\text{-CO}_2^-})(\sigma_{m\text{-Cl}} - \sigma_{m\text{-H}}) \tag{3-19}$$

and

$$\log \frac{K_{m\text{-NO}_2}}{K_{\text{H}}} = \tau_m(\sigma_{m\text{-CO}_2\text{H}} - \sigma_{m\text{-CO}_2^-})(\sigma_{m\text{-NO}_2} - \sigma_{m\text{-H}}) \tag{3-20}$$

In fact, for any *m*-X-benzoic acid we may write

$$\log \frac{K_{m\text{-X}}}{K_{\text{H}}} = \tau_m(\sigma_{m\text{-CO}_2\text{H}} - \sigma_{m\text{-CO}_2^-})(\sigma_{m\text{-X}} - \sigma_{m\text{-H}}) \qquad (3\text{-}21)$$

Inasmuch as the expression for the ionization constant of any benzoic acid in water at 25° C will contain the same factor, $\tau_m(\sigma_{m\text{-CO}_2\text{H}} - \sigma_{m\text{-CO}_2^-})$, it is useful to have an abbreviation for this factor. Since this factor is a characteristic of the reaction it is denoted ρ, called the reaction constant, and perhaps given a subscript to remind us what reaction it refers to, for example, $\rho_{\text{ArCO}_2\text{H}}$. The remaining term on the right-hand side of the equation is the difference between two substituent constants. Since the substituent constants appear in our equations only as differences, their absolute values are undefined. That is, if we added a constant to each substituent constant, the ability of the set of σ's to correlate K and k values would be unaffected. For this reason we are free to choose the absolute value of any one substituent constant arbitrarily. It is convenient to do this by setting the substituent constant for hydrogen equal to zero. With this definition and the introduction of the reaction constant, Eq. 3-21 becomes

$$\log \frac{K_{m\text{-X}}}{K_{\text{H}}} = \rho_{\text{ArCO}_2\text{H}}\sigma_{m\text{-X}} \qquad (3\text{-}22)$$

Although the similarity of this expression to the Hammett equation is clear, it is necessary to add one more definition before it becomes identical to the Hammett equation for meta-substituted compounds. Any set of ρ's and σ's used for correlating rate and equilibrium constants by equations like 3-22 could be replaced by an equally good new set by multiplying each σ and dividing each ρ by some constant. For this reason we are free to assign arbitrarily an absolute value to one more substituent constant or to one reaction constant. To make Eq. 3-22 identical to the Hammett equation for meta-substituted compounds we assign the value 1.000 to ρ for the ionization of benzoic acids in water at 25° C.

The relation

$$\rho_{\text{ArCO}_2\text{H}} = \tau_m(\sigma_{m\text{-CO}_2\text{H}} - \sigma_{m\text{-CO}_2^-}) \qquad (3\text{-}23)$$

introduced as a definition in the present derivation, can be obtained algebraically by assuming that the Hammett equation is generally applicable.[14] It is a special case of the more general relationship that the reaction constant for any equilibrium involving meta-substituted aromatic compounds is proportional to the difference in substituent constants between the two groups being interchanged by the equilibrium, and the proportionality constant is τ_m for the set of reaction conditions in question, that is, for

$$\mathrm{ArB} \rightleftharpoons \mathrm{ArC}$$

$$\rho = \tau_m(\sigma_{m\text{-B}} - \sigma_{m\text{-C}}) \tag{3-24}$$

This equation thus provides a relationship between substituent constants and reaction constants. This relationship, though useful, is no more general than the assumptions (e.g., the generality of the Hammett equation) from which it may be derived.

A different approach may be based on a treatment by Leffler and Grunwald.[4] Let us assume that the free energy change for a reaction is a function of a number of separable factors such as polar (p), resonance (r), steric (s), etc., factors.

$$-\Delta G^\circ = f(p, r, s, \ldots) \tag{3-25}$$

When a change is made in reactant structure, in solvent, etc., ΔG° will change by an amount $\Delta\Delta G^\circ$ as a result of the change in polar, resonance, steric, etc., factors.

$$-(\Delta G^\circ + \Delta\Delta G^\circ) = f(p + \Delta p, r + \Delta r, s + \Delta s, \ldots) \tag{3-26}$$

A Taylor's series expansion gives

$$\begin{aligned}
-(\Delta G^\circ + \Delta\Delta G^\circ) &= f(p, r, s, \ldots) + \Delta p \frac{\partial [f(p, r, s, \ldots)]}{\partial p} \\
&+ \frac{\Delta p^2}{2} \frac{\partial^2 [f(p, r, s, \ldots)]}{\partial p^2} + \cdots + \Delta r \frac{\partial [f(p, r, s, \ldots)]}{\partial r} \\
&+ \frac{\Delta r^2}{2} \frac{\partial^2 [f(p, r, s, \ldots)]}{\partial r^2} + \cdots + \Delta s \frac{\partial [f(p, r, s, \ldots)]}{\partial s} \\
&+ \cdots + \Delta p\, \Delta r \frac{\partial^2 [f(p, r, s, \ldots)]}{\partial p\, \partial r} + \cdots
\end{aligned} \tag{3-27}$$

If the changes in p, r, s, etc., are small enough, we can drop all but the first derivative terms. This and subtraction of Eq. 3-25 give

$$-\Delta\Delta G^\circ = \Delta p \frac{\partial f}{\partial p} + \Delta r \frac{\partial f}{\partial r} + \Delta s \frac{\partial f}{\partial s} + \cdots \tag{3-28}$$

where f is $f(p, r, s, \ldots)$. If only one of the terms on the right-hand side of Eq. 3-28 is significant, for example, the Δp term, we have

$$-\Delta\Delta G^\circ = \Delta p \frac{\partial f}{\partial p} \tag{3-29}$$

In the case where the introduction of the substituent X into the reference compound changes the polar effect by the amount Δp_X, we may write

$$-\Delta\Delta G_X^\circ = \Delta p_X \frac{\partial f}{\partial p} \tag{3-30}$$

or

$$\log \frac{K_X}{K_0} = \frac{\Delta p_X}{2.3RT} \frac{\partial f}{\partial p} \tag{3-31}$$

If $\Delta p_X/(2.3RT)$ is defined as the substituent constant σ_X, and $\partial f/\partial p$ as the reaction constant ρ, we have

$$\log \frac{K_X}{K_0} = \rho\sigma_X \tag{3-32}$$

Thus an equation in the form of the Hammett equation should be applicable if ΔG° can be expressed as a function of various factors p, r, s, etc., if only one of those factors is important in the case at hand, and if the change in that factor is not too large. However, these are not the only circumstances under which an equation of the form of Eq. 3-32 should be applicable. If the changes in one factor are for some reason proportional to the changes in another factor for all the structural changes under consideration, then an equation like Eq. 3-32 may hold even if both factors are important. For example, suppose that the change in the resonance effect is proportional to the change in the polar effect for all the substituents studied.

$$\Delta r_X = C\Delta p_X \tag{3-33}$$

Also suppose that only polar and resonance effects are important, so that Eq. 3-28 assumes the form

$$-\Delta\Delta G_X^\circ = \Delta p_X \frac{\partial f}{\partial p} + \Delta r_X \frac{\partial f}{\partial r} \tag{3-34}$$

Then substitution of Eq. 3-33 into 3-34 and replacement of $-\Delta\Delta G_X{}^\circ$ by $2.3RT \log (K_X/K_0)$ gives

$$\log \frac{K_X}{K_0} = \frac{\Delta p_X}{2.3RT} \left(\frac{\partial f}{\partial p} + C \frac{\partial f}{\partial r}\right) \tag{3-35}$$

in which σ_X may be defined as it was in the transformation of Eq. 3-31 to 3-32, and

$$\rho = \frac{\partial f}{\partial p} + C \frac{\partial f}{\partial r} \tag{3-36}$$

In many cases, however, it appears that the simplifying assumptions required to give equations of the form of 3-32 are not satisfactory approximations. For example, in cases where there are two important mechanisms of interaction between the substituent and reaction center, an equation of the following type might be applicable.

$$\log \frac{K_X}{K_0} = \rho\sigma + \rho'\sigma' \tag{3-37}$$

In other cases, when the change in a given factor is too large it may be necessary to include the second derivative term from Eq. 3-27 and use correlation equations of the form

$$\log \frac{K_X}{K_0} = \rho_1\sigma + \rho_2\sigma^2 \tag{3-38}$$

or

$$\log \frac{K_X}{K_0} = \rho_1\sigma + \rho'\sigma + \rho''\sigma\sigma' \tag{3-39}$$

In most cases at present, however, the number and range of available experimental data are not large enough to use equations like 3-39 with any confidence that agreement between experiment and equation would be anything more than a trivial result of the fact that a large number of disposable parameters were available.

3-3. SUBSTITUENT CONSTANTS FOR THE HAMMETT AND RELATED EQUATIONS

3-3a. Substituent Constants from the Acidities of Benzoic Acid Derivatives. Inasmuch as the Hammett ρ is defined as 1.000 for the ionization of benzoic acids in water at 25° C, a reaction whose equilibrium constants are increased by electron-withdrawing substituents, it follows that substituents with positive σ constants are electron withdrawers and those with negative σ constants are electron donors. Hammett σ constants for a large number of substituents have been determined from the defining relationship, Eq. 3-4, that is, from the ionization constant of the appropriately substituted benzoic acid in water at 25° C. Many of these values are collected in Table 3-1 together with values calculated from ionization constants of benzoic acids determined under other conditions, most commonly in aqueous ethanol mixtures where ρ is larger than it is in pure water. Most of the values are from the compilation of McDaniel and Brown[15] or the pK_a determinations of Sheppard[16] or Exner[17,18] and Lakomý.[18] A σ value

Table 3-1. Hammett Substituent Constants[a]

Substituent	σ Meta	σ Para	Substituent	σ Meta	σ Para
Me	−0.07	−0.17	$NHCO_2Et$	0.09[c]	−0.13[c]
Et	−0.07	−0.15	NHAc	0.21	0.00
i-Pr	−0.07[b]	−0.15	NHCHO	0.22[c]	0.05[c]
t-Bu	−0.10	−0.20	$NHCOCF_3$	0.35[c]	0.14[c]
CH_2Ph	−0.18[c]	−0.11[c]	$NHSO_2Me$	0.21[c]	0.05[c]
C≡CH	0.20[d]	0.23[d]	OH	0.12	−0.37
Ph	0.06	−0.01	OMe	0.12	−0.27
Picryl	0.27[e]	0.31[e]	OEt	0.1	−0.24
CH_2SiMe_3	−0.16	−0.21	OPh	0.25	−0.32[n]
CH_2OMe	0.02[c]	0.03[c]	OCF_3	0.40[h]	0.35[h]
CH_2OPh	0.03[c]	0.07[c]	OAc	0.39	0.31
CHO	0.36[f]	0.44[f]	OSO_2Me	0.39[c]	0.37[c]
Ac	0.38	0.50	F	0.34	0.06
COPh	0.36[c]	0.44[c]	$SiMe_3$	−0.04	−0.07
CO_2^- [g]	−0.1	0.0	PO_3H^- [g]	0.2	0.26
CO_2Et	0.37	0.45	SH	0.25	0.15
CN	0.56	0.66	SMe	0.15	0.00
CH_2CN	0.15[c]	0.17[c]	SCF_3	0.40[h]	0.50[h]
CH_2I	0.07[c]	0.09[c]	SAc	0.39	0.44
CH_2Br	0.11[c]	0.12[c]	SOMe	0.52	0.49
CH_2Cl	0.09[c]	0.12[c]	SO_2Me	0.60	0.72
CF_3	0.43	0.54	SO_2CF_3	0.79[h]	0.93[h]
$CF(CF_3)_2$	0.37[h]	0.53[h]	SO_2NH_2	0.46	0.57
NH_2	−0.16	−0.66	SO_3^- [g]	0.05	0.09
NMe_2	−0.15[i]	−0.83	SMe_2^+ [g]	1.00	0.90
$N(CF_3)_2$	0.40[j]	0.53[j]	SF_5	0.61	0.68
NMe_3^+ [g]	0.88	0.82	Cl	0.37	0.23
N_3	0.37[k]	0.08[k]	Br	0.39	0.23
N=NPh	0.30[l]	0.35[l]	I	0.35	0.28
N_2^+ [g]	1.76[m]	1.91[m]	IO_2	0.70	0.76
NO_2	0.71	0.78			

[a] These substituent constants are based on the ionization costants of benzoic acids taken from the compilation of McDaniel and Brown[15] and rounded off, unless otherwise stated.

[b] Ref. 2.

[c] Based on a p*K* value in 50% aqueous ethanol and the corresponding ρ value (1.52).

[d] J. A. Landgrebe and R. H. Rynbrandt, *J. Org. Chem.*, **31,** 2585 (1966).

[e] D. J. Glover, *J. Org. Chem.*, **31,** 1660 (1966).

[f] A. A. Humffray, J. J. Ryan, J. P. Warren, and Y. H. Yung, *Chem. Commun.*, 610 (1965).

may also be calculated from the equilibrium constant or rate constant for any reaction of an appropriately substituted reactant if the reaction has been shown to obey the Hammett equation. Values obtained in this way and additional values based on the ionization constants of benzoic acids may be found in various larger compilations of σ values such as that of Hansch and co-workers,[19] who give data on 236 different substituents.

The values of the substituent constants for various groups on the whole fit in fairly well with the qualitative theory of substituent effects in aromatic systems that was developed in the 1920s. There are a number of substituents in which the atom attached to the ring has unshared electrons and is fairly electronegative. Such substituents, of which the halogens, alkoxy, hydroxy, and amino are examples, were said to remove electrons from all parts of the rings by a polar (inductive plus field) effect and feed electrons to the ortho and para positions by a resonance effect involving the unshared pair(s) of electrons.* Thus in the meta position such substituents were

*These effects are sometimes represented, especially by British chemists, by capital letter symbols. The inductive effect is called the *I* effect and the resonance effect called the *T* (for tautomeric) or *K* (for *Konjugationseffekt*). The inductive effect has been divided into the static or polarization effect I_s and the dynamic or polarizability effect I_d. The static resonance effect has been called the *M* (for mesomeric) effect and the dynamic effect called the *E*(for electromeric) effect. In the Ingold convention substituents with $+I$ and $+K$ effects are electron donors, and $-I$ and $-K$ groups are electron withdrawers. The Robinson convention, which is also still in use, employs plus and minus signs in the opposite sense. The *I* effect may be symbolized by replacing the single bond by an arrow aimed in the direction the effect displaces the electrons. The *K*, *M*, etc., effects may be symbolized similarly using curved arrows leading from the pi electrons (the "second" bond of a double bond) to the atom or bond toward which the effect displaces the electrons. Thus a combined electron-withdrawing *I* effect and electron-donating *K* effect can be represented as shown in the following formulas.

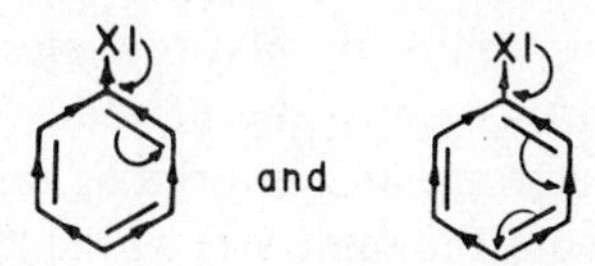

[g] Values of σ for electrically charged groups are relatively unreliable.
[h] Ref. 16.
[i] J. C. Howard and J. P. Lewis, *J. Org. Chem.*, **31**, 2005 (1966).
[j] F. S. Fawcett and W. A. Sheppard, *J. Amer. Chem. Soc.*, **87**, 4341 (1965).
[k] P. A. S. Smith, J. H. Hall, and R. O. Kan, *J. Amer. Chem. Soc.*, **84**, 485 (1962).
[l] M. Syz and H. Zollinger, *Helv. Chim. Acta*, **48**, 383 (1965).
[m] E. S. Lewis and M. D. Johnson, *J. Amer. Chem. Soc.*, **81**, 2070 (1959).
[n] This value is almost implausibly small, much smaller than the value 0.14 reported in 50% ethanol. See footnote *c*.

thought to exert only an electron-withdrawing polar effect and should have positive substituent constants. In the para position the greater distance from the reaction center will tend to make the electron-withdrawing polar effect smaller. However, most substituents have the positive ends of their dipoles aimed more directly at the reaction center when they are para to it than when they are meta. There is no general agreement on which of these two factors is more important. In any event, in the para position the electron-withdrawing polar effect is opposed by an electron-donating resonance effect. This is probably the principal reason why para-substituent constants for groups of the kind we are considering tend to be less positive than the corresponding meta-substituent constants, and why they may even be negative (when the resonance effect is large enough to overwhelm the polar effect).

The agreement with this generalization is good but not perfect. For all the halogens, all the substituents attached via an oxygen atom, and the amino, dimethylamino, acetamino, sulfhydryl, and methylthio substituents, σ_m is larger than σ_p. Unexpectedly, however, σ_p is larger than σ_m for the CF_3S, CH_3COS, and $N(CF_3)_2$ substituents. These are certainly substituents in which the resonance electron-donating ability of the atom attached to the ring would be expected to be reduced to a low level. It is hard to say to what extent the larger value of σ_p arises from polar effects being larger in the para than in the meta position,[9,17,18] from nonbonded interactions between the pi cloud of the aromatic ring and certain atoms in the substituents,[16] from resonance electron withdrawal involving *d* orbitals in the case of the sulfur substituents, from experimental errors, or from some other source.

The fact that $\sigma_m - \sigma_p$ is larger for fluorine than for the other halogens, larger for hydroxy than for sulfhydryl, etc., is attributed to greater resonance electron-donating ability by first-row elements than by elements in the same family but further down the Periodic Table. First-row elements have their unshared electron pairs in orbitals with a large amount of $2p$ character. These are about the same size as the $2p$ orbitals in the aromatic ring. Since the first-row elements are also attached to the ring by relatively short bonds there is efficient overlap between the pi system of the ring and the orbital containing the unshared electron pair.

If the meta-substituent constant were a reflection of the polar effect only, it would be expected that $\sigma_{m\text{-F}} > \sigma_{m\text{-OH}} > \sigma_{m\text{-NH}_2} > \sigma_{m\text{-CH}_3}$ and that $\sigma_{m\text{-F}} > \sigma_{m\text{-Cl}} > \sigma_{m\text{-Br}} > \sigma_{m\text{-I}}$. The observed deviations from these sequences suggest that resonance effects are *significant* in the meta position even though they are not as large as they are in the para position. Thus perhaps the amino substituent donates electrons so strongly to the positions ortho and para to it that these positions exert an electron-donating polar effect on the meta position between them.

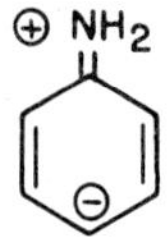

3-3b. σ^+, σ^-, and σ^n Constants. Of the deviations from the Hammett equation in the simple form that we have described, there are some so large and systematic that they came immediately to Hammett's attention. These were found in reactions in which there was at the reaction center (in the reactants, the products, or both) a group capable of strong resonance electron donation to the ring. In particular, it was the amino or dialkylamino group and the O^- group, which figure in such reactions as the acidity of phenols and anilinium ions,

$$ArOH \rightleftharpoons ArO^- + H^+$$

$$ArNH_3^+ \rightleftharpoons ArNH_2 + H^+$$

that were associated with the deviations. These deviations were not observed with meta-substituted compounds, which obeyed the Hammett equation satisfactorily, but with para-substituted compounds in which the para substituent was a group like nitro, cyano, or acetyl, capable of withdrawing electrons strongly by a resonance effect. The deviations were all in such a direction as to indicate that the presence of a strongly resonance electron-withdrawing group and a strongly resonance electron-donating group para to each other on an aromatic ring led to a species that was more stable than would be expected from the simple Hammett equation. This is explained in terms of the contribution of valence bond structures such as **1**, in which there may be said to be direct resonance interaction between the two groups. It is argued that structure **2** cannot contribute to explaining

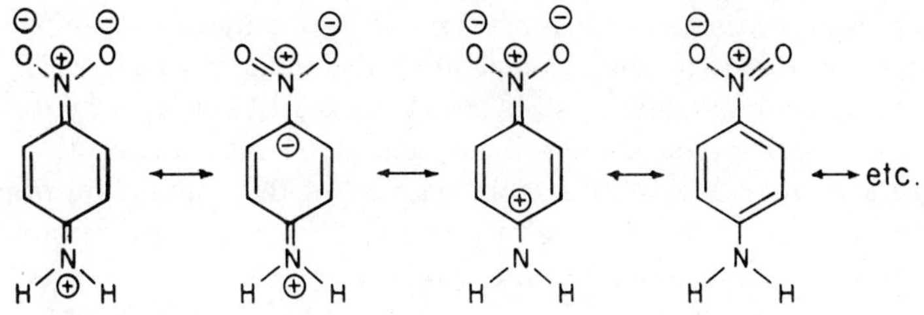

1 ⟷ **2** ⟷ **3** ⟷ **4** ⟷ etc.

the enhanced acidity of *p*-nitroanilinium ions since such structures may be written for any aniline, and structure **3** cannot contribute to an explanation since such structures may be written for any nitro compound and hence have already been taken into account when one uses a substituent constant for the *p*-nitro group that is based on the ionization constant of *p*-nitrobenzoic acid.

To cover cases of the type just described, Hammett assigned to various strongly resonance electron-withdrawing para substituents like nitro special substituent constants for use in reactions in which there is a group capable of strong resonance electron donation at the reaction center. Such substituent constants later became known as sigma minus (σ^-) values.

If substituent A is undergoing direct resonance interaction with substituent B, it is obvious that B is undergoing direct resonance interaction with A. Hence it is not surprising that deviations from the Hammett equation were later found in cases where there is a group capable of strong resonance electron withdrawal at the reaction center and a para substituent capable of strong resonance electron donation. For example, in the reaction

$$XC_6H_4{-}\underset{\displaystyle OH}{\overset{\displaystyle Ph}{C}}{-}Ph + H^+ \rightleftharpoons XC_6H_4{-}\underset{\displaystyle \oplus}{\overset{\displaystyle Ph}{C}}{-}Ph + H_2O$$

the equilibrium constant for the case in which X is the *p*-methoxy substituent is larger than one would expect from a Hammett equation correlation using a ρ calculated from data on meta substituents and para substituents that are relatively incapable of resonance electron donation and a σ for the *p*-methoxy group calculated from the ionization constant of *p*-methoxybenzoic acid. To cover deviations of this kind, a set of substituent constants for para substituents capable of resonance electron donation was devised for use in reactions in which the reaction center is capable of withdrawing electrons strongly from the ring by a resonance effect.[20–22] Such substituent constants are called sigma plus (σ^+) values.

The use of two sigmas for a given para substituent, a regular sigma and a sigma plus or a regular sigma and a sigma minus, may give the impression that there are only two kinds of reactions: reactions in which there is no direct resonance interaction at all between the substituent and the reaction center, and those in which there is an amount of resonance sufficient to change log K by a certain invariable fraction of the amount by which it would be changed in the absence of this direct resonance interaction. The truth appears to be nearer what one would have expected in advance; the fraction contributed to Δlog K by direct resonance interaction between the substituent and the reaction center may vary quite widely. Strong evidence for this point of view was provided by van Bekkum, Verkade, and Wepster,[2] who showed that the values of sigma that would be required for the exact correlation of the data for the ρ-nitro derivative in 34 selected reactions varied continuously from 0.77 to 1.38 instead of falling in two clusters, one around σ (0.78) and one around σ^- (1.24). Nevertheless, the use of dual sigmas, σ and σ^- or σ and σ^+, is often quite convenient.

It was also pointed out by van Bekkum, Verkade, and Wepster that the ordinary sigma values calculated from data on the ionization of benzoic acids are, in the case of resonance electron-donating para substituents, significantly "contaminated" by the effects of direct resonance interaction between the substituent and the carboxy group. By avoiding reactions in which direct resonance interaction between substituent and reaction center seemed plausible, they calculated a set of substituent constants known as normal substituent constants or sigma *n* (σ^n) values. Taft and co-workers calculated a similar set of contants, which they refer to as σ° values.[23]

In the early calculations of σ° and σ^n values the ionization constants of phenylacetic acids, β-phenylpropionic acids, etc., were assumed to be unaffected by direct resonance interaction of any substituents with the reaction center. Subsequently, however, Wepster and co-workers have pointed to the evidence that groups that are electron withdrawers by the resonance effect do give direct resonance interactions with the reaction center in such cases.[24] (Note that the point for the *p*-nitro substituent in Fig. 3-1 lies above the best straight line that could be drawn through the points for other meta and para substituents.) Like the methyl group, the carboxymethyl group can act as a resonance electron-donating group and the $CH_2CO_2^-$group must be an even better resonance electron donor. A recent set of σ^n values[25] was calculated by uşing as "primary substituent constants" the σ values for 10 meta substituents (including 3,5-dimethyl and 3,5-dinitro, which have the values -0.098 and 1.379) determined from the ionization constants of benzoic acids in water at 25° C. These were used to calculate ρ constants for reaction series in which the reaction center was attached to the aromatic ring via a saturated carbon atom. Values from this set have been combined with values from the older set based on reactions of derivatives of phenol, aniline, and thiophenol[2] in order to get the more broadly based set of σ^n values listed in Table 3-2. Note that each σ^n value is less negative or more positive than the corresponding σ_p value listed in Table 3-1. The differences are particularly large for the amino and dimethylamino substituents, for which direct resonance interaction with a *p*-carboxy group should be particularly important.

Also listed in Table 3-2 are σ^+ constants calculated largely from rates of solvolysis of *t*-cumyl chlorides ($ArCMe_2Cl$) using a ρ calculated from data on substituents for which direct resonnance interaction with the reaction center seemed implausible.[22,25] This reaction involves the rate-controlling formation of a tertiary carbonium ion that may be markedly stabilized by

Table 3-2. Values of σ^n and σ^+ for Para Substituents

Substituent	$\sigma^{n\ a}$	$\sigma^{+\ b}$
Me	−0.12	−0.32
Et	−0.12	−0.31
i-Pr	−0.14	−0.29
t-Bu	−0.14	−0.27
Ph	0.10	−0.21
NH_2	−0.27	−1.47
NMe_2	−0.22	−1.67
NHAc	0.14	−0.58
NHCHO	0.19	
OH	−0.16	−0.91
OMe	−0.13	−0.79
OPh	0.09	−0.5[c]
O^-	−0.50[d]	−2.3[d]
F	0.15	−0.08
SMe	0.13	−0.62
Cl	0.26	0.11
Br	0.28	0.14
I	0.30	0.13

[a] Values of σ^n based on reactions of derivatives of phenol, aniline, and thiophenol in those cases[2] where ρ was calculated from at least four data and in which the correlation coefficient was greater than 0.990 were averaged with the more recently published σ^n values.[25]
[b] Ref. 25.
[c] Ref. 22.
[d] Substituent constants for electrically charged groups are relatively unreliable.

resonance electron-donating substituents, as suggested by structure **5** for the *p*-methoxy case. Since the $-CMe_2^+$ group is a much stronger electron withdrawer than the carboxy group, all the σ^+ values in Table 3-2 are more negative or less positive than the corresponding σ_p constants based on benzoic acid acidities listed in Table 3-1. Although σ^n and σ^+ constants may also be calculated for substituents that are incapable of resonance electron donation, they tend not to differ significantly from the constants given in Table 3-1 and will not be listed here.

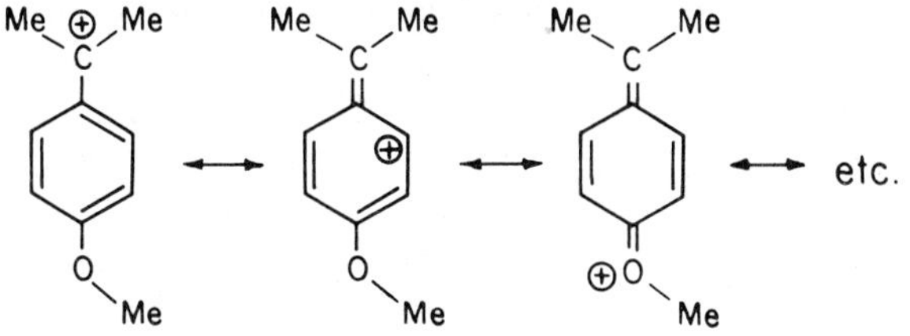

5

Table 3-3 contains σ^- values, largely calculated from the acidities of the appropriate phenols.[9] As in the case of σ^+ values, σ^- values for para substituents are essentially the same as σ_p values when there is no direct resonance interaction between the substituent and the reaction center. Comparisons between Tables 3-1 and 3-3 show that such resonance is probably not significant for the p-CF_3 and p-SF_5 derivatives of the phenoxide ion, even though such no-bonded structures as **6** may be written.[26]

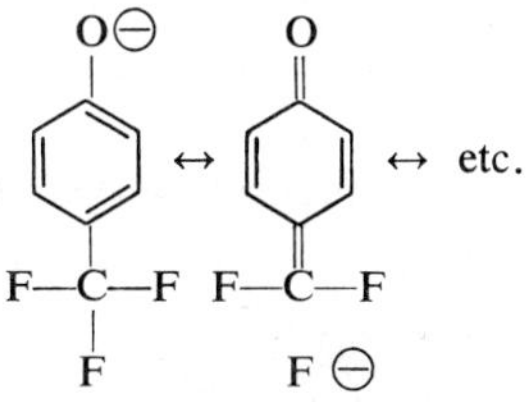

6

Table 3-3. Values of σ^- for Para Substituents[a]

Substituent	σ^-	Substituent	σ^-
C≡CPh	0.36	N_2^+	3.2[d,e]
CHO	0.98[b]	NO_2	1.23[b]
Ac	0.82[b]	$SiMe_3$	0.17
CO_2H	0.73	SOMe	0.73
CO_2Me	0.74[b]	SO_2Me	1.05[b]
CO_2Et	0.74[b]	SO_2CF_3	1.36
$CONH_2$	0.62[b]	SO_2NH_2	0.94
CN	0.99[b]	SO_2F	1.32
CF_3	0.56[c]	SF_5	0.70
N=NPh	0.69	SMe_2^+	1.16[d,f]

[a] From Ref. 9 unless otherwise noted.
[b] Ref. 25.
[c] Ref. 26.
[d] Substituent constants for electrically charged groups are relatively unreliable.
[e] E. S. Lewis and M. D. Johnson, *J. Amer. Chem. Soc.*, **81**, 2070 (1959).
[f] F. G. Bordwell and P. J. Boutan, *J. Amer. Chem. Soc.*, **78**, 87 (1956); **79**, 717 (1957).

3-3c. Resolution of Meta- and Para-Substituent Constants into Polar and Resonance Components. We have already discussed the total electronic effects of certain substituents qualitatively in terms of a superposition of polar and resonance effects. It is natural to consider a quantitative treatment. For example, if we assume that any group X attached to an aromatic ring has a polar substituent constant P_X and a resonance substituent constant R_X, we might use equations like 3-40 and 3-41 to try to correlate equilibrium (or rate) constants. A set of

$$\log \frac{K_{p\text{-X}}}{K_0} = \rho_P{}^p P_X + \rho_R{}^p R_X \tag{3-40}$$

$$\log \frac{K_{m\text{-X}}}{K_0} = \rho_P{}^m P_X + \rho_R{}^m R_X \tag{3-41}$$

reaction series for which the simple Hammett equation (3-5) is obeyed perfectly would be a set in which the *relative* magnitudes of the four ρ constants in Eqs. 3-40 and 3-41 were unchanged. When this is true we could call $\rho_P{}^p$ the ρ constant for any given reaction series and the substituent constants would then have the meanings given in Eqs. 3-42 and 3-43. Correlation of the data obtained in such a set of reaction series in terms of

$$\sigma_{p\text{-X}} = P_X + \frac{\rho_R{}^p}{\rho_P{}^p} R_X \tag{3-42}$$

$$\sigma_{m\text{-X}} = \frac{\rho_P{}^m}{\rho_P{}^p} P_X + \frac{\rho_R{}^m}{\rho_P{}^p} R_X \tag{3-43}$$

Eqs. 3-40 and 3-41 is equivalent to correlating the Hammett meta- and para-substituent constants in terms of Eqs. 3-42 and 3-43, in which the three ratios $\rho_R{}^p/\rho_P{}^p$, $\rho_P{}^m/\rho_P{}^p$, and $\rho_R{}^m/\rho_P{}^p$ are constants. Such a treatment is simplified by the permissibility of certain arbitrary assignments. For example, if we have a set of values for the three ratios and the various P_X and R_X values that fits the observed σ values with a certain precision, we can get another set that fits with exactly the same precision by multiplying each R_X value by the constant c and dividing $\rho_R{}^p/\rho_P{}^p$ and $\rho_R{}^m/\rho_P{}^p$ by the same constant. Therefore we can multiply each R_X by the constant $\rho_R{}^p/\rho_P{}^p$ and divide the two ratios by this constant to get the simplified Eqs. 3-44 and 3-45. Equations 3-44 and 3-45 for n different substituents

$$\sigma_{p\text{-X}} = P_X + R_X \tag{3-44}$$

$$\sigma_{m\text{-X}} = \frac{\rho_P{}^m}{\rho_P{}^p} P_X + \frac{\rho_R{}^m}{\rho_R{}^p} R_X \tag{3-45}$$

comprise a set of $2n$ equations with $2n + 2$ unknowns (n P's, n R's, $\rho_P{}^m/\rho_P{}^p$, and $\rho_R{}^m/\rho_R{}^p$). Hence there are an infinite number of sets of solutions, and more information, assumptions, or assignments are needed to get a unique set. Furthermore, even if we have a set of values of P and R that fits the data, how do we know that the ones we call P are measures of polar effects and the R's measures of resonance effects? (Exchanging every P in the original Eqs. 3-40 and 3-41 for an R and every R for a P leaves the overall equations unchanged.)

Taft and co-workers resolved Hammett substituent constants into polar and resonance components by assuming that the polar effects produced by substituents on aromatic rings, as measured by the P constant in Eqs. 3-40–3-45 were proportional to the Taft substituent constants (σ^*) of the same substituents.[23,27–32] These σ^* constants fairly adequately describe the effects of substituents on log K and log k for polar reactions in many cases in which the substituent is attached to saturated carbon and in which significant steric effects seem unlikely (Section 3-6). Such substituent constants should be fairly clean measures of the polar character of substituents. In addition it was assumed that polar effects were transmitted from the meta and para positions with equal efficiency. That is, $\rho_P{}^m/\rho_P{}^p$ was assumed to be 1.0 (after trying other values and learning that values outside the range 0.8–1.2 gave significantly poorer results). This assumption, the use of Taft's symbols σ_R and α for R and $\rho_R{}^m/\rho_R{}^p$, respectively, and the assumption that P is proportional to σ^* (i.e., $P = c\sigma^*$) transform Eqs. 3-44 and 3-45 to 3-46 and 3-47. Elimination of σ_R between these two equations

$$\sigma_p = c\sigma^* + \sigma_R \tag{3-46}$$

$$\sigma_m = c\sigma^* + \alpha\sigma_R \tag{3-47}$$

gives Eq. 3-48. A plot of the right-hand side for 22 pairs of σ_m and σ_p

$$c\sigma^* = \frac{\sigma_m - \alpha\sigma_p}{1 - \alpha} \tag{3-48}$$

values versus σ^* gave a moderately good straight line of slope 0.45 when an α value of $\frac{1}{3}$ was used.[28] Hence 0.45 was used as the value of c to put the σ^* constants on the appropriate scale for aromatic polar effects. The product $c\sigma^*$ or $0.45\sigma^*$ was denoted the inductive substituent constant σ_I.

Substitution of the definition of σ_I into Eqs. 3-46 and 3-47 and the resulting relationships into the simple Hammett equation gives Eqs. 3-49 and 3-50. Application of these equations to rate and equilibrium data that did

$$\log\left(\frac{K}{K_0}\right)_{\text{para}} = \rho(\sigma_I + \sigma_R) \tag{3-49}$$

$$\log\left(\frac{K}{K_0}\right)_{\text{meta}} = \rho(\sigma_I + \alpha\sigma_R) \quad (3\text{-}50)$$

not fit the simple Hammett equation very well showed that greatly improved fits could be obtained if α was permitted to vary with the nature of the reaction. Most of the values of α were in the range 0.1–0.6.[28,29,33]

Exner reached a significantly different conclusion concerning the contribution of resonance effects to Hammett substituent constants.[9,17] He plotted values of $\log (K/K_0)_{\text{para}}$ versus $\log (K/K_0)_{\text{meta}}$ for a number of substituted benzoic acids including a large number with substituents of the type CH_2Y. If the points for electrically charged substituents, for substituents like methoxy and halogens, in which the atom attached to the ring has unshared electrons, and the point for $B(OH)_2$, in which the atom attached to the ring has an incomplete octet, are excluded, the remaining points fall near a straight line. The line (of slope 1.14) based on data obtained in 80% methyl Cellosolve gave an excellent fit to the data except for the points for the carboxy and benzoyl substituents, whose $\log (K/K_0)_{\text{para}}$ values were slightly too large to fit. The line (of slope 1.15) based on data obtained in 50% ethanol was almost as good. Other lines of similar slope, based on other reactions, were also in reasonable agreement with the data. The large deviations by the excluded points that referred to substituents with unshared electrons on the atom attached to the ring were in the proper direction to be explained by the electron-donating resonance effect of such substituents, which is felt much more strongly in the para than in the meta position. To the extent to which the other points are fit perfectly by a straight line of slope λ, we may say that $\sigma_{p\text{-X}}$ is equal to $\lambda\sigma_{m\text{-X}}$, or, on the basis of Eqs. 3-44 and 3-45 we may write Eq. 3-51. Since the ratios $\rho_P{}^m/\rho_P{}^p$ and $\rho_R{}^m/\rho_R{}^p$ are taken to be constant

$$P_X + R_X = \lambda\left(\frac{\rho_P{}^m}{\rho_P{}^p} P_X + \frac{\rho_R{}^m}{\rho_R{}^p} R_X\right) \quad (3\text{-}51)$$

for a given reaction series, it follows that Eq. 3-51 can be fit perfectly only if R_X for all the substituents in question is zero or if R_X is proportional to P_X. Although it is perfectly plausible that the resonance effects of a series of substituents of the type CH_2Y might be proportional to the polar effects and that there might be a rough correlation between polar and resonance effects for a random assortment of substituents, it is very hard to understand why there should be an excellent correlation between P_X and R_X for the wide variety of substituents in Exner's correlation, which included CF_3, CN, NO_2, SO_2Y, and other groups. Hence it would seem preferable to assume that resonance effects are zero for the type of substituents under consideration if the straight line plots were perfect. However, since we can never be sure that any relationship involving experimental values is perfect, it is more reasonable to assume merely that these resonance effects are small. Exner concluded that electron-withdrawing resonance contributions to σ_m and σ_p are probably negligible except perhaps in the case of such carbonyl substituents as benzoyl, carboxy, etc. If this were true treatments of aromatic rate and equilibrium data that assume equal effectiveness of transmittal of polar effects from the meta and para positions should yield resonance substituent constants that are rather precisely correlated with the corresponding polar substituent constants. They do not. Hence, although resonance does not contribute as much to meta- and para-substituent effects with known electrically neutral electron-withdrawing substituents (in the absence of conjugation with the reaction center) as it does with many resonance electron-donating substituents, it may well contribute significantly. The fact that a 4-nitro substituent is an appreciably weaker electron-withdrawing substituent when there are two methyl groups ortho to it [24] is direct evidence for such a contribution of a resonance effect. The *o*-methyl groups keep the nitro group from lying in the plane of the ring and thereby minimize the contribution of structures like **7** to the total structure of the compound.

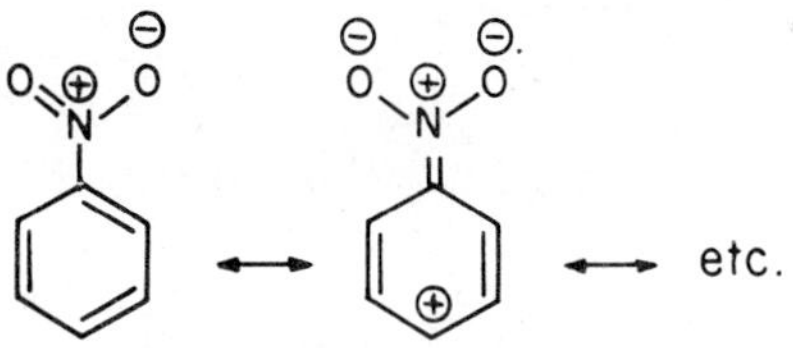

etc.

7

3-3d. The Yukawa-Tsuno Equations. As already pointed out, the use of the Hammett equation with σ^n, σ, σ^+, or σ^- values has the disadvantage of giving discontinuities. For example, one may use σ^- to allow for a certain extent of direct resonance interaction of an electron-withdrawing substituent with an electron-donating reaction center, or may use σ and allow for none. There is no "in between." Yukawa and Tsuno devised equations to provide a continuous variation in the amount of resonance interaction between substituent and reaction center.[25,34–38] These equations use the ordinary substituent constant for meta substituents. In one form they permit the effective para-substituent constant to vary between σ_p and σ^+ or σ^-, and in another form they permit it to vary between the Taft σ° and σ^+ or σ^-. Hoefnagel and Wepster have rewritten the latter form in terms of the similar σ^n constant and have calculated a set of optimum substituent constants.[25] In this form we have Eq. 3-52 for electron-withdrawing reaction centers and 3-53 for electron-donating reaction centers. The extent of

$$\log\left(\frac{K}{K_0}\right)_{\text{para}} = \rho(\sigma^n + r^+ \Delta\sigma_R^+) \quad (3\text{-}52)$$

$$\log\left(\frac{K}{K_0}\right)_{\text{para}} = \rho(\sigma^n + r^- \Delta\sigma_R^-) \quad (3\text{-}53)$$

resonance interaction between substituents and reaction center is measured by r^+ or r^-. Since $\Delta\sigma_R^+$ is defined as $\sigma^+ - \sigma^n$ and $\Delta\sigma_R^-$ as $\sigma^- - \sigma^n$, when r^+ or r^- is 1.0 the right-hand side of Eq. 3-52 or 3-53 becomes $\rho\sigma^+$ or $\rho\sigma^-$. That is, direct resonance interaction between substituent and reaction center comprises as large a fraction of the total substituent effect as it does in the reaction used to define σ^+ or σ^-. When r^+ or r^- is zero the right-hand side of Eq. 3-52 or 3-53 becomes $\rho\sigma^n$; that is, there is no direct resonance interaction between substituents and the reaction center. A set of $\Delta\sigma_R^+$ and $\Delta\sigma_R^-$ values is listed in Table 3-4.

Table 3-4. Values of $\Delta\sigma_R^+$ and $\Delta\sigma_R^-$

Substituent	$\Delta\sigma_R^+$ [a]	Substituent	$\Delta\sigma_R^-$
Me	−0.20	CHO	0.55[b]
Et	−0.19	Ac	0.32[b]
i-Pr	−0.15	CO_2Me	0.28[b]
t-Bu	−0.13	$CONH_2$	0.31[b]
Ph	−0.31	CN	0.29[b]
NH_2	−1.20	CF_3	0.02[c]
NMe_2	−1.45	N=NPh	0.34[c]
NHAc	−0.72	N_2^+	1.3[c,d]
OH	−0.75	NO_2	0.45[b]
OMe	−0.66	$SiMe_3$	0.24[c]
OPh	−0.59	SOMe	0.24[c]
O^-	−1.8	SO_2Me	0.32[b]
SMe	−0.75	SO_2CF_3	0.43[c]
F	−0.23	SO_2NH_2	0.37[c]
Cl	−0.15	SF_5	0.02[c]
Br	−0.14	SMe_2^+	0.26[c]
I	−0.17		

[a] Calculated from the σ^n and σ^+ values in Table 3-2.

[b] Ref. 25.

[c] Calculated from the σ^- in Table 3-3 and σ^n taken as σ_p from Table 3-1.

[d] Substituent constants for electrically charged groups are relatively unreliable.

3-3e. σ_R(BA), $\sigma_R{}^\circ$, $\sigma_R{}^+$, and $\sigma_R{}^-$ Constants. Ehrenson and co-workers have separated aromatic meta- and para-substituent effects into polar and resonance components in an approach involving fewer assumptions than those we have described so far.[5,31,32] First it was assumed that Eqs. 3-40 and 3-41 are applicable; that is, that the overall substituent effect may be expressed as the sum of a purely polar term and a purely resonance terms. Then it was assumed that the polar term may be expressed as the product of the σ_I value for the given substituent (see Section 3-7) and a ρ parameter that depends only on the nature of the reaction and the position of the substituent. It had been hoped that each substituent could then be assigned a unique resonance substituent parameter just as it was given a unique polar substituent parameter. Such treatments of a broad range of data on aromatic substituent effects have been carried out and are useful for many purposes.[39] However, all those proposed to date give large and systematic deviations from the experimental data in certain types of cases and therefore inspire attempts at improved correlations. To obtain what they regarded as satisfactory agreement with the data, Ehrenson and co-workers divided all reactions into four categories (and toyed with the possibility of a fifth category) and used a different set of resonance substituent constants for each category. One category, for which the resonance substituent constants are called σ_R(BA) is the benzoic acid category, consisting of the ionization and other reactions of benzoic acids, hydrolysis of benzoate esters, and other reactions with similar possibilities for resonance interactions with substituents. The $\sigma_R{}^\circ$ category of reactions consists largely of processes in which there is a methylene group between the aromatic ring and the reaction center. The $\sigma_R{}^+$ reactions are those in which there is a strong resonance electron-withdrawing reaction center conjugated with the aromatic ring (the same reactions as those that σ^+ values were designed for). Last is the $\sigma_R{}^-$ category—reactions in which there is a strong resonance electron-donating reaction center conjugated with the aromatic ring. These constants are sometimes called $\sigma_R{}^-$(A) constants since they are based on the acidity of anilinium ions. Thus the relations used were Eqs. 3-54 and 3-55, where the σ_R in the para equation may be σ_R(BA), $\sigma_R{}^\circ$, $\sigma_R{}^+$, or $\sigma_R{}^-$, depend-

$$\log\left(\frac{K}{K_0}\right)_{\text{para}} = \rho_P{}^p\,\sigma_I + \rho_R{}^p\,\sigma_R \tag{3-54}$$

$$\log\left(\frac{K}{K_0}\right)_{\text{meta}} = \rho_P{}^m\,\sigma_I + \rho_R{}^m\,\sigma_R \tag{3-55}$$

ing on the nature of the reaction. For meta-substituted compounds, which were studied in less detail, the resonance effect is so much less important that any set of σ_R values may be used, but the $\sigma_R{}^\circ$ values are preferred.

To select from among the infinite number of equivalent sets of parameters, ρ_P^p and ρ_R^p were each assigned the value 1.0 for the ionization of benzoic acids in water at 25° C. Then a least squares optimization of data in the σ_R(BA) category gave the set of σ_R(BA) values. The scales for the other sets of σ_R values were established by assigning certain σ_R's the same values they have in the σ_R(BA) set and by various other assignments. The resulting values are listed in Table 3-5.

The numerical values of σ_R on the various scales illustrate the difficulties that are involved in trying to fit all the data with any one set of σ_R values. To give but one of a number of possible examples, note that when there is no possibility of conjugation with the reaction center (the $\sigma_R°$ scale) fluorine is as strong a resonance electron-donating substituent as phenoxy and

Table 3-5. Values of σ_R(BA), $\sigma_R°$, σ_R^+, and σ_R^-[32]

Substituent	σ_R(BA)	$\sigma_R°$	σ_R^+	σ_R^-
NMe_2	−0.83	−0.52	−1.75	−0.34
NH_2	−0.82	−0.48	−1.61	−0.48
NHAc	−0.36	−0.25	−0.86	
OMe	−0.61	−0.45	−1.02	−0.45
OPh	−0.58	−0.34	−0.87	
SMe	−0.32	−0.20	~−0.7[a]	−0.14
Me	−0.11	−0.11	−0.25	−0.11
Ph	−0.11	−0.11	−0.30	0.04
F	−0.45	−0.34	−0.57	−0.45
Cl	−0.23	−0.23	−0.36	−0.23
Br	−0.19	−0.19	−0.30	−0.19
I	−0.16	−0.16	−0.25	−0.11
H	0.00	0.00	0.00	0.00
SOMe	0.00	0.00	0.00	
SCF_3	0.04	0.04	0.04	0.14
$SiMe_3$	0.06	0.06	0.06	0.14
SF_5	0.06	0.06	0.06	0.20
$SOCF_3$	0.08	0.08	0.08	
CF_3	0.08	0.08	0.08	0.17
SO_2Me	0.12	0.12	0.12	0.38
CN	0.13	0.13	0.13	0.33
CO_2R	0.14	0.14	0.14	0.34
NO_2	0.15	0.15	0.15	0.46
Ac	0.16	0.16	0.16	0.47

[a] Values ranging from −0.48 to −0.98 were needed to fit the data used.

stronger than acetamino; however, when conjugated with an electron-withdrawing reaction center (the σ_R^+ scale) fluorine is a much weaker electron donor than phenoxy or acetamino.

Equations 3-54 and 3-55 and the set of parameters in Table 3-5 probably cover more data on more aromatic substituent effects more reliably than any other treatment that has been carried out. Nevertheless, this treatment is unsatisfying in several ways. Its empirical success owes much to the large number of parameters used. It is not always clear in advance which set of σ_R constants should be used. The practice of using the one that works best can be a way of hiding experimental error. One feels that if σ_R cannot be a constant it should at least change continuously rather than jumping from σ_R° to σ_R^+ or σ_R^-, etc.

Blagdon has suggested an interesting alternative treatment in which polar effects are handled in the usual way but resonance effects follow a hyperbolic rather than a linear function.[40] Each substituent has two resonance substituent constants, one that may be thought to be based on the extent of resonance electron donation or withdrawal that the substituent enters into in the absence of interaction with other groups, and a second that may be thought of as a measure of the resonance polarizability of the group—the ease with which its resonance effect changes upon interaction with another group. This treatment has the virtue of being applied in the same way to reactions in which there is direct resonance interaction between the substituents and the reaction center as to reactions in which there is not. Flachskam has experimented with the use of such hyperbolic functions and also with the use of parabolic functions in the calculation of energies of resonance interaction in an approach analogous to that followed in Eqs. 3-7 through 3-13.[41]

3-3f. Variable Substituent Constants. Reactions of compounds with electrically charged substituents do not in general give as good agreement with the Hammett equation as the reactions of compounds with dipolar substituents. The energy of interaction between two charges is proportional to $1/r$, whereas the energy of interaction of a charge and a dipole is proportional to $1/r^2$ (cf. Eqs. 2-1 and 2-3). Hence the coulombic interaction of a charged reaction center with a dipolar substituent would be decreased more as the reaction center was moved further away (as on going from benzoic acids to cinnamic acids) than the coulombic interaction with a charged substituent would be. Inasmuch as most of the data used in correlations deal with dipolar substituents, ρ constants change from reaction to reaction largely in such a manner as to give agreement with data on dipolar-substituted compounds, at the expense of disagreement by the smaller group of data on compounds with charged substituents.

Another factor that should cause data on compounds with electrically charged substituents to deviate from the Hammett equation is the strong interaction that would exist between such a substituent and the solvent in many cases. The O^- group in a hydroxylic solvent, where it would be strongly hydrogen bonded to several surrounding solvent molecules, would be expected to be a considerably different substituent from what it would be in a nonhydroxylic solvent where such hydrogen bonding could not occur.

As an example of the behavior of charged substituents let us compare the positively charged trimethylammonio substituent with the electrically neutral nitro substituent. The substituent effect of the *m*-trimethylammonio group on the pK_a of benzoic acid in aqueous solution at 25° C is only 15% larger than that of the *m*-nitro group (Table 3-1). The effect of *m*-trimethylammonio substitution on the pK_a of β-phenylpropionic acid in 50% or 75% ethanol—water is about 2.7 times as large as that of *m*-nitro.[24] There may be a solvent effect component in this change in relative substituent effects but it probably arises largely from differences in distances between substituent and reaction center.

Solvent effects should also be capable of causing deviations from the Hammett equation in the cases of electrically neutral substituents, especially when there is a particularly strong interaction, such as a hydrogen bond, between the solvent and substituent.[42] Another way of looking at the matter is to consider the substituent to include the surrounding solvation shell. To implement this approach one could then let σ constants be solvent dependent, at least in those cases where the solvent-substituent interactions are large. This would complicate the applications of the Hammett equation and is rarely done.

3-3g. Relative Magnitudes of Meta- and Para-Substituent Constants. The first derivation given in Section 3-2 led to the conclusion (Eq. 3-24) that for reactions of the type

$$\text{ArB} \rightleftharpoons \text{ArC} \tag{3-56}$$

where Ar is a meta-substituted phenyl group, the reaction constant, which will be denoted ρ_m for clarity, may be expressed

$$\rho_m = \tau_m(\sigma_{m\text{-B}} - \sigma_{m\text{-C}}) \tag{3-57}$$

The same type of derivation could be used for para-substituted compounds to give

$$\rho_p = \tau_p(\sigma_{p\text{-B}} - \sigma_{p\text{-C}}) \tag{3-58}$$

In the Hammett equation only one reaction constant is used, and it is applied to data on both meta- and para-substituted compounds. Thus, in terms of the derivation that leads to Eqs. 3-57 and 3-58, the Hammett equation sets ρ_m and ρ_p equal to each other. From this it follows that

$$\tau_m(\sigma_{m\text{-B}} - \sigma_{m\text{-C}}) = \tau_p(\sigma_{p\text{-B}} - \sigma_{p\text{-C}}) \tag{3-59}$$

If this relationship is to be general, it must apply to the case where B is hydrogen and hence $\sigma_{m\text{-B}}$ and $\sigma_{p\text{-B}}$ are zero. When this is so, Eq. 3-59 may be rearranged to give

$$\sigma_{p\text{-C}} = \frac{\tau_m}{\tau_p} \sigma_{m\text{-C}} \tag{3-60}$$

Equations 3-59 and 3-60 show, as McDaniel pointed out,[43] that if the Hammett equation is generally applicable to equations of the type of Eq. 3-56, then a plot of σ_p versus σ_m for all possible substituents will give a straight line of slope τ_m/τ_p passing through the origin. In order to avoid the complications of resonance interaction between substituent groups and a reaction center, $\sigma_p{}^n$ values (or no values at all) have been used instead of benzoic acid-based σ_p values for substituents capable of resonance-electron donation. The plot in Fig. 3-2 of data from Tables 3-1 and 3-2 shows a clear but rough correlation between $\sigma_p{}^n$ and σ_m. As noted in Section 3-3c in regard to a similar plot (using the equivalent of σ_p instead of $\sigma_p{}^n$), when attention is restricted to electrically neutral substituents in which the atom attached to the ring has a filled octet and no unshared electrons (i.e., to all the unlabeled open circles plus all other unlabeled points for which σ_m is larger than 0.4) there is a much more nearly linear correlation. Even better is the fit of the points for electrically neutral substituents that are attached to the ring by a carbon atom (open circles) to the line shown in the figure. Hence in the absence of direct resonance between substituents and reaction center, we would expect reactions involving interconversion of electrically neutral carbon groups (e.g., $ArCHO \rightleftharpoons ArCH_2OH$) to give good agreement with the Hammett equation using the same ρ for meta and para substituents. Since the points for sulfur substituents approximate another straight line and the nitrogen points still another one, there are similar expectations for such reaction series as $ArSMe \rightleftharpoons ArSO_2Me$ and $ArNH_2 \rightleftharpoons ArNHCHO$. (The fact that the sulfur and nitrogen lines miss the origin appreciably means that the equilibration could not involve hydrogen as a reaction center, as in a reaction like $ArH \rightleftharpoons ArSO_3H$.)

The deviations by the points for charged substituents may arise from the greater uncertainty with which these substituent constants are known. If so, the general success of the procedure of using the same ρ for both meta and para substituents becomes more understandable. However, the

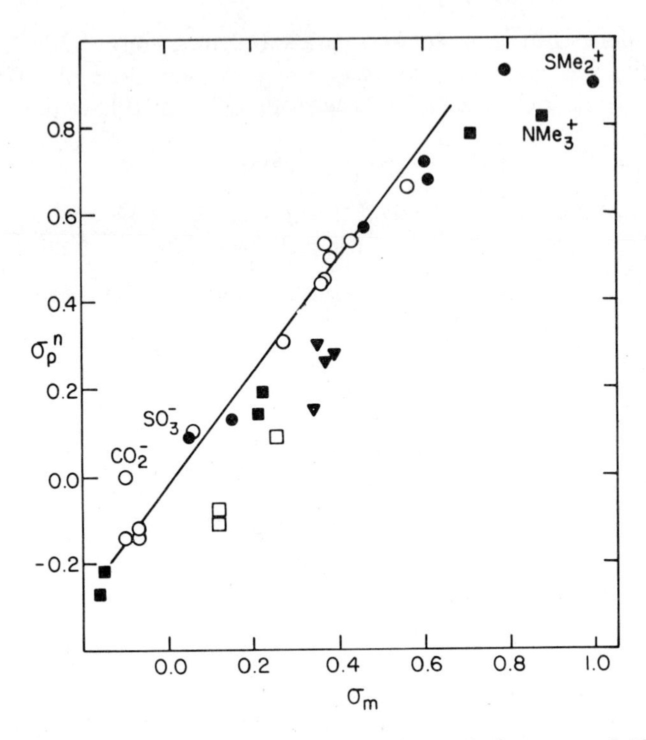

Figure 3-2. Plot of σ_p (or σ^n) versus σ_m. ○ Carbon substituents. ● Sulfur substituents. □ Oxygen substituents. ■ Nitrogen substituents. ▼ Halogen substituents. All electrically charged groups are labeled.

reliability of some of the experimental work and the fact that both points for negatively charged substituents lie above and both points for positively charged substituents lie below the best line through the points for the element in question suggest that the deviations are real. If this is so it may give rise to cases in which correlations are significantly improved by using a separate ρ_p and ρ_m. To investigate this point Jaffé used separate ρ_m and ρ_p values in correlating the data on 336 reaction series.[44] He reported that the number of reactions in which ρ_m and ρ_p were significantly different was larger than would be expected purely by chance. However, the improvement in the precision of the correlation brought about by using two ρ's instead of one was not very great. Any detailed interpretation of the nature of the reactions in which two ρ's are better than one would be complicated by the fact that the substituent constants used are based almost entirely on reactions in which an electrically neutral group is changed to a charged one or vice versa.

The tendency of the points for substituents attached by a given element to fall near a straight line provides a reason why reactions in which there is no change in the atom by which the reaction center is attached to the ring are not usually correlated much better by a dual ρ treatment. It was such reactions—aromatic *side-chain* reactions—for which the Hammett equation was devised. Reaction series in which the two reaction centers being equilibrated are attached to the ring by different elements are rare (Jaffé's set of reactions included none), but the correlation of the rather scanty data on the reaction $ArH \rightleftharpoons ArBr$[45] would be improved by use of a separate ρ_m and ρ_p.

3-4. ORTHO-SUBSTITUENT EFFECTS

In his original work on the $\rho\sigma$ relationship, Hammett noted that although it was rather generally applicable to the reactions of meta- and para-substituted benzene derivatives, it often gave poor results when applied to the reactions of ortho-substituted compounds. For example, both *o*-hydroxy and *o*-methyl substituents increase the acidity of benzoic acid although both groups would be expected to be electron donors. The inapplicability of the $\rho\sigma$ relationship to ortho-substituted compounds is believed to be usually the result of steric interactions between the substituent and reaction center. The *o*-methyl substituent in *o*-toluic acid prevents the carboxy group from being coplanar with the benzene ring. This diminishes direct resonance interaction between the carboxy group and the aromatic ring, an interaction in which the ring donates electrons to the carboxy group. (This follows from the fact that the carboxy group attached to an aromatic ring acts as an electron withdrawer by a resonance effect.) In some cases direct bonding interactions of some sort between substituent and reaction center appear to be significant. Stabilization of the salicylate anion by internal hydrogen bonding is probably an important factor in causing salicylic acid to be stronger than benzoic acid.

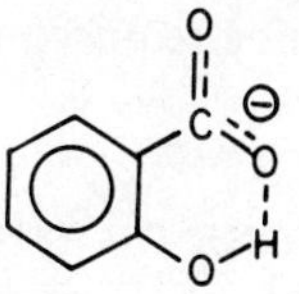

Direct resonance interactions between substituent and reaction center are often possible in the case of ortho-substituted compounds just as they are in the case of the analogous para-substituted compounds. Therefore there are often significant contributions of bonding, resonance, or steric terms to G_{ABN} in Eq. 3-10. Even if these terms are negligible so that

$$G_{ABN} = G_{polar}^{ABN}$$

there are still reasons why the $\rho\sigma$ relationship should not work as well for ortho- as for meta- and para-substituted aromatic compounds. The angular orientation of a dipole (cf. Eqs. 2-3 and 2-4) in an ortho substituent with respect to a charge in the reaction center would vary more with the nature of the reaction center than it would in the case of a meta or para substituent. Also, the effective dielectric constant of the medium separating the dipole and charge would vary more in the case of an ortho substituent. Even when there is no direct contact between the ortho substituent and the reaction center each may interfere with the solvation of the other.

Taft used the assumption that steric effects are the same in acidic ester hydrolysis as in basic ester hydrolysis (cf. Section 3-6) to separate ortho-substituent effects on ester hydrolysis into steric and electronic components.[21,46] The resulting steric substituent constants are certainly qualitatively plausible; they have given quantitative correlations of steric effects in a few cases but not in most of the reactions of ortho-substituted aromatic compounds. The resulting electronic substituent constants tend to run parallel to the corresponding para-substituent constants. They too have been found to be of only limited utility in correlating ortho-substituent effects. The *o*-hydrogen substituent, that is, the unsubstituted compound, often deviates from attempted correlations, and, in fact, does not fit the correlation used in defining the substituent constant (in which *o*-methyl was used as the reference substituent).

Charton treated a large number of ortho-substituent effects in terms of an equation analogous to Eq. 3-54, in which there are separate terms for polar and resonance effects.[41,48] A number of the correlations seem satisfactory but in most cases the number of data does not exceed the number of parameters by as large an amount as would be desired, and in some cases steric factors were accidentally kept fairly constant. In most cases the point for hydrogen did not fit the correlation, or to be more precise, the substituent "constant" for hydrogen was not required to be constant.

3-5. STRUCTURAL EXTENSIONS OF THE HAMMETT EQUATION

Although the Hammett equation was initially applied very largely to side-chain reactions of benzene derivatives with one meta or para substituent, a few other types of compounds were included. Subsequently more and more new types of compounds were added. One structural extension of the simplest form of the Hammett equation consists of correlations of data on polysubstituted aromatic compounds. In the simplest cases of this type, one substituent is held constant and a second substituent varied. Thus Roberts and Yancey found that the pK_a values for 11 4- and 5-substituted 2-methylbenzoic acids in 50% aqueous ethanol at 25° C gave a satisfactory Hammett equation correlation with a ρ of 1.67.[49] This value may or may not be significantly larger than the ρ for monosubstituted benzoic acids in the same solvent, for which a value of 1.46 was reported by one group of workers[50] but for which a value of 1.60 may be calculated[1] from the data of another group.[51]

A fairly common procedure in treating data on compounds with several meta and para substituents is simply to add their substituent constants, as in Eq. 3-61. To include various 3-nitro-5-X-benzoic acids, for example,

$$\log \frac{K}{K_0} = \rho \sum_i \sigma_i \tag{3-61}$$

with monosubstituted benzoic acids in a correlation of log K values using Eq. 3-61 is equivalent to assuming that ρ for 3-nitro-5-X-benzoic acids is the same as for monosubstituted benzoic acids. It happens that this assumption gives a pretty good fit to the data but not as good as when 3-nitro-5-X-benzoic acids are permitted to have a ρ that is different from that for monosubstituted benzoic acids. However, if we are trying to devise a correlation that will fit all 3,5-disubstituted benzoic acids as well as all 3- and 4-monosubstituted benzoic acids, we cannot let the ρ for all 3-nitro-5-X-benzoic acids (including the ones where X is hydrogen and methoxy) and the ρ for all 3-methoxy-5-X-benzoic acids (including the ones where X is hydrogen and nitro) have no relation to each other or to the ρ for monosubstituted benzoic acids. The use of the Hammett equation in correlating data on polysubstituted compounds is a subdivision of the use of linear free energy relationships in correlating data when there are multiple structural variations (including possibly variations in solvent, coreactant, etc., as well as substituent). Miller treated this problem in general terms.[52] Application of his treatment to the acidities of 3-X-5-Y-benzoic acids gives Eq. 3-62, in which K_{XY} is the ionization constant of the X,Y-sub-

$$\log \frac{K_{XY}}{K_{00}} = \rho(\sigma_X + \sigma_Y) + \rho'\sigma_X\sigma_Y \tag{3-62}$$

stituted compound and K_{00} is that of the completely unsubstituted compound, in which both X and Y are hydrogen. When one substituent, say Y, is always hydrogen, this relationship reduces to Eq. 3-63, showing that ρ

in Eq. 3-62 is the reaction constant for monosubstituted benzoic acids.

$$\log \frac{K_{X0}}{K_{00}} = \rho\sigma_X \tag{3-63}$$

When Y is held constant, but is not hydrogen, Eq. 3-62 may be transformed to Eq. 3-64 inasmuch as log (K_{0Y}/K_{00}) is equal to $\rho\sigma_Y$. Equation 3-64

$$\log \frac{K_{XY}}{K_{0Y}} = (\rho + \rho'\sigma_Y)\sigma_X \tag{3-64}$$

shows that the reaction constant for the acidity of 3-X-5-Y-benzoic acids, where X is varied and Y is constant, varies linearly with σ_Y. The fact that Eq. 3-61 is a good approximation for 3-X-5-Y-benzoic acids shows that $\rho'\sigma_Y$ is never very large compared with ρ. This would not be expected to be the case if we were correlating equilibrium constants for the carbonium-ion formation process shown in Eq. 3-65. If Eq. 3-61 were strictly appli-

$$Ar_2CHOH + H^+ \rightleftharpoons Ar_2CH^+ + H_2O \tag{3-65}$$

cable to these equilibrium constants, the reaction constant for 3- or 4-X-benzhydrols would be the same as for 3- or 4-X-4′-dimethylaminobenzhydrols. However, the interaction of a *p*-dimethylamino substituent with a benzyl-type carbonium-ion center is so much larger than the interaction of any known meta substituent with a carboxy or carboxylate anion grouping that $\rho'\sigma_Y$ may well be comparable in magnitude to ρ in the case of reaction 3-65 even though it is not for 3-X-5-Y-benzoic acids.

The same kind of treatment that Hammett applied to benzene derivatives may also be applied to derivatives of polynuclear aromatic hydrocarbons. Wells, Ehrenson, and Taft described correlations of a large amount of data on naphthalene derivatives in terms of their equation, which is analogous to Eqs. 3-54 and 3-55.[31]

The preceding discussion of multiply substituted compounds involved using Hammett substituent constants to characterize the behavior of substituents that were ortho neither to the reaction center nor to each other. In applying Eq. 3-61 to 3,4-disubstituted, 3,4,5-trisubstituted, etc., compounds, the location of substituents ortho to each other may cause significant deviations. A nitro substituent may be kept out of coplanarity with the ring and hence made less electron withdrawing, or a dimethylamino group may be kept out of coplanarity and hence made more electron withdrawing. One substituent could hydrogen bond to another. One might be conjugated with the other. With innocuous substituents like methyl and halogen, however, Eq. 3-61 is likely to be relatively reliable.

Many of the structural features that make linear free energy relationships particularly applicable to aromatic compounds are also found in heterocyclic compounds, and hence it is not surprising that much of the existing data on heterocyclic compounds may be correlated by extension of the Hammett equation. Applications of the Hammett equation to heterocyclic compounds have been reviewed by Jaffé and Jones.[53] As an example of one type of extension of the Hammett equation to heterocyclic species, the replacement of a CH group in a benzene ring by a nitrogen atom (thus getting a pyridine ring), may be regarded as introducing a substituent. In this case, however, there seem to be no substituent constants that give good agreement for a variety of reactions. Better correlations are obtained with reactions at the heterocyclic nitrogen atom of 3- and 4- substituted pyridines. The transmission of substituent effects through a heterocyclic ring was correlated by Hammett, who found a ρ of 1.39 for the ionization of 5-substituted 2-furoic acids

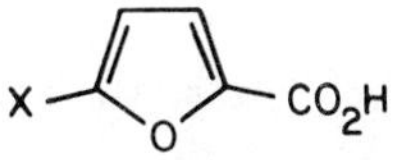

X—(furan ring, O)—CO_2H

using para-substituent constants for the 5-substituents.[54]

Hammett substituent constants have even been found to be applicable in certain reactions of acyclic compounds. For example, Charton and Meislich obtained a ρ of 2.23 in a good correlation of the ionization constants of *trans*-3-substituted acrylic acids with σ_p values.[55] Evidently in this and the other cases in which σ_p values may be used for nonpara substituents, the field, inductive, resonance, etc., effects of the substituents are mixed in about the same proportion that they are for para substituents in a benzene ring.

Charton has described a large number of other correlations of data on nonaromatic unsaturated compounds using the Hammett equation and related relationships in which the relative contribution of resonance and polar effects is a disposable parameter.[56]

3-6. THE TAFT EQUATION[3–8,27,46,57]

3-6a. σ^* Constants. Taft estimated the magnitude of polar effects in ester hydrolysis reactions by a method that may be expressed as follows. Let us assume that the rate constant for the basic hydrolysis of an ester in the series RCO_2R', where R varies but R′ is the same for all members of the series, relative to that of the ester chosen as the reference compound may be expressed

$$\log\left(\frac{k_R}{k_0}\right)_B = (E_R^{polar})_B + (E_R^{resonance})_B + (E_R^{steric})_B \qquad (3\text{-}66)$$

where k_0 is the rate constant for the reference ester and the E's are polar, resonance, and steric effects, as labeled. If it is further assumed that the polar effect may be expressed in terms of a rho-sigma relationship (using the symbols ρ^* and σ^*), we may write

$$\log\left(\frac{k_R}{k_0}\right)_B = \rho_B{}^*\sigma_R{}^* + (E_R^{resonance})_B + (E_R^{steric})_B \qquad (3\text{-}67)$$

The same type of equation may be written for the rate constants for acid hydrolysis of the same two esters:

$$\log\left(\frac{k_R}{k_0}\right)_A = \rho_A{}^*\sigma_R{}^* + (E_R^{resonance})_A + (E_R^{steric})_A \qquad (3\text{-}68)$$

Ingold had suggested earlier that steric effects in the acid hydrolysis of esters are of about the same magnitude as those in basic hydrolysis. Taft assumed that resonance effects were also equal in the basic and acidic hydrolysis of esters (or at least that steric plus resonance effects were equal in the two cases), so that when Eq. 3-68 is subtracted from 3-67 the last two terms in each equation disappear. For application of Eq. 3-69, the resulting relationship, the methyl group was chosen as the reference sub-

$$\log\frac{(k_R/k_0)_B}{(k_R/k_0)_A} = (\rho_B{}^* - \rho_A{}^*)\sigma_R{}^* \qquad (3\text{-}69)$$

stituent; that is, $\sigma^*_{CH_3}$ is zero by definition, and $\rho_B{}^* - \rho_A{}^*$ for the ester hydrolysis reactions for which data were available was assigned the value 2.48 (an average value for $\rho_B - \rho_A$ for basic and acidic hydrolysis of esters of *m*- and *p*-substituted benzoic acids). From these assignments and the available data on ester hydrolysis, a number of substituent constants could be calculated. When these substituent constants were applied to a number of other rate and equilibrium processes, such as the acidities of acids of the type RCO_2H, plots of log K_R versus $\sigma_R{}^*$ were found to give good approaches to straight lines. Such reactions were then used as secondary sources for the calculation, via Eq. 3-70, of substituent constants for

$$\log\frac{K}{K_0} = \rho^*\sigma^* \qquad (3\text{-}70)$$

groups for which the appropriate ester hydrolysis data were not available

Table 3-6. Taft Substituent Constants[a]

Substituent	σ^*	Substituent	σ^*
CCl_3	2.65	CH=CHPh	0.41
CHF_2	2.05	$CHPh_2$	0.40
CO_2Me	2.00	CH_2CH_2Cl	0.38
$CHCl_2$	1.94	CH=CHMe	0.36
Ac	1.65	$CH_2CH_2CF_3$	0.32
C≡CPh	1.35	CH_2Ph	0.22
CH_2SO_2Me	1.32	CH_2CH=CHMe	0.13
CH_2CN	1.30	$(CH_2)_3CF_3$	0.12
CH_2F	1.10	CH_2CH_2Ph	0.08
CH_2Cl	1.05	Me	0.000
CH_2Br	1.00	Et	−0.10
CH_2CF_3	0.92	*n*-Pr	−0.12
CH_2I	0.85	*n*-Bu	−0.13
CH_2OPh	0.85	Cyclohexyl	−0.15
CH_2OMe	0.64[b]	CH_2Bu-*t*	−0.16
CH_2Ac	0.60	*i*-Pr	−0.19
Ph	0.60	Cyclopentyl	−0.20
CH_2OH	0.56	$CHEt_2$	−0.22
$CH_2CH_2NO_2$	0.50	CH_2SiMe_3	−0.26
H	0.49	*t*-Bu	−0.30

[a] From Ref. 27 unless otherwise noted.

[b] P. Ballinger and F. A. Long, *J. Amer. Chem. Soc.*, **82,** 795 (1960).

(and in some cases would probably be unobtainable). More than half the σ^* values listed in Table 3-6 are derived from ester hydrolysis data; most of the rest come from data on other carboxylic acid derivatives, especially from ionization constants of carboxylic acids (for which ρ^* is 1.721 in water at 25° C).

To point out a number of ways in which the relative magnitudes of Taft substituent constants agree with our qualitative ideas concerning polar effects, we may note that $\sigma^*_{Cl_3C}$ (2.65) > $\sigma^*_{Cl_2CH}$ (1.94) > $\sigma^*_{ClCH_2}$ (1.05) > $\sigma^*_{CH_3}$ (0.00); $\sigma^*_{FCH_2}$ (1.10) > $\sigma^*_{ClCH_2}$ (1.05) > $\sigma^*_{BrCH_2}$ (1.00) > $\sigma^*_{ICH_2}$ (0.85); $\sigma^*_{PhC\equiv C}$ (1.35) > $\sigma^*_{PhCH=CH}$ (0.41) > $\sigma^*_{PhCH_2CH_2}$ (0.08); $\sigma^*_{F_3CCH_2}$ (0.92) > $\sigma^*_{F_3C(CH_2)_2}$ (0.32) > $\sigma^*_{F_3C(CH_2)_3}$ (0.12).

Unlike Hammett substituent constants, which have been used successfully only when the substituent is attached to sp^2-hybridized carbon, Taft substituent constants have been applied to cases in which the substituent is attached to a number of different kinds of atoms. This is probably one reason why such applications have revealed a larger fraction of deviant data and larger deviations than are usually observed in the application of the Hammett equation to the much smaller body of data to which it is generally felt to be applicable. Many of these deviations from Eq. 3-70, which is considered to be a measure of polar effects only, seem obviously attributable to the intrusion of resonance effects, hydrogen bonding between substituent and reaction center, or steric effects, including conformational effects that change the relative orientations of the substituent and reaction center. Other deviations probably arise from imperfections in the σ^* constants listed in Table 3-6. It is too much to hope that the cancellation of resonance and steric effects assumed in the derivation of Eq. 3-69 was perfect (and also that there were no errors in the experimental data to which this equation was applied). In fact, Chapman, Shorter, and coworkers found direct evidence for differences between steric effects on rates of acidic and basic ester hydrolysis.[57–59] Brown discussed the probable effects of polarizability on σ^* values.[60] Since some of the alkyl groups are the largest and have the greatest total polarizability of all the groups in Table 3-6, they should be among those for which σ^* is the least reliable measure of the polar substituent effect. In view of the relatively nonpolar character of alkyl groups such unreliability must be particularly great when expressed as a *percentage* of the total σ^*.

3-6b. Limits on the Generality of the Taft Equation. The preceding paragraph provides a partial explanation for the results of Ritchie's analysis of the possible generality of Eq. 3-70 and the σ^* values obtained from the defining reactions (Eq. 3-69).[61] This analysis starts with the assumption that the Taft equation is generally applicable to reactions of the type shown in Eq. 3-71, where A is the substituent, C is the reaction center in the

$$\text{A–N–C} \rightleftharpoons \text{A–N–D} \qquad (3\text{-}71)$$

reactant form, D is the reaction center in the product form, and N is the connecting group, if any. Equation 3-72 may be considered to be another reaction belonging to the same reaction series, for which the reaction

$$\text{ACH}_2\text{–N–C} \rightleftharpoons \text{ACH}_2\text{–N–D} \qquad (3\text{-}72)$$

constant will be denoted ρ_0^*. On this basis we may write

$$\log \frac{K_{\text{ACH}_2\text{NC}}}{K_{\text{C}_2\text{H}_5\text{NC}}} = \rho_0^* \left(\sigma^*_{\text{ACH}_2} - \sigma^*_{\text{CH}_3\text{CH}_2}\right) \qquad (3\text{-}73)$$

Alternatively, Eq. 3-72 may be considered to belong to a series of its own in which the substituent is not ACH_2 but A, and the connecting group is not N but CH_2N. If the reaction constant for this series is called $\rho^*_{CH_2}$, we may write

$$\log \frac{K_{ACH_2NC}}{K_{C_2H_5NC}} = \rho^*_{CH_2}\sigma_A{}^* \tag{3-74}$$

since $\sigma^*_{CH_3}$ is defined as zero. Elimination of the log term between Eqs. 3-73 and 3-74 gives

$$\sigma^*_{ACH_2} = \frac{\rho^*_{CH_2}}{\rho_0{}^*}\sigma_A{}^* + \sigma^*_{CH_3CH_2} \tag{3-75}$$

according to which a plot of $\sigma^*_{ACH_2}$ versus $\sigma_A{}^*$ should give a straight line whose intercept is $\sigma^*_{CH_3CH_2}$ and whose slope is $\rho^*_{CH_2}/\rho_0{}^*$. Since the definition of terms that go into Eq. 3-70 requires that all σ^* values be universal constants, whose values do not depend on solvent, temperature, etc., the ratio $\rho^*_{CH_2}/\rho_0{}^*$ must also be a universal constant—the factor by which ρ^* always changes when the substituents are moved one methylene group further from the reaction center. The plot of $\sigma^*_{ACH_2}$ versus $\sigma_A{}^*$ in Fig. 3-3 shows that there are two lines, one described by the points for hydrogen and all the saturated alkyl groups and the second described by all the other points. We have already given reasons why the σ^* values for alkyl groups should be relatively unreliable, There is also evidence that the value of $\sigma_H{}^*$ listed in Table 3-6 is often a poor measure of the polar character of the hydrogen substituent and that the polar character of hydrogen, relative to that of alkyl groups, for example, varies with the nature of the atom to which the substituents are attached. Not only is hydrogen frequently found to be an unusually deviant substituent in Taft equation correlations, but also data on the dipole moments of hydrocarbons attest to the variability of its relative polar character. All ordinary alkanes have dipole moments that are essentially zero. Therefore, replacing a hydrogen atom in an alkane by an alkyl group, which simply gives another alkane, does not change the dipole moment. Hence by this measure the polar character of a hydrogen atom is the same as that of an alkyl group when both are attached to saturated carbon. Benzene, ethylene, and acetylene also have dipole moments of zero, but replacing a hydrogen atom in these compounds with an alkyl group gives a product with a significant dipole moment (with the sp^3-hybridized carbon at the positive end and the sp^2- or sp-hybridized carbon at the negative end of the dipole).[62] Therefore, by this measure alkyl groups are electron-donating substituents relative to hydrogen when attached to sp^2- or sp-hybridized carbon. Since $\sigma_A{}^*$ for hydrogen in Fig. 3-3 refers to compounds in which hydrogen is attached to sp^2-hybridized carbon and $\sigma^*_{CH_2A}$ to compounds in which it is attached to

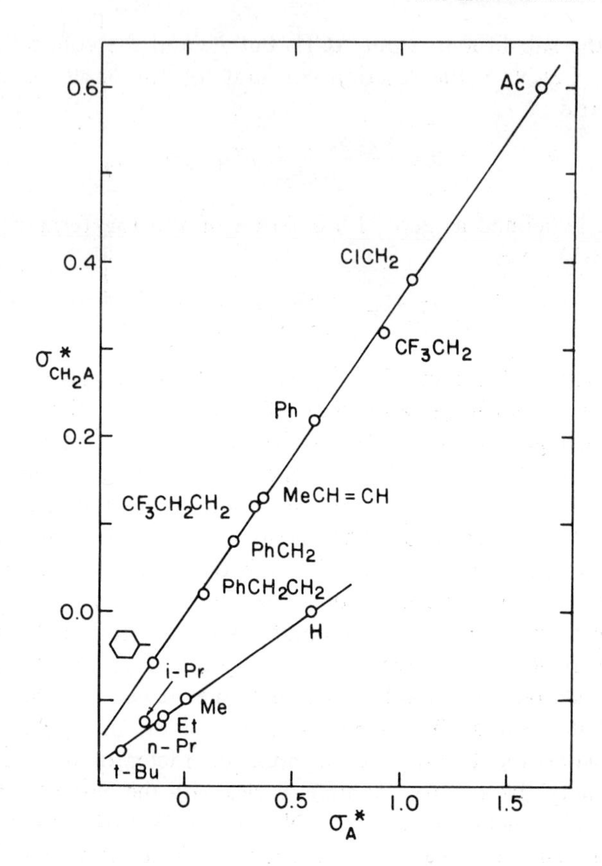

Figure 3-3. Plot of σ_{CH_2A} versus σ_A for all the points for which data were available in Ref. 27.

saturated carbon, it is not surprising that σ_A* is too large relative to $\sigma^*_{CH_2A}$ for the point for hydrogen to fall on the line that describes most of the points in Fig. 3-3. For these reasons it seems more reasonable to assume that the points for alkyl groups and hydrogen and the line through them are misleading than to make such an assumption about the other points and the line through them.

As Ritchie points out, the plot becomes satisfactory if σ* for hydrogen and all saturated alkyl groups are defined as zero. He further points out that although these values of zero give poorer agreement with experimental data on ester hydrolysis and a number of other reactions, there seems to be a comparable number of reactions in which they give improved correlations. An alternative method of obtaining a more self-consistent set of σ^* constants would be to give σ_H^* a considerably smaller value and to calculate substituent constants for alkyl groups from Eq. 3-75, obtaining the value of $\rho^*_{CH_2}/\rho_0^*$ from the slope of a pilot like that in Fig. 3-3. However, a considerably reduced σ_H^* value could not fit the data that are satisfactorily correlated by the σ_H^* value listed in Table 3-6. These data tend to be reactions in which the substituent is attached directly to the carbonyl carbon atom of a carboxylic acid derivative. For this reason, the use of the present σ_H^* should probably be restricted to such cases.

Although it is possible to get a set of σ^* values that fits Eq. 3-75 satisfactorily and therefore gives a reasonable approach to a straight line in a plot like that shown in Fig. 3-3, there is another reason why Eq. 3-70 must not be completely general. All we need is the assumption that Eq. 3-70 is applicable in order to derive an equation equivalent to 3-75. (If the reference substituent were Y instead of methyl, the term $\sigma^*_{CH_3CH_2}$ in Eq. 3-75 would have to be replaced by $\sigma^*_{YCH_2}$.) This equation tells us that the polar substituent effect of any substituent in any reaction changes by the factor $\rho^*_{CH_2}/\rho_0^*$, which is a universal constant, when an added methylene group is interposed between the substituent and the reaction center. Experimentally this is not found to be true, however. In studying substituent effects on acidity, for example, the Taft equation has been most commonly applied in cases where there are only 2–4 atoms between the acidic proton and the substituent dipole. To the extent to which the agreement has been satisfactory this means that the substituent effects are decreasing by a factor near 2.8 (the reciprocal of the slope of the major line in Fig. 3-3) every time they are moved one methylene group further from the reaction center. In comparing such reaction series as the meta- and para-substituted phenylacetic acids and β-phenylpropionic acids, however, where the distance between substituent dipole and acidic proton is significantly larger, the factor is smaller. (The ratio of Hammett ρ values is 1.9 in the specific pair of reactions referred to.[25]) This is another way of saying that although σ^* values are not ordinarily calculated for $ArCH_2$ and $ArCH_2CH_2$ groups, if they were their values would deviate significantly from Eq. 3-75. Such deviations are inconsistent with a strictly inductive picture of polar substituent effects in which the amount of change in charge induced by a polar substituent on a chain carbon atom differs by a constant fraction from that induced on the next carbon atom closer to the substituent. They are just what is expected from a polar effect in which a direct electrostatic or field effect is a major component.

Consider the following crude picture of the effect of increasing the number of methylene groups between the acidic functional group AH and the dipolar substituent X–Y.

$$\text{Y—X—(CH}_2)_n\text{—AH} \rightleftharpoons \text{H}^+ + \text{Y—X—(CH}_2)_n\text{—A}^- \qquad (3\text{-}76)$$

Assume that the substituent effect of XY (in terms of ΔpK) is proportional to the energy of electrostatic interaction between the XY dipole and the negative A^- group. Take the distance between XY and A to be one unit when the number of methylene groups n is zero and to be $n + 1$ units in general. Let the effective dielectric constant be the same in all cases. Neglect any changes in the angular orientation of XY with respect to A. Then, according to Eq. 2-3, ΔpK_n may be expressed as shown in Eq. 3-77, where C contains the various proportionality constants relating to the

$$\Delta pK_n = \frac{C}{(n + 1)^2} \tag{3-77}$$

dielectric constant, angular orientation, conversion of energy to log K, etc. From Eq. 3-77 it follows that $\Delta pK_n/\Delta pK_{n+1}$, the factor by which the substituent effect changes when an additional methylene group is inserted between substituent and reaction center, must follow Eq. 3-78. According to

$$\frac{\Delta pK_n}{\Delta pK_{n+1}} = \left(\frac{n + 2}{n + 1}\right)^2 \tag{3-78}$$

this equation $\Delta pK_n/\Delta pK_{n+1}$ decreases from 4.0 to 2.25 to 1.78 to 1.56 to 1.44 as n increases from 0 through 4. Equation 3-78 makes it clear that the "fall-off" factor per added methylene group" should decrease with increasing separation of substituent and reaction center. But why did $\Delta pK_n/\Delta pK_{n+1}$ remain near 2.8 over a wide enough range of cases for the Taft equation with the σ^* values plotted in Fig. 3-3 to be as useful as it has been? There are probably a number of factors that contribute to the answer to this question. There is probably a significant contribution of a classical inductive effect when the substituent is particularly close to the reaction center. In most reaction series increasing the number of methylene groups makes the species longer and relatively slender, thus increasing the effective dielectric constant. This increases $\Delta pK_n/\Delta pK_{n+1}$ for intermediate n values. Then before n has become very large the substituent effect may become too small to measure reliably (or often the appropriate compounds were not studied). Finally, there must be a certain amount of coincidence in the unexpectedly high quality of the main line in Fig. 3-3.

The preceding discussion shows that deviations from the Taft equation will be expected when we use two or more substituents that differ markedly in the distance between their dipoles and points of attachment in two or more reaction series that differ markedly in the distances between the reaction center and the point of attachment of substituents.

3-7. INDUCTIVE SUBSTITUENT CONSTANTS (σ_I)

As a measure of the polar effect of substituents, σ^* has been replaced in common usage to a considerable extent by the "inductive" substituent constant σ_I. Several different sets of reactions as well as such physical measurements as nuclear magnetic resonance shifts have been used to obtain σ_I values. The equivalent of Eq. 3-79 was once used as a definition,[30] the factor 0.45 being that needed to give values that could be taken as the

$$\sigma_I = 0.45\sigma^* \tag{3-79}$$

polar component of σ_m and σ_p values (so that σ_p could be equated to $\sigma_I + \sigma_R$, for example; cf. Section 3-3c). In Table 3-7 is a collection of σ_I values from the publications of Taft and co-workers,[27–32,42,63] who have carried out most of the basic research in this area. In evolving from a set of values based on Eq. 3-79 to that listed in the table, the scale has been changed to make hydrogen the reference substituent, whose σ_I constant is zero by definition. Among other changes, the difference between the values for methyl and hydrogen has been shrunk from the 0.22 that would result from Eq. 3-79 and the σ^* values in Table 3-6 to the 0.04 seen in Table 3-7.† This was probably done to get better agreement with data on reactions in which the substituents were attached directly to an aromatic ring or to a saturated carbon atom. However, a difference of zero between σ_I for methyl and hydrogen would probably fit the available data about as well as a difference of 0.04. Bicyclo[2.2.2]octane derivatives (including quinuclidine, the 1-aza derivative) with a substitutent at one bridgehead and an acidic or basic group at the other appear to be especially suitable for determining the polar effects of substituents attached to saturated carbon. Relative to the unsubstituted compound, a 4-methyl substituent decreases the acidity of the 1-carboxylic acid by 0.01,[65] decreases the acidity of the dibenzo-1-carboxylic acid by 0.03,[65] increases the acidity of the $\Delta^{2,3}$-1-carboxylic acid by 0.04,[65] and increases the acidity of the quinuclidinium ion by 0.08.[66–68] These methyl substituent effects correspond to σ_I values ranging from -0.02 to 0.02 on the basis of a σ_I value of zero for hydrogen. It would probably be advisable to use a σ_I value of zero for alkyl groups attached to saturated carbon and the value -0.04 in Table 3–7 for cases in which the substituent is attached to aromatic carbon.

The σ_I values shown by footnotes to have been obtained from ^{19}F nmr chemical shifts are of reduced reliability. Several differences of 0.10 or

†Charton published a list of more than 100 σ_I values calculated on a somewhat different basis from that used for the values in Tables 3-7.[64]

Table 3-7. Values of σ_I[a]

Substituent	σ_I	Substituent	σ_I	Substituent	σ_I
NMe_3^+	0.92[b]	COF	0.42[c]	CH_2Cl	0.17
SO_2Cl	0.86[c]	I	0.39	$NHNH_2$	0.15
SO_2F	0.86	OAc	0.39	NH_2	0.12
NO_2	0.65	OPh	0.38	CH_2OH	0.10
$SOCF_3$	0.64	$CH{=}CHNO_2$	0.38	Ph	0.10
SO_3Et	0.63[c]	NO	0.37[c]	NMe_2	0.06
NH_3^+	0.60[b]	CHO	0.31[c]	$CH{=}CH_2$	0.05
SO_2Me	0.59	CO_2Et	0.30	CH_2NH_2	0.00
SF_5	0.57	Ac	0.28	H	0.00
CN	0.56	OMe	0.27	Me	−0.04
COCN	0.55[c]	NHAc	0.26	Et	−0.05
OCF_3	0.55	$CH_2NH_3^+$	0.25	*i*-Pr	−0.06[d]
SOMe	0.50	OH	0.25	*t*-Bu	−0.07[d]
F	0.50	SH	0.25	$B(OH)_2$	−0.08
SO_2NH_2	0.46[c]	SO_3^-	0.25[b,c]	$SiMe_3$	−0.10
Cl	0.46	N=NPh	0.25[c]	CH_2SiMe_3	−0.11
CF_3	0.45	SMe	0.23	O^-	−0.12[b]
Br	0.44	CH_2CN	0.23	CO_2^-	−0.35[b,c]
SCF_3	0.42	$CONH_2$	0.21	$B(OH)_3^-$	−0.36[b,c]

[a] Unless otherwise noted these values were obtained from Refs. 32 and 42, with preference being given to the former source in cases of disagreement. They are generally taken to be useful in solvents no more acidic than formic acid.

[b] Substituent constants for electrically charged groups are relatively unreliable.

[c] Obtained from ^{19}F chemical shifts. Such values are listed only when values based on chemical reactions were not given.

[d] Ref. 63.

more between σ_I values obtained in this way and those obtained from rate and equilibrium data have been found.[42]

When resonance and steric effects are negligible inductive substituent constants may be used in Eq. 3-80, which is analogous to the Taft equation

$$\log \frac{K}{K_0} = \rho_I \sigma_I \tag{3-80}$$

(Eq. 3-70) and whose generality is subject to some of the same limitations discussed in Section 3-6b. The σ_I constants listed in Table 3-7 are based on reactions of compounds in which the substituents are attached to saturated carbon and sp^2-hybridized carbon. It is not clear how well they will work when attached to other types of atoms. In fact, as mentioned for the case of the hydrogen substituent, it is possible that there will be systematic deviations depending on what else is bonded to the carbon atom to which the substituents are attached and on whether the carbon is sp^2- or sp^3-hybridized.

When the appropriate equation analogous to Eq. 3-75 is derived for σ_I values and the data in Table 3-7 used in a plot analogous to that shown in Fig. 3-3, the $\sigma_I(CH_2A)$ values fit the best straight line with an average deviation of 0.014. The slope of the line is 0.5, corresponding to a fall-off factor of 2.0 per methylene group instead of 2.8 as noted with σ^* values. This probably indicates the strong reflection in σ_I values of data on side-chain reactions of meta- and para-substituted aromatic compounds, where the fall-off factors are smaller than they are in cases where the dipole of the substituent is nearer the reaction center. Values of σ_I defined in terms of basicities of 4-substituted quinculidines[66–68] would give larger fall-off factors.

PROBLEMS

1. If ρ for the ionization of phenols in water at 25° C is 2.23 and the K's of phenol and p-nitrophenol are 1.0×10^{-10} and 1.0×10^{-7}, respectively, by how much has direct resonance interaction between substituent and reaction center changed the free energy of ionization?

2. For each of the following equilibrium processes tell whether you would expect ρ to be positive or negative and give your reasons.
   a. $ArCHO + 2MeOH \rightleftharpoons ArCH(OMe)_2 + H_2O$
   b. $ArNMe_2 + MeI \rightleftharpoons ArNMe_3^+ + I^-$
   c. $ArI + Cl_2 \rightleftharpoons ArICl_2$
   d. $ArOH + AcOH \rightleftharpoons ArOAc + H_2O$

3. Tell whether you would expect the Taft ρ^* (or ρ_I) for the following reaction series (equilibria) to be positive or negative and why.

$$RNH_2 + HCO_2Me \rightleftharpoons RNHCHO + MeOH$$

Tell whether you would expect K for R = Ph to fit the correlation well, to be too large, or to be too small, and why.

4. Estimate the value of Hammett's ρ for the reaction

$$ArSCH_3 + CH_3I \rightleftharpoons ArS(CH_3)_2^+ + I^-$$

5. Name two substituents for which it would probably be impossible (for two different reasons) to obtain the Hammett substituent constant by determination of

the strength of the appropriately substituted benzoic acid, and suggest ways in which their substituent constants could be determined from rate or equilibrium data.

6. Name two substituents whose Taft substituent constants could probably not be determined from ester hydrolysis data and suggest how these substituent constants could be determined from rate or equilibrium data.

REFERENCES

1. H. H. Jaffé, *Chem. Rev.*, **53,** 191 (1953).
2. H. van Bekkum, P. E. Verkade, and B. M. Wepster, *Rec. Trav. Chim. Pays-Bas*, **78,** 815 (1959).
3. J. Hine, *Physical Organic Chemistry*, 2nd ed., McGraw-Hill, New York, 1962, Chap. 4.
4. J. E. Leffler and E. Grunwald, *Rates and Equilibria of Organic Reactions*, Wiley, New York, 1963, Chaps. 6 and 7.
5. S. Ehrenson, *Progr. Phys. Org. Chem.*, **2,** 195 (1964).
6. P. R. Wells, *Linear Free Energy Relationships*, Academic Press, New York, 1968.
7. C. K. Ingold, *Structure and Mechanism in Organic Chemistry*, 2nd ed., Cornell University Press, Ithaca, N.Y., 1969, Chap. XVI.
8. L. P. Hammett, *Physical Organic Chemistry*, 2nd ed., McGraw-Hill, New York, 1970, Chap. 11.
9. O. Exner, in *Advances in Linear Free Energy Relationships*, N. B. Chapman and J. Shorter, Eds., Plenum Press, New York, 1972, Chap. 1.
10. Cf. J. W. McBain and O. C. M. Davis, *Z. Phys. Chem.*, **78,** 369 (1912); A. E. Bradfield and B. Jones, *J. Chem. Soc.*, 1006, 3073 (1928); 2903 (1931).
11. L. P. Hammett, *Chem. Rev.*, **17,** 125 (1935).
12. Cf. G. N. Burkhardt, *Nature*, **136,** 684 (1935).
13. Cf. W. F. Sager and C. D. Ritchie, *J. Amer. Chem. Soc.*, **83,** 3498 (1961).
14. J. Hine, *J. Amer. Chem. Soc.*, **81,** 1126 (1959).
15. D. H. McDaniel and H. C. Brown, *J. Org. Chem.*, **23,** 420 (1958).
16. W. A. Sheppard, *J. Amer. Chem. Soc.*, **85,** 1314 (1963); **87,** 2410 (1965).
17. O. Exner, *Collect. Czech. Chem. Commun.*, **31,** 65 (1966).
18. O. Exner and J. Lakomý, *Collect. Czech. Chem. Commun.*, **35,** 1371 (1970).
19. C. Hansch, A. Leo, S. H. Unger, K. H. Kim, D. Nikaitani, and E. J. Lien, *J. Med. Chem.*, **16,** 1207 (1973).
20. D. E. Pearson, J. F. Baxter, and J. C. Martin, *J. Org. Chem.*, **17,** 1511 (1952).
21. H. C. Brown, J. D. Brady, M. Grayson, and W. H. Bonner, *J. Amer. Chem. Soc.*, **79,** 1897 (1957).

22. H. C. Brown and Y. Okamoto, *J. Amer. Chem. Soc.*, **80,** 4979 (1958).
23. R. W. Taft, Jr., S. Ehrenson, I. C. Lewis, and R. E. Glick, *J. Amer. Chem. Soc.*, **81,** 5352 (1959).
24. A. J. Hoefnagel, J. C. Monshouwer, E. C. G. Snorn, and B. M. Wepster, *J. Amer. Chem. Soc.*, **95,** 5350 (1973).
25. A. J. Hoefnagel and B. M. Wepster, *J. Amer. Chem. Soc.*, **95,** 5357 (1973).
26. C. L. Liotta, D. F. Smith, Jr., H. P. Hopkins, Jr., and K. A. Rhodes, *J. Phys. Chem.*, **76,** 1909 (1972).
27. R. W. Taft, in *Steric Effects in Organic Chemistry*, M. S. Newman, Ed., Wiley, New York, 1956, Chap. 13.
28. R. W. Taft and I. C. Lewis, *J. Amer. Chem. Soc.*, **80,** 2436 (1958).
29. R. W. Taft and I. C. Lewis, *J. Amer. Chem. Soc.*, **81,** 5343 (1959).
30. R. W. Taft, *J. Phys. Chem.*, **64,** 1805 (1960).
31. P. R. Wells, S. Ehrenson, and R. W. Taft, *Progr. Phys. Org. Chem.*, **6,** 147 (1968).
32. S. Ehrenson, R. T. C. Brownlee, and R. W. Taft, *Progr. Phys. Org. Chem.*, **10,** 1 (1973).
33. J. L. Roberts and H. H. Jaffé, *J. Amer. Chem. Soc.*, **81,** 1635 (1959).
34. Y. Yukawa and Y. Tsuno, *Bull. Chem. Soc. Japan*, **32,** 971 (1959).
35. Y. Yukawa, Y. Tsuno, and M. Sawada, *Bull. Chem. Soc. Japan*, **39,** 2274 (1966).
36. Y. Yukawa, Y. Tsuno, and M. Sawada, *Bull. Chem. Soc. Japan*, **45,** 1198 (1972).
37. Cf. J. Hine, *J. Amer. Chem. Soc.*, **82,** 4877 (1960).
38. B. M. Wepster, *J. Amer. Chem. Soc.*, **95,** 102 (1973).
39. Cf. C. G. Swain and E. C. Lupton, Jr., *J. Amer. Chem. Soc.*, **90,** 4328 (1968).
40. D. E. Blagdon, personal communication, 1973.
41. R. L. Flachskam, personal communication, 1974.
42. R. W. Taft, E. Price, I. R. Fox, I. C. Lewis, K. K. Andersen, and G. T. Davis, *J. Amer. Chem. Soc.*, **85,** 709, 3146 (1963).
43. D. H. McDaniel, *J. Org. Chem.*, **26,** 4692 (1961).
44. H. H. Jaffé, *J. Amer. Chem. Soc.*, **81,** 3020 (1959).
45. J. Hine and H. E. Harris, *J. Amer. Chem. Soc.*, **85,** 1476 (1963).
46. R. W. Taft, *J. Amer. Chem. Soc.*, **74,** 3120 (1952).
47. M. Charton, *J. Org. Chem.*, **30,** 3341 (1965).
48. M. Charton, *Progr. Phys. Org. Chem.*, **8,** 235 (1971).
49. J. D. Roberts and J. A. Yancey, *J. Amer. Chem. Soc.*, **73,** 1011 (1951).
50. J. D. Roberts, E. A. McElhill, and R. Armstrong, *J. Amer. Chem. Soc.*, **71,** 2923 (1949).
51. W. L. Bright and H. T. Briscoe, *J. Phys. Chem.*, **37,** 787 (1933).
52. S. I. Miller, *J. Amer. Chem. Soc.*, **81,** 101 (1959).

53. H. H. Jaffé and H. L. Jones, *Advan. Heterocycl. Chem.*, **3,** 209 (1964).
54. L. P. Hammett, *Physical Organic Chemistry*, 1st ed., McGraw-Hill, New York, 1940, p. 189.
55. M. Charton and H. Meislich, *J. Amer. Chem. Soc.*, **80,** 5940 (1958).
56. M. Charton, *Progr. Phys. Org. Chem.*, **10,** 81 (1973).
57. J. Shorter, in *Advances in Linear Free Energy Relationships*, N. B. Chapman and J. Shorter, Eds., Plenum Press, New York, 1972, Chap. 2.
58. N. B. Chapman, J. Shorter, and K. J. Toyne, *J. Chem. Soc.*, 2543 (1961).
59. N. B. Chapman, A. Ehsan, J. Shorter, and K. J. Toyne, *J. Chem. Soc. B*, 570 (1967); 178 (1968).
60. T. L. Brown, *J. Amer. Chem. Soc.*, **81,** 3229 (1959).
61. C. D. Ritchie, *J. Phys. Chem.*, **65,** 2091 (1961).
62. L. E. Sutton, in *Determination of Organic Structures by Physical Methods*, E. A. Braude and F. C. Nachod, Eds., Academic Press, New York, 1955, Chap. 9.
63. R. W. Taft and I. C. Lewis, *Tetrahedron*, **5,** 210 (1959).
64. M. Charton, *J. Org. Chem.*, **29,** 1222 (1964).
65. F. W. Baker, R. C. Parish, and L. M. Stock, *J. Amer. Chem. Soc.*, **89,** 5677 (1967).
66. W. Eckhardt, C. A. Grob, and W. D. Treffert, *Helv. Chim. Acta*, **55,** 2432 (1972).
67. E. Ceppi, W. Eckhardt, and C. A. Grob, *Tetrahedron Lett.*, 3627 (1973).
68. J. Paleček and J. Hlavatý, *Collect. Czech. Chem. Commun.*, **38,** 1985 (1973).

4

Equilibrium in Conformational and Cis-Trans Isomerization

4-1. CONFORMATIONAL EQUILIBRIA IN ACYCLIC COMPOUNDS[1-3]

4-1a. Rotation Around Bonds between *sp*³-Hybridized Carbon Atoms. Rotation around the carbon-carbon bond in species of the type XCH_2CH_2Y involves passing over potential energy minima when the bonds are staggered and maxima when they are eclipsed. One of the minima corresponds to the trans conformer (**1**) and the other two to the gauche conformers (**2** and **3**). The stabilities of **2** and **3** will be identical if X and Y are achiral

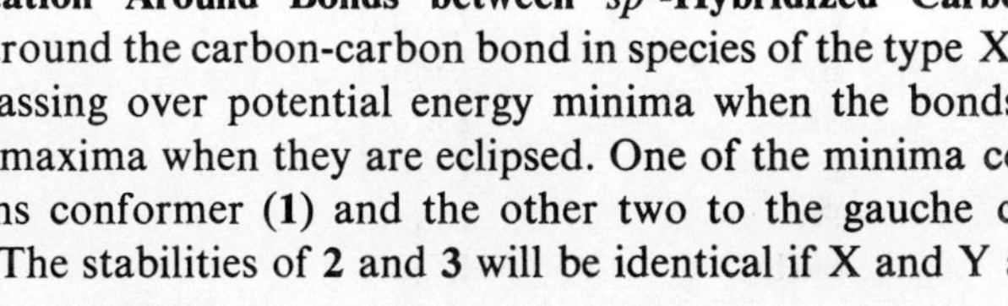

X, H, H, H, H, Y — **1**; X, H, Y, H, H, H — **2**; X, Y, H, H, H, H — **3**

(superposable on their mirror images). Data on how structural and environmental changes influence the relative stabilities of gauche and trans forms are listed in Table 4-1.

Table 4-1. Relative Conformational Stabilities for Compounds of the Type XCH_2CH_2Y[a]

X	Y	State	ΔH^{0b} (kcal/mole)
F	F	Gas	0.0
F	F	Liquid	−0.9
Cl	Cl	Gas	1.2
Cl	Cl	Liquid	0.0
Br	Br	Gas	1.8
Br	Br	Liquid	0.7
Br	Br	Hexane soln	0.9
Br	Br	$MeNO_2$ soln	0.4
Cl	F	Gas	0.5
Br	Cl	Gas	1.4
CN	CN	Liquid	−0.4
F	OH	CCl_4 soln	−2.1
Cl	OH	CCl_4 soln	−1.2
Br	OH	CCl_4 soln	−1.2
I	OH	CCl_4 soln	−0.8
Me	Me	Gas	0.6[c]
Me	F	Gas	−0.05
Me	Cl	Gas	−0.5
Me	Br	Gas	−0.3
Me	C≡CH	Gas	−0.1[d]
t-Bu	CN	Liquid	0.8[e]
t-Bu	CO_2Me	Liquid	1.0[e]
t-Bu	Cl	Liquid	1.0[e]
t-Bu	Br	Liquid	1.3[e]
t-Bu	SPh	Liquid	1.3[e]
t-Bu	I	Liquid	1.6[e]
t-Bu	Ph	Liquid	1.7[e]
t-Bu	$SiMe_3$	Liquid	2.3[e]

[a] Data from Ref. 2 unless otherwise noted.

[b] H^0(gauche) − H^0(trans). In some cases the method used gives ΔE, but this differs from ΔH by less than the experimental uncertainty.

[c] ΔG_{chem}.

[d] F. J. Wodarczyk and E. B. Wilson, *J. Chem. Phys.*, **56,** 166 (1972).

[e] G. M. Whitesides, J. P. Sevenair, and R. W. Goetz, *J. Amer. Chem. Soc.*, **89,** 1135 (1967).

Some of the changes in relative conformational stabilities seem clearly to result from repulsions between large groups in the gauche conformation. Thus, for example, when X is *t*-butyl, the energy content of the gauche relative to the trans form drops from 1.6 to 1.3 to 1.0 to <0.6 (too small to measure) to 0.0 kcal/mole as Y goes from iodine to bromine to chlorine to fluorine to hydrogen. When X is methyl, however, the situation is more complicated. For example, the gauche form is more stable (relative to the trans) when Y is chlorine than when it is fluorine or hydrogen (for which ΔH is zero) even though fluorine and hydrogen substituents are smaller than chlorine. Some of the factors to which attention should be given in explaining these results illustrate the general complexity of structural effects on conformational equilibria. When X is methyl, if the gauche form has perfect staggering and tetrahedral bond angles, the carbon-Y bond is parallel to the nearest carbon-hydrogen bond of the methyl group, as shown in **4**. Therefore when Y is changed from hydrogen to fluorine to chlorine

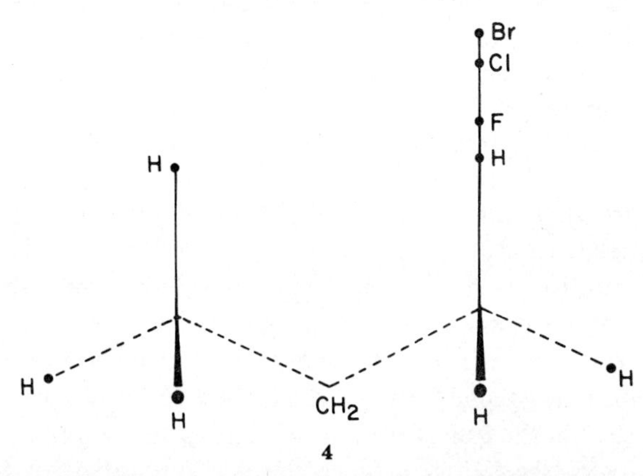

4

to bromine it moves somewhat further from the nearest methyl hydrogen atom and considerably further from the methyl carbon atom. When X is *t*-butyl, as shown in **5**, changing Y from hydrogen to fluorine to chlorine to bromine moves it somewhat nearer the nearest methyl hydrogen atoms and does not greatly change its distance from the carbon. Thus when X is methyl, steric destabilization resulting from increases in the van der Waals radii of Y may be offset by increases in the carbon-Y bond lengths. When X is *t*-butyl the changes in bond lengths and van der Waals radii act in the same direction.

It is also significant that when X is *t*-butyl all the methyl-Y distances are in the destabilizing region of the van der Waals potential energy curve. Hence the energy content of the gauche conformer tends to increase with increasing size of the Y group. When X is methyl some of the methyl-Y

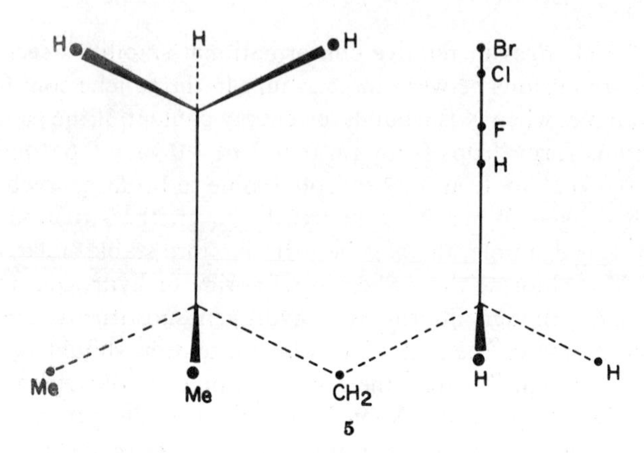

5

distances in the gauche form are less than 11% smaller than the sums of the van der Waals radii. According to the Lennard-Jones equation (Section 2-3) with the usual value for n, this means that the interactions are in the destabilizing region of the van der Waals curve. Relatively small distortions in molecular geometry can increase the net stabilizing interaction between the methyl group and Y, which results from dipole-induced dipole interactions (Section 2-2a), London forces (Section 2-2c), etc. Thus in some cases the gauche conformers are more stable than the trans. Hydrogen bonding is another factor that can stabilize the gauche relative to the trans form; it may do so in the case of the β-haloethanols.

If there are no distortions in bond angles or lengths, the carbon-carbon and carbon-halogen distances in the gauche form will be the same as those in the trans form of an ethylene dihalide. Hence electrostatic interactions should destabilize the gauche form, in which the halogen-halogen distance is much smaller than in the trans form. Such interactions should increase with electronegativity of the halogen. Since the energy content of the gauche ethylene dihalide relative to the trans is seen in Table 4-1 to increase in the sequence F $<$ Cl $<$ Br in the gas phase, it appears that steric effects overwhelm the differences in electrostatic interactions. The reality of the electrostatic interactions may be seen in the changes in the relative stabilities of the two conformers that accompany changes in the dielectric constant (ϵ) of the medium in which measurements were made. With ethylene dibromide the energy content of the gauche (relative to the trans) conformer decreases from 1.8 kcal/mole in the gas phase (ϵ = 1.0) to 0.9, 0.7, and 0.4 kcal/mole in hexane (ϵ = 1.9), liquid ethylene dibromide (ϵ = 4.8), and nitromethane (ϵ = 36), respectively.

Some of the results obtained with more highly substituted ethanes seem predictable from the results on 1,2-disubstituted ethanes shown in Table

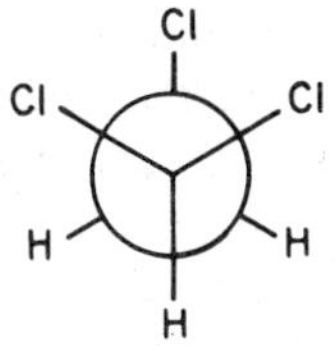

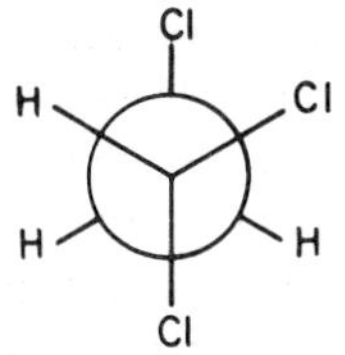

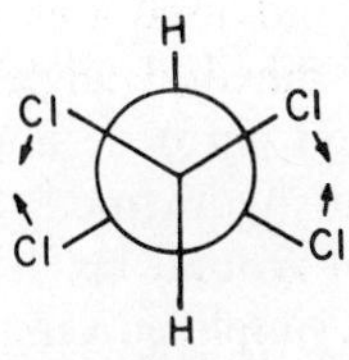

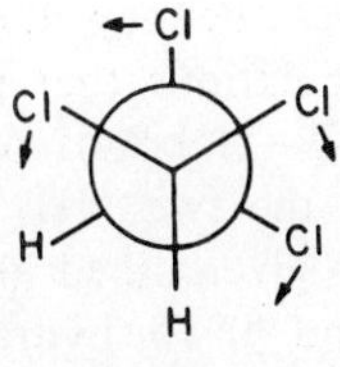

4-1. In the case of 1,1,2-trichloroethane there is one conformer (**6**) with one chlorine gauche to two other chlorine atoms and two conformers (**7** and its mirror image) with a chlorine gauche to one chlorine and trans to another chlorine atom. Since the gauche form of ethylene dichloride is about 1.2 kcal/mole less stable than the trans form in the gas phase, we would expect the doubly gauche form **6** to be less stable than the gauche-trans form **7**. The difference in stabilities between **6** and **7** would be expected to be larger than the difference between the two conformers of ethylene dichloride. The strain between the two chlorine atoms in *gauche*-ethylene dichloride may be relieved by a little rotation around the carbon-carbon bond. In fact, the dihedral angle between the two carbon-chlorine bonds has been found to be $71 \pm 5°$ instead of the 60° that would be characteristic of "perfect" staggering.[4] Strain energy in **6** may not be relieved in this manner, however; rotation in the direction that would decrease one gauche repulsion would increase the other. Therefore it is not surprising that **6** is about 2.6 kcal/mole less stable than **7** in the gas phase.[2] That conformers **8** and **9** of 1,1,2,2-tetrachloroethane are about equally stable in the gas phase[2] seems less predictable, since **8** has only two gauche chlorine interactions whereas **9** has three. The unexpected relative stability of conformers like **9** has been rationalized in terms of the interaction of each halogen atom with the halogen atom attached to the same carbon atom.[5] This interaction, which causes the Cl—C—Cl bond angle in methylene chloride to be almost 112°,[6] will tend to move the chlorine atoms in the directions shown by the arrows in **8** and **9**. The result in **8** will be to aggravate both the chlorine-chlorine gauche interactions, whereas in **9** it will certainly decrease the magnitude of one of the three destabilizing interactions. The greater stability of gauche conformers is common enough that it has been postulated as a general rule that the most stable structure will be the one with the maximum number of gauche interactions of polar bonds.[7,8] However, this rule gives no explanation for why the conformer analogous to **8** in the case of 1,1,2,2-tetrafluoroethane is more stable than the one analogous to **9** by 1.2 kcal/mole in the gas phase.[2] The facts may again be explained in terms of the interactions of pairs of halogen atoms attached to the same carbon atom. The F—C—F angle of 108.3° observed in methylene fluoride[9] shows that there is a net attraction between two fluorine atoms attached to the same carbon atom. To envision the result of such an attraction the arrows in **8** and **9** should be reversed. This should relieve any gauche repulsions in **8** and increase at least one, if not all three, of them in **9**.

4-1b. Rotation Around Bonds to Oxygen and Nitrogen Atoms. Rotation around the carbon-oxygen bonds in alcohols and dialkyl ethers and the carbon-nitrogen bonds in alkyl amines is analogous to rotation around bonds between sp^3-hybridized carbon atoms in that the energy minima correspond to staggered and the energy maxima to eclipsed conformations. These results can be rationalized qualitatively in terms of the same bond-bond (or atom-atom) repulsions that qualitatively explain the preference for staggered conformations in bonds between saturated carbon atoms. When both of the atoms joined by a bond have unshared electron pairs, however, destabilizing interactions between the electron pairs may be important. Such interactions appear to be a major factor controlling the conformations of hydrogen peroxide, hydrazine, hydroxylamine, and their simple derivatives. Penney and Sutherland used this argument in semi-quantitative calculations that predicted dihedral angles of about 90° between the two O—H bonds of hydrogen peroxide and between the two planes bisecting the two NH_2 groups of hydrazine.[10] Subsequent experimental work has given dihedral angles of around 111° for hydrogen peroxide[11,12] and around 90° for hydrazine.[13] A simplified version of the argument may be stated as follows. Figure 4-1 is a projection of the oxygen atom in the compound YOH looking down the O—Y bond. The group Y (not shown) is behind the plane of the paper. The angle Y—O—H is taken to be between 90 and 109.5°. Thus the hydrogen atom is not only to the left of the oxygen but also somewhat above the plane of the paper. For convenience the p_x orbital on oxygen has been located so that its axis is in the Y—O—H plane. The axis of the p_y orbital is taken as the O—Y bond. For this reason the p_y orbital, like the s orbital, is cylindrically symmetrical with respect to the O—Y bond and has no influence on rotation around this bond. Hence, to avoid cluttering, neither the s nor the p_y orbital is shown. The assignments of the locations of the p_x and p_y orbitals require

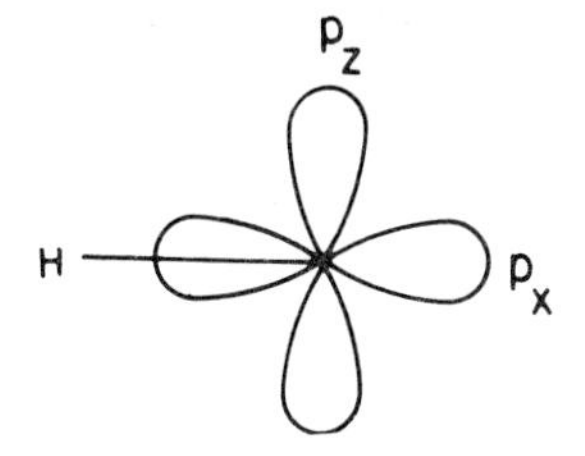

Figure 4-1. A projection down the O—Y bond in the compound YOH.

that the p_z orbital be perpendicular to the YOH plane, as shown in the figure. The orbitals used by the oxygen atom for bonding to Y and H will be *s-p* hybrids. If the bond angle is 109.5° they will be sp^3 hybrids; that is, they will be $\frac{1}{4}$ *s* and $\frac{3}{4}$ *p*, and the *p* orbital used in forming the bond to hydrogen will be p_x. At the other extreme, if the bond angle is 90° the contribution of *s* will be zero; the p_y orbital will be used for bonding to Y and the p_x for bonding to hydrogen. In either case the p_x orbital is at least three-fourths used for bonding and the p_z orbital is used entirely for unshared electrons. Thus, in the case where Y is another hydroxy group, that is, in the case of hydrogen peroxide, the repulsions between unshared electrons will be minimized if the p_z orbitals are placed at right angles to each other, as in projection **10**. The hydrogen atoms must repel each

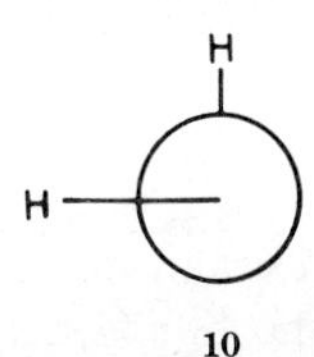

10

other, both by a van der Waals interaction and by a dipole-dipole interaction. (With O—O and O—H bond distances of 1.475 and 0.950 Å, respectively, and an O—O—H bond angle of 94.8°,[14] they would be only 2.1 Å apart if the dihedral angle were 90°; the van der Waals radius for hydrogen is 1.2 Å.) Hence a dihedral angle somewhat larger than 90° is not surprising.

In the case of hydrazine, if the bonding orbitals used by nitrogen were purely *p*, so that the bond angles were all 90°, the unpaired electrons will be in *s* orbitals so that the conformation would not be affected by lone pair interactions. In the real case, however, the bonding orbitals have appreciable *s* character since the bond angles are around 110°.[13] This case may be analyzed in terms of Fig. 4-2, which is a projection down the N—Y bond in the compound YNH_2. Again the *s* and p_y orbitals are not

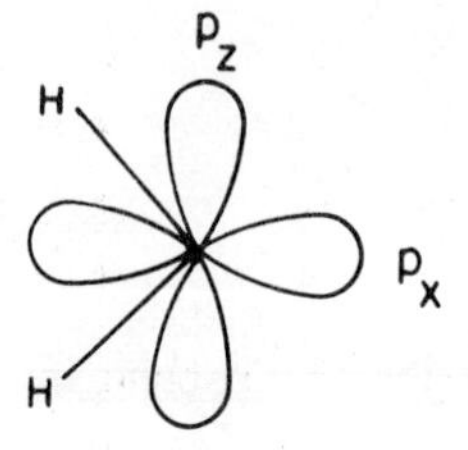

Figure 4-2. A projection down the N—Y bond in the compound YNH_2.

shown since they are cylindrically symmetrical with respect to the bond around which rotation may occur and therefore will not influence such rotation. Calculations of the coefficients of the hybrid orbitals[15] show that all the p_z orbital but only part of the p_x orbital must be used in constructing the bonding orbitals. If the bonding orbitals are taken as sp^3 hybrids, then one-third of the p_x orbital is used, leaving the other two-thirds for unshared electrons. Thus in hydrazine, where two amino groups are attached directly to each other, the lone pair repulsions are minimized when there is a 90° dihedral angle between the two p_x orbitals, as in projection **11**.

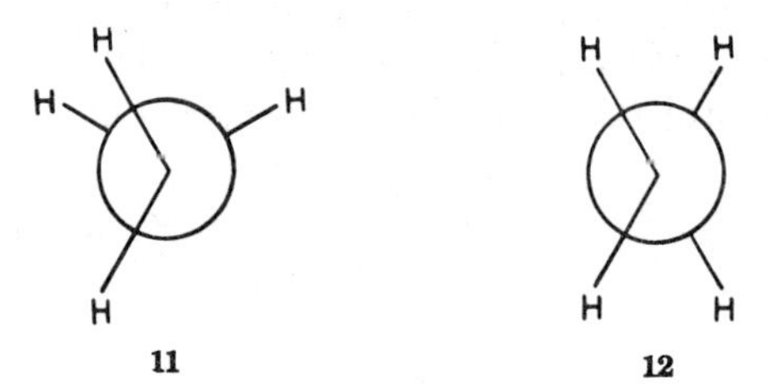

The bond-bond repulsions in **11** could be decreased by decreasing the dihedral angle so as to stagger the bonds better. They would also be decreased in conformer **12** (where the unshared pair repulsions would be at a maximum).

4-1c. Rotation Around Single to sp^2-Hybridized Carbon Atoms. In both propene and acetaldehyde one of the carbon-hydrogen bonds of the methyl group is eclipsed with the double bond to the adjacent carbon atom in the most stable conformer of the compound. One way of rationalizing the relative stability of this conformer is to represent the double bond as two curved bonds corresponding to overlap of two of the four sp^3 orbitals of each of the double-bonded atoms. In these terms a Newman projection down the single bond of propene (**13**) shows that staggering with respect to the two components of the double bond results in eclipsing of the double bond as a whole. Propionaldehyde, its *N*-methyl imine, and 1-butene exist in conformations analogous to **13**. The conformer in which

13

the double bond is eclipsed by a methyl group is the more stable conformer by about 1 kcal/mole for the aldehyde[2] and by about 0.2 kcal/mole for the imine.[16] In the case of 1-butene, the two conformers are of about the same stability (within 0.15 kcal/mole[17]).

4-2. CONFORMATIONAL EQUILIBRIA IN CYCLOHEXANE DERIVATIVES[1]

Sachse recognized long ago that with perfect tetrahedral bond angles and equal carbon-carbon bond lengths there are two possible forms of cyclohexane.[18] One is a rigid form called the *chair* form (**14**), which cannot be

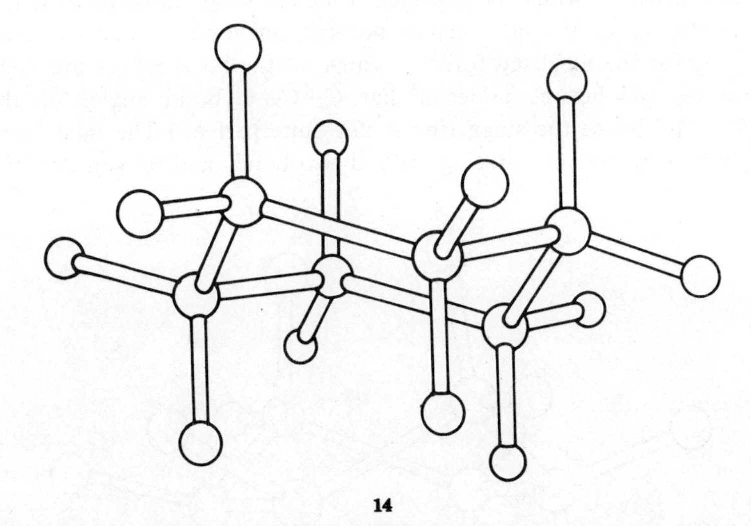

14

distorted by rotation around the carbon-carbon bonds without also changing the bond angles and/or bond lengths. The other is a flexible form, which, by rotations around the carbon-carbon bonds without change in bond angles or lengths, can pass through an infinite number of geometric forms including six different (but distinguishable only if the atoms are made distinguishable, as by isotopic labeling) *twist* forms (**15**) and six dif-

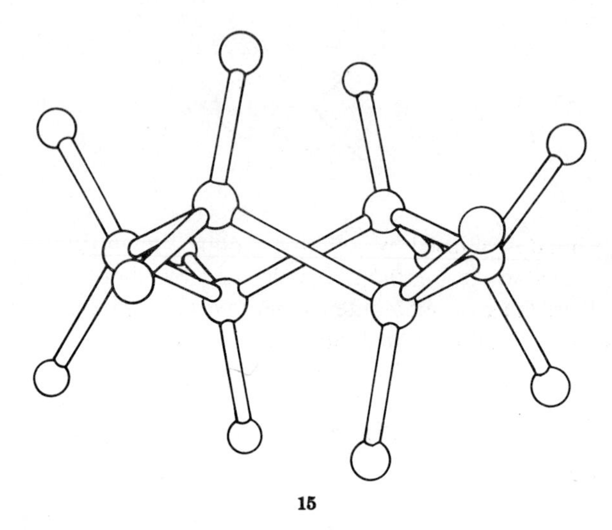

15

ferent *boat* forms (**16**). Hassel showed by electron diffraction measurements that cyclohexane actually has the chair structure,[19] and Pitzer pointed out that this structure would be expected to be the most stable because it has perfect staggering around every carbon-carbon bond.[20,21] (This is exactly true only for the idealized form in which all the bond angles are 109.5°; the actual cyclohexane molecule has C—C—C bond angles of about 111.5°[21] and hence the staggering is not quite perfect.) The boat form is destabilized by perfect eclipsing around two bonds and by van der Waals

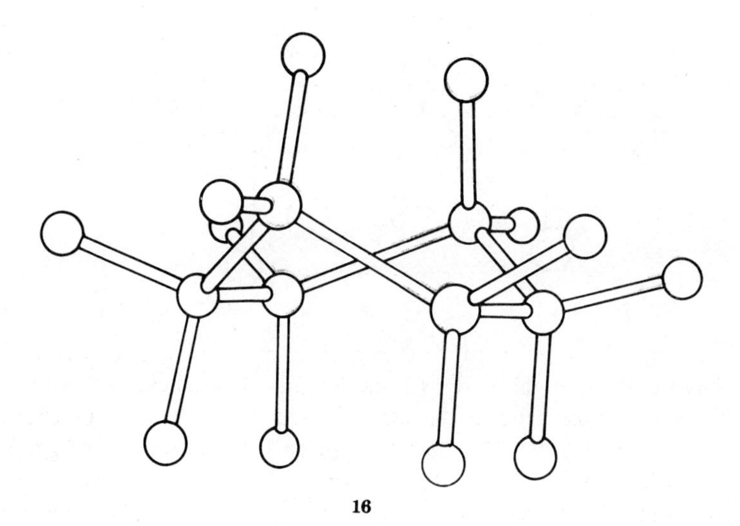

16

repulsions between the two "flagpole" hydrogen atoms. In the twist form (also called the twist-boat or skew-boat form) there is about half as much eclipsing around twice as many bonds and there is considerably less hydrogen-hydrogen repulsion. The twist form is believed to represent the energy minimum for the flexible form and the boat form to represent the energy maximum.[22,23]

The carbon-hydrogen bonds in the chair form of cyclohexane are divided into two categories. There are six, referred to as *axial* bonds, that are parallel to the threefold symmetry axis of the molecule and six, referred to as *equatorial* bonds, that are roughly perpendicular to the threefold symmetry axis. Each carbon atom has one axial and one equatorial bond. The transformation of chair cyclohexane to twist cyclohexane is a rapid process at room temperature, although the equilibrium constant is so small that the twist form has not been detected directly. When chair cyclohexane is re-formed in such an equilibrium there is a 50% probability that a new chair form will be produced in which every hydrogen atom that had been axial is now equatorial, and every one that had been equatorial is now axial. Such a "new" form would not be distinguishable from the old form in the case of cyclohexane itself, but it would be with a substituted cyclohexane (if the substitution is not too symmetrical). With a monosubstituted cyclohexane the result is to equilibrate the equatorial and axial forms of the compound, as shown in Eq. 4-1. Equilibrium constants for the

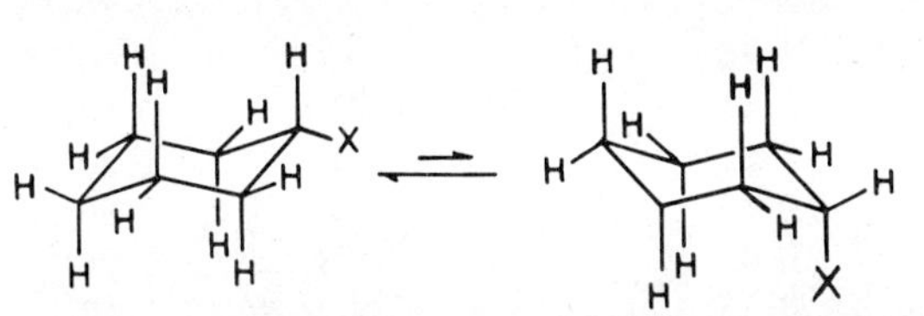

(4-1)

equatorial-axial interconversion of a number of monosubstituted cyclohexanes have been determined or estimated. (In this, as in the case of most other evaluations of physical constants, it would be difficult or impossible to draw a sharp line between determinations and estimates.) Average values for the free energy change for several substituents,[24,25] usually in an organic solvent near room temperature, are listed in Table 4-2. In some cases the value appears to be relatively independent of the nature of the solvent, but in others, where there is a marked interaction between the solvent and the substituent, there is a significant solvent dependence. All the groups listed except —HgOAc are seen to be more stable in the equatorial than in the axial position. This is generally attributed to repulsions between the axial substituent and carbon atoms 3 and 5 and the axial hydrogen atoms on these carbons (although, as pointed out in more detail in the next section,

$\Delta G°$ is a function of *all* the interactions and distortions that are present). These hydrogens are nearer to axial X (for any X except hydrogen) than are the hydrogen atoms on the adjacent carbons 2 and 6. If there were no distortion from perfect tetrahedral bond angles, perfect staggering of bonds, etc., the interactions of the chlorine atom in axial cyclohexyl chloride with carbon and hydrogen atoms at position 3 or 5 would be expected to be essentially like the interaction between the chlorine atom and the methyl group in *gauche-n*-propyl chloride. Yet the interaction in the cyclohexane case is destabilizing (relative to that in the alternative equatorial conformer), whereas the interaction in *gauche-n*-propyl chloride was seen (Table 4-1) to be stabilizing (relative to that in *trans-n*-propyl chloride). The most plausible explanation for this difference is that the much greater flexibility of the propyl chloride molecule makes much easier its distortion

Table 4-2. Standard Free Energy Changes for Equatorial to Axial Changes in Cyclohexyl X[a]

X	$\Delta G°$ [b]	X	$\Delta G°$ [b]
Me	1.70	OH	0.52[f]
Et	1.75	OH	0.87[e]
i-Pr	2.15	OMe	0.60[f]
t-Bu	>4	OCHO	0.59[e,f]
CH=CH_2	1.35	OAc	0.60[f]
C≡CH	0.41[c]	OTs	0.50
Ph	3.0	NH_2	1.20[f]
F	0.15	NH_2	1.60[e]
Cl	0.43	NH_3^+	1.9[e]
Br	0.38	NO_2	1.10
I	0.43	—NC	0.21[c]
CO_2H	1.35[d]	—NCO	0.51[c]
CO_2^-	1.92[e]	—NCS	0.28[c]
CO_2Et	1.20	—SCN	1.23[c]
COCl	1.25	SH	0.9
—CN	0.17	SMe	1.07[c]
$SiCl_3$	0.61[c]	HgOAc	0.0[c]

[a] From Ref. 24 unless otherwise noted.
[b] $G°$ (axial) − $G°$ (equatorial).
[c] Ref. 25.
[d] In oxygen-containing solvents.
[e] In hydroxylic solvents.
[f] In aprotic solvents.

to a geometry permitting strongly stabilizing van der Waals interactions. Thus, for example, in *gauche-n*-propyl chloride the dihedral angle between the carbon-chlorine and carbon-methyl bond has been increased from the 60° characteristic of perfect staggering to about 70°.[26] Such an increase in one of the two relevant dihedral angles of cyclohexyl chloride would decrease the other. Because atoms with large van der Waals radii tend also to have large covalent radii the $\Delta G°$ values in Table 4-2 do not always increase with the size of the group. The situation is the same as that discussed in connection with conformation **4**.

The data in Table 4-2 may also be used as the basis for conclusions about polysubstituted cyclohexanes. For example, since equatorial-axial change in *trans*-1,4-dimethylcyclohexane transforms two equatorial methyl groups to axial methyl groups,

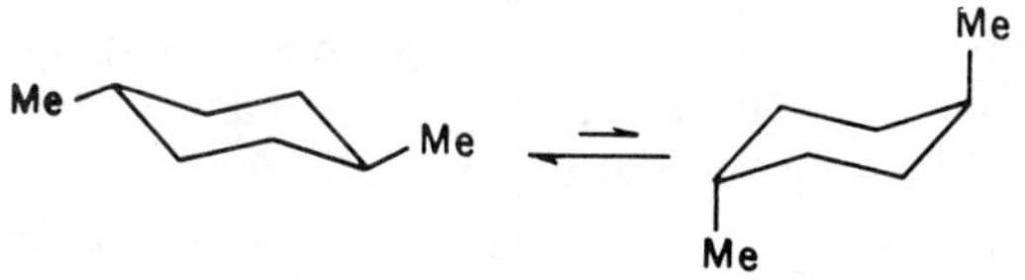

Me Me ⇌ Me Me

$\Delta G°$ would be expected to be around 3.4 kcal/mole, twice the value for methylcyclohexane. With *cis*-1,3-dimethylcyclohexane, transformation of the diequatorial to the diaxial conformer

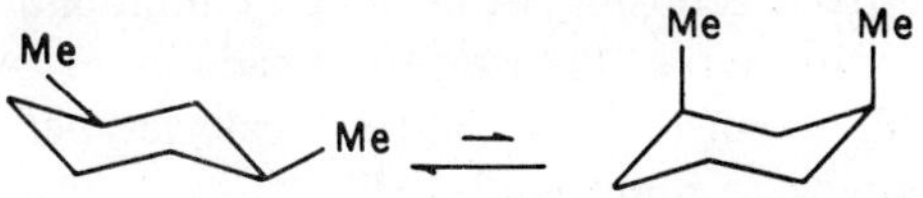

Me Me ⇌ Me Me

should be uphill by more than 3.4 kcal/mole since the two axial methyl groups are now considerably more severely crowded because they are now repelling each other.

4-3. CONFORMATIONAL EQUILIBRIA IN OTHER CYCLIC COMPOUNDS[1]

The eclipsing strain present in the coplanar conformation of cyclobutane may be decreased by torsion around the carbon-carbon bonds in such a way as to raise and lower the carbon atoms alternately. Unlike the case of cyclohexane, however, such torsion increases the strain in the C—C—C bond angles. Apparently the decrease in eclipsing strain is initially greater than the increase in angle strain; cyclobutane is a nonplanar molecule.[27] The angle between the C1—C2—C3 plane and the C1—C4—C3 plane is around 20° for cyclobutane and several of its derivatives.[27–29] Both cyclobutyl bromide[28] and cyclobutyl chloride[29] were found to have the halogen atom predominantly in an equatorial position, as in **17**.

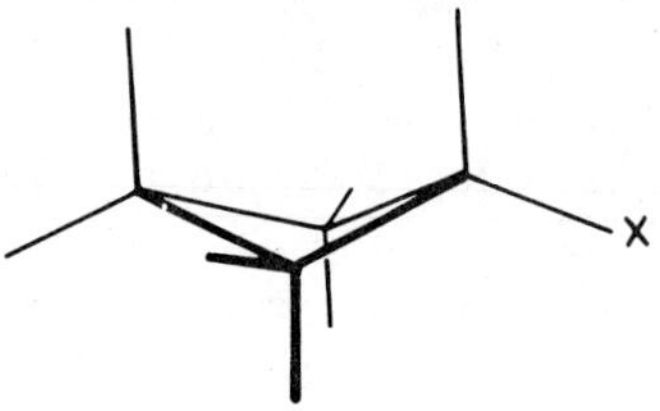

X

17

With cyclopentane and other compounds with odd numbers of atoms in the ring the eclipsing strain of the planar form cannot be as simply relieved by alternately raising and lowering atoms around the ring as may be done with even-ring compounds. Cyclopentane exists as a flexible nonplanar conformer; the carbon atoms vibrate back and forth across the average plane of the ring without ever being in the same plane at the same time and without passing over any significant energy barrier for the molecule as a whole.

Unlike cyclobutane and cyclopentane, cycloalkanes with more than six ring atoms may exist in conformations in which all the valence angles are 109.5°. However, for the cycloalkanes with 7–13 ring atoms all such conformations suffer from eclipsing strain and/or nonbonded repulsions between hydrogen atoms across the ring from each other. Accordingly these compounds are all strained by several kilocalories per mole relative to cyclohexane or to acyclic compounds.[30,31]

With heterocyclic compounds there may be additional factors that influence the conformational equilibrium. As an example we shall consider the "anomeric effect," which is reflected in the tendency of the α anomer of various glucose derivatives to be more stable than might be expected from the usual tendency of substituents larger than hydrogen to avoid axial positions.[32] It seems likely that electrostatic effects are important in this case.[33] Newman projection **18** looks down the bond from carbon-1 to the ring oxygen atom of the pyranose form of a sugar and shows that an atom attached to carbon-1 is nearer carbon-5 when it is α (axial) than when it is

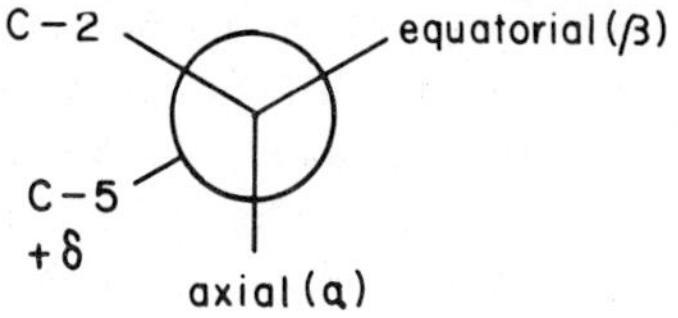

C−2
equatorial (β)
C−5
+δ
axial (α)

18

β (equatorial). When the atom attached to carbon-1 is highly electronegative (as has been the case for the observed examples of the anomeric effect) an α position permits a larger stabilizing electrostatic interaction with carbon-5, which is partially positive because of its attachment to the ring oxygen atom. This electrostatic explanation of the anomeric effect (which is also important with acyclic compounds) is supported by the observation that the magnitude of the effect decreases with increasing dielectric constant of the solvent.[34]

The same factors that operate to produce the anomeric effect in certain heterocyclic compounds also have major influences on conformational equilibria for acyclic compounds in which another electronegative atom is attached to the same carbon atom as an oxygen or amino nitrogen atom.

4-4. CALCULATION OF RELATIVE CONFORMATIONAL STABILITIES[35,36]

A method based on classical mechanics was developed some time ago for the estimation of the preferred atomic positions and energy contents of the various conformers of a compound. In spite of the many approximations implicit in the method, it involved so much tedious computation that it was not widely used until high-speed computers became available to organic chemists. It would be desirable to take polar and resonance effects into account, and some efforts in this direction have been made, but ordinarily only steric effects have been considered. For simplicity we shall discuss here how the method may be applied to a compound like a saturated hydrocarbon, for which the interactions may be reasonably treated as purely steric in nature. In this treatment it is assumed that there is a "normal" bond length for any bond of a given type, a "normal" bond angle for a given pair of bonds to a given atom, and a "normal" dihedral angle for a given pair of bonds separated by a given third bond. Deviations from these normal values may be achieved only by the expenditure of energy. The energy required for stretching a given bond depends on the magnitude of the stretch and on the force constant for stretching the bond. The energy required for bending similarly depends on the magnitude of the bend and the bending force constant. The energy required to change the dihedral angle, which is referred to as the torsional energy, is usually assumed to be a function of the cosine of the dihedral angle. The remaining energy term that is evaluated is the van der Waals energy, which is calculated from equations like those described in Section 2-3. No van der Waals energies are calculated for interactions between pairs of atoms that are bonded to each other since such an interaction is taken to be covered by the normal bond length and the stretching force constant. Ordinarily no energies of interaction of pairs of atoms attached to a common atom are calculated since these are taken to be covered by the normal bond angle and the bending force constant. The van der Waals interactions between the other pairs of atoms may, of course, be either stabilizing or destabilizing. The total energy E may thus be expressed as shown in Eq. 4-2, where E_c is the sum

$$E = E_c + E_b + E_t + E_v \qquad (4\text{–}2)$$

of the energies of compressing and stretching the bonds, E_b is the energy of bond bending, E_t is the torsional energy, and E_v is the van der Waals energy.

With an aliphatic compound the calculation may be started by giving each of the atoms coordinates such that all the bond lengths, bond angles, and dihedral angles are normal. The total energy, which would simply be E_v in this case, is then calculated. Ordinarily there are small changes in the atomic positions that may be made that decrease E_v more than they increase $E_c + E_b + E_t$. The computer program adjusts the atomic positions so as to minimize E. This occurs when even the most favorable further changes in atomic position increase $E_c + E_b + E_t$ as fast as they decrease E_v. Energy minimization procedures have been reviewed.[36] If the energy is minimized with respect to all the coordinates, that is, if any combination of small changes in coordinates produces an increase in the energy, the corresponding set of atomic positions is the optimum set for the given conformer. Usually such a minimum will be only one of several possible minima. By changing the starting geometry to that near the optimum for another conformer we may repeat the minimization procedure and learn the energy content and optimum atomic positions for a second conformer. From the energies of all the possible conformers in their optimum geometries we can tell which is the most stable conformer for the compound. In the case of cyclic compounds there is usually no set of atomic positions such that the bond distances, bond angles, and dihedral angles are all essentially normal. (Cyclohexane, for example, would be an exception if the normal angles were taken to be 109.5°.) But after some moderately plausible initial geometry is fed to the computer, the energy minimization for various conformers is carried out in about the same way as for acyclic compounds.

There are a number of unsatisfactory aspects to the procedure. The idea that the total energy may be expressed as the sum of independent terms is an approximation that may be greatly in error for highly deformed molecules. The equations used for calculating E_c, E_b, E_t, and E_v are only approximate in form, and furthermore many of the constants needed for application of these equations are not available from experimental sources. In fact, there is more than one approach to the problem of arriving at the appropriate equations and constants. In one approach, for example, the normal carbon-hydrogen bond length may be taken as that actually found in a certain compound. In another approach the normal carbon-hydrogen bond length is a disposable parameter whose value is taken so that the energy minimization procedure will give the best agreement with experimental atomic positions and conformational energy content in the case of some known compound or compounds. The normal carbon-hydrogen bond length can be given a unique value for use with any carbon-hydrogen bond or, alternatively, we can use one value for the carbon-hydrogen bonds in carbon-bound methyl groups, another for a methylene group attached to a carbon and an oxygen, etc. The use of a smaller number of parameters simplifies the calculations. The use of a larger number of parameters and the empirical determination of their values certainly permit better agreement with experimental data to be obtained, but the probability of obtaining comparable agreement when treating significantly different types of compounds is decreased. The point is the familiar one that interpolations are usually better than extrapolations.

In the first substantial application of a high-speed computer to conformational analysis Hendrickson neglected bond stretching and compression.[23] This neglect probably did not contribute any major uncertainty to his results. It is so much harder to move atoms by bond stretching and compression than it is by bond bending and rotation around bonds that bond distances of a given type are remarkably independent of the nature of the molecules in which they appear. The chair form of cyclohexane was calculated to be 5.7 kcal/mole more stable than the twist form, in excellent agreement with the values 5.9 and 5.5 kcal/mole,[37,38] which were arrived at from experimental data in a somewhat indirect manner. The value 1.0 kcal/mole, calculated for the difference in energies between the axial and equatorial forms of methylcyclohexane,[39] does not agree as well with the experimental value of about 1.7 kcal/mole. Methods of calculating the various energy terms were developed further by several groups of workers. The steric energy calculation was combined with 10 additive parameters for structural features to calculate the enthalpies of formation of more than 40 saturated hydrocarbons containing widely varied amounts of strain with the agreement usually within the experimental uncertainty.[40] In this calculation, by Allinger and co-workers, the bending force constants were chosen to fit molecular geometry rather than spectra, the van der Waals interactions of hydrogen were taken as centered around a point

8% of the way from the hydrogen nucleus to the nucleus of the atom to which it was bonded, allowance was made for interaction of bending and stretching bonds, and different normal bond angles were chosen depending on the nature of the surrounding atoms.

Attention has also been given to computer conformational analysis of nonhydrocarbons. Allinger and co-workers made calculations on aldehydes, ketones, chlorides, alcohols, sulfoxides, and nitriles.[41] Kitaigorodsky and co-workers studied aromatic compounds.[42] In their calculations on polypeptides in solution, Scheraga and co-workers used constant bond lengths and constant bond angles, but they allowed for electrostatic interactions, hydrogen bonding, and interactions with the solvent.[43]

The method has also been used for the calculation of transition state stabilities, in which case the energy is not minimized with respect to *all* the coordinates. It must be a maximum with respect to the reaction coordinate.

4-5. EQUILIBRIUM IN CIS-TRANS ISOMERIZATION

4-5a. Cis-Trans Equilibria in Cyclohexane Derivatives. Data on cis-trans isomerism in the cyclohexane series may often be explained satisfactorily in terms of free energy changes for the transformation of various substituents from axial to equatorial positions, such as those listed in Table 4-2. For example, isomerization of *cis*-1,4-dimethylcyclohexane to *trans*-1,4-dimethylcyclohexane changes a compound that must have one methyl group axial to one that has available a conformation with both methyl groups equatorial. According to Table 4-2 an axial methyl group is destabilized by 1.70 kcal/mole. If the trans isomer is assumed to consist purely of the diequatorial conformer, there will also be a ΔS_σ contribution of $-R \ln 2$ or -1.38 eu resulting from the fact that the cis isomer has a symmetry number of 1 and the trans isomer a symmetry number of 2. Combination of these two factors gives a K value of 3.37 for isomerization of the cis to the diequatorial trans isomer at 175° C. A similar treatment gives a K value of 0.07 for isomerization of the cis to the diaxial trans isomer at this temperature. The total equilibrium constant of 3.44 for isomerization of *cis*- to *trans*-1,4-dimethylcyclohexane at 175° C is not far from the experimental value of 4.1.[44] (Treatment of the 1.70 kcal/mole contribution as temperature-independent is an implicit assumption that it is purely an enthalpy factor.) With the 1,3-dimethylcyclohexanes the situation is similar except that it is now the trans isomer that must have one methyl group axial. Therefore the equilibrium constant for the formation of diequatorial *cis*-1,3-dimethylcyclohexane from the trans isomer would be calculated to be 3.37 at 175° C. In this case the diaxial conformer is too unstable to add significantly to this value. The experimental value is 4.3.[44]

With the 1,2-dimethylcyclohexanes the situation is different in that each conformer of each isomer is strained. The trans isomer has available to it the highly strained diaxial conformer and the diequatorial conformer, which has a gauche methyl-methyl interaction. The cis isomer exists as an axial-equatorial conformer that also has a gauche methyl-methyl interaction. Assuming one gauche methyl-methyl interaction to be the same size as the other, the isomerization of the cis to the trans isomer would be calculated to have the same equilibrium constant as for the 1,4-dimethylcyclohexanes. The experimental value, 5.8,[44] is considerably larger, probably partly because the gauche methyl-methyl interaction in the cis isomer is larger than that in the diequatorial conformer of the trans isomer. In the cis isomer the axial methyl group is involved in three different gauche interactions simultaneously; this gives it much less freedom for distortion to minimize the interactions.

Equilibration of the cis and trans isomers of a number of 3- and 4-substituted derivatives of *t*-butylcyclohexane has been studied. In most of these cases it seemed reasonable to assume that the *t*-butyl group would be so crowded in the axial position that conformers with axial *t*-butyl groups may be neglected. If this is true, $\Delta G°$ for the isomerization of *cis*-3-X-*t*-butylcyclohexane to its trans isomer or of *trans*-4-X-*t*-butylcyclohexane to its cis isomer should just be the free energy of destabilization of X in the axial position. Sodium ethoxide in ethanol was found to transform the ethyl 4-*t*-butylcyclohexanecarboxylates into an equilibrium mixture containing 84.7% of the trans isomer at 78° C.[45] This result gives a $\Delta G°$ value of 1.2 kcal/mole for the carbethoxy group, in agreement with the value listed in Table 4-2.

4-5b. Cis-Trans Equilibria in Other Cyclic Compounds. Seigler and Bloomfield made a systematic study of the effect of ring size on equilibrium in cis-trans isomerism in their investigation of five dimethyl cycloalkane-1,2-dicarboxylates.[46] As shown in Table 4-3 the equilibrium constant is larger in the cyclopropane case than any other. This is the case in which there is most nearly perfect eclipsing of bonds to adjacent carbon

H
C—CO_2Me
$(CH_2)_{n-2}$
C—CO_2Me
H

NaOMe / MeOH ⇌

H
C—CO_2Me
$(CH_2)_{n-2}$
C—H
CO_2Me

(4-3)

Table 4-3. Equilibrium in the Cis-Trans Isomerization of Dimethyl Cycloalkane-1,2-dicarboxylates[a]

Cycloalkane	K	$\Delta G°$
Cyclopropane	99	−2.62
Cyclobutane	8.58	−1.27
Cyclopentane	6.08	−1.06
Cyclohexane	11.7	−1.45
Cycloheptane	3.99	−0.82

[a] For the cis → trans reaction in methanol at 50° C.[46]

atoms. Thus dimethyl *cis*-cyclopropane-1,2-dicarboxylate is the compound with the largest destabilizing interaction between carbomethoxy groups. The other four equilibrium constants do not differ greatly from each other. Apparently there are several factors at work, and they do not all operate in the same direction. For example, the even number of atoms in the cyclobutane and cyclohexane rings tends to facilitate staggering of the bonds around the ring, but it also requires one of the carbomethoxy groups to occupy an axial position in the trans isomer.

A qualitative analogy to the cyclohexane case may be seen in the observation that the equilibrium constants for the formation of the trans from the cis isomer in the case of 3-*t*-butylcyclobutanol (0.21) and methyl 3-*t*-butylcyclobutanecarboxylate (0.46) at 100° C are both less than 1.0.[47] Only in the more stable cis isomers can both substituents be in axial positions at the same time. Similarly, the equilibrium constants for the formation of *trans*-1,3-dihalocyclobutanes from the cis isomers at 124° C are 0.69, 0.48, and 0.46 for the chloro, bromo, and iodo compounds, respectively.[48] In these cases the steric factors favoring the cis isomers must overwhelm unfavorable dipole-dipole repulsions between the carbon-halogen dipoles.

4-5c. Cis-Trans Equilibria in Alkenes. Steric effects can make the equilibrium in the cis-trans isomerization of alkenes lie almost entirely on the side of the cis isomer or almost entirely on the side of the trans isomer. For example, although the trans isomers of large-ring cycloalkenes are somewhat more stable than the cis isomers (just as in the case of acyclic alkenes), with smaller rings the cis isomers become more stable. The equilibrium constants for cis to trans isomerism in acetic acid at 80° C are 1.87, 2.49, 0.062, and 0.0034 for cyclododecene, cycloundecene, cyclodecene, and cyclononene, respectively.[49] The equilibrium constants for smaller rings do not appear to have been measured, but the observation that the hydrogenation of *trans*-cyclooctene (the smallest *trans*-cycloalkene that has been isolated) is 9.3 kcal/mole more exothermic than that of the cis isomer shows that the cis to trans equilibrium constant must be very small.[50] On the other hand, with the 1,2-di-*t*-butylethylenes[51] it is the cis isomer whose hydrogenation is more exothermic (by 9.3 kcal/mole).[52]

A number of equilibrium constants for cis to trans isomerization by propenyl compounds[53–62] and other compounds of the type XCH=CHY[55,58,62–70] are listed in Tables 4-4 and 4-5, respectively. The former table shows that even the smallest alkyl groups interact in a destabilizing manner when they are cis to each other across a carbon-carbon double bond. The cis interaction between methyl and halogen or cyano substituents appears to be stabilizing (relative to the corresponding trans interaction, at least). It might be argued that this stabilization results from electrostatic attraction of the methyl group, made positive by electron donation to the carbon-carbon double bond, and the halogen or cyano groups, which are negative because of their electron-withdrawing ability. However, such an argument cannot explain why several of the 1,2-dihaloethylenes, as shown in Table 4-5, are more stable in the cis than in the trans form. Statistical mechanical calculations of entropies, heat capacities, zero-point vibrational energies, etc., (cf. Section 1-5) show that the greater stability of the cis isomer in the case of the difluoro, chlorofluoro, and dichloro compounds, at least, is a potential energy effect rather than a kinetic energy effect.[67] The most plausible explanation for the greater stabilities of the cis isomers is that they arise from London forces (Section 2-2c) between the two halogen atoms. Certainly the energy of interaction between two atoms or groups will commonly be stabilizing if the distance between them is not too small. Prediction of when the interaction will be stabilizing is made difficult by the fact that the critical distance where the interaction changes from stabilizing to destabilizing must depend on the direction of approach, hybridization of the bonding to the atom, and various other factors. Thus, although *cis*-1,2-difluoroethylene is considerably more stable than its trans isomer, with the 1,2-difluorocyclopropanes, where the fluorine-fluorine distances must be similar (but not identical), the trans isomer is more stable—K is 12 at 313° C.[67] On the other hand, the fact that the gauche conformer of 1,2-difluoroethane is as stable as the trans isomer, or more so (Table 4-1), shows that there is probably a significant stabilizing interaction from London forces to counteract the unfavorable dipole-dipole repulsion.

Table 4-4. Equilibrium Constants for Reactions of the Type *cis*-MeCH=CHX ⇄ *trans*-MeCH=CHX at 25° C.

X	*K*	Medium	Ref.
Me	3.0	Gas	53
Et	4.6	Gas	53
n-Pr	4.0[a]	Gas	54
n-Bu	4.0[a]	Gas	54
i-Pr	5.5[a]	Gas	54
cis-CH_2=CHCH=CH	1.8[a]	Gas	55
trans-CH_2=CHCH=CH	3.0[a]	Gas	55
CH_2CN	2.0	*t*-BuOH	56
CO_2H	12[b]	Liquid	57
CO_2Me	11[b]	$(Me_2N)_3PO$	58
CN	0.62[a]	Gas	59
NO_2	30	Heptane	60
OMe	1.2	Cyclohexane	58
F	0.28[a]	Gas	61
Cl	0.27[a]	Gas	61
Br	0.26[a]	Gas	61
SO_2Me	~40[b]	Et_3N	62

[a] Calculated from values of ΔH^0 and ΔS^0 determined at higher temperatures.

[b] Assuming that $\Delta G°$ is the same at 25° C as at the higher temperature at which the experiment was carried out.

The rather large values of *K* listed in Table 4-4 for crotonic acid, methyl crotonate, and 1-nitropropene might at first seem surprising. If the plane of the nitro or CO_2R group were turned perpendicular to the plane of the carbon-carbon double bond, the interaction with the methyl group in the cis isomer should be no larger than the interaction of a halogen with the methyl group. However, such a perpendicular orientation would deprive the cis isomer of the stabilizing resonance interaction between the nitro or CO_2R group and the carbon-carbon double bond. Such an interpretation is supported by the *K* values listed for 1,2-diphenylethylene and 1-phenyl-2-mesitylethylene in Table 4-5. In spite of the greater size of the mesityl substituent the mesityl compound exists as the cis isomer to a much greater extent than its phenyl analogue does at equilibrium. This must be because the methyl groups in the mesityl compound prevent the mesityl ring from being coplanar with the carbon-carbon double bond even in the trans isomer (**19**). Hence the nonplanarity of the cis isomer gives a smaller loss

Table 4-5. Equilibrium Constants for Reactions of the Type *cis*-XCH═CHY ⇄ *trans*-XCH═CHY at 25° C

X	Y	*K*	Medium	Ref.
Et	*n*-Pr	6.1[a]	Gas	63
Ph	Ph	480	Cyclohexane	64
Ph	Mesityl[b]	13	Cyclohexane	64
Et	OMe	1.8	Cyclohexane	58
$PhCH_2$	OMe	0.56	Me_2SO	65
n-$C_{10}H_{21}$	SMe	1.4	Me_2SO	62
CO_2Me	OMe	420[c]	Cyclohexane	58
trans-MeCH═CH	CH_2═CH	21[c]	Gas	55
CH_2═CH	F	0.54[c]	Benzene	66
F	F	0.22[a]	Gas	67
F	Cl	0.30[a]	Gas	67
F	I	0.4[c]	Liquid	68
Cl	Cl	0.39[a]	Gas	67
Br	Br	0.85[c]	Liquid	69
Br	I	1.4[c]	Liquid	69
I	I	3.1[c]	Liquid	69
PhS	PhS	0.8[c]	*p*-Xylene	70

[a] Calculated from values of $\Delta H°$ and $\Delta S°$ determined at higher temperatures.

[b] 2,4,6-Trimethylphenyl.

[c] Assuming that $\Delta G°$ is the same at 25° C as at the higher temperature at which the experiment was carried out.

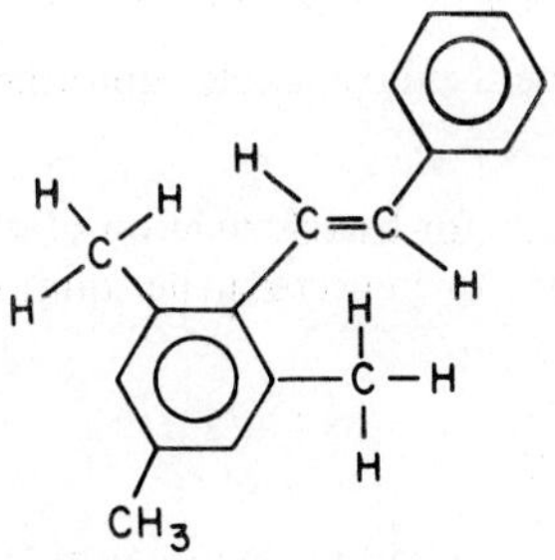

19

of resonance interactions between the double bond and the aromatic rings than in the case of 1,2-diphenylethylene.

The large value of *K* for methyl 3-methoxypropenoate shown in Table 4-5 presumably reflects the large amount of resonance energy—that of the two substituents with each other as shown in contributing structure **20** as

$$\overset{\oplus}{\text{MeO}}{=}\text{CH}{-}\text{CH}{=}\text{C}\begin{matrix}\diagup \text{O}^{\ominus} \\ \diagdown \text{OMe}\end{matrix}$$

20

well as that of each substituent with the double bond—that is present in the trans isomer. It is not clear how much of this resonance stabilization is lost by the deviations from coplanarity present in the cis isomer, but the loss appears to be greater than with methyl crotonate or crotonic acid (Table 4-4).

PROBLEMS

1. Would you expect ethylene dichloride to have a larger dipole moment in the gaseous or in the liquid state (at a given temperature)? Why?

2. Draw a Newman projection of the most stable conformer of hydroxylamine.

3. The pK_a values for the conjugate acids of pyridine, pyridazine (1,2-diazine), pyrimidine (1,3-diazine), and pyrazine (1,4-diazine) are 5.2, 2.2, 1.2, and 0.5, respectively.[71] Thus replacing any CH group of pyridine by nitrogen decreases its basicity. Why does the largest decrease take place when the most distant CH group is replaced and the smallest decrease take place when the nearest CH group is replaced?

4. Use the values of $\Delta G°$ listed in Table 4-2 to estimate the equilibrium constant for cis-trans isomerism in 1,4-dibromocyclohexane at 25° C. In which direction should electrostatic effects influence this equilibrium constant?

5. What do you expect to be the most stable conformer of dimethoxymethane and why?

6. Is the equilibrium constant for transformation of *cis*-1,2-dichloroethylene to its trans isomer larger in pentane or in acetonitrile solution? Why?

REFERENCES

1. E. L. Eliel, N. L. Allinger, S. J. Angyal, and G. A. Morrison, *Conformational Analysis*, Wiley, New York, 1966.
2. J. P. Lowe, *Progr. Phys. Org. Chem.*, **6,** 1 (1968).
3. E. B. Wilson, *Chem. Soc. Rev.*, **1,** 293 (1972).
4. L. E. Sutton, Ed., *Tables of Interatomic Distances and Configuration in Molecules and Ions*, Special Publication No. 11, The Chemical Society, London, 1958.

5. I. Miyagawa, T. Chiba, S. Ikeda, and Y. Morino, *Bull. Chem. Soc. Japan*, **30,** 218 (1957).
6. D. Chadwick and D. J. Millen, *Trans. Faraday Soc.*, **67,** 1539 (1971).
7. S. Wolfe, A. Rauk, L. M. Tel, and I. G. Csizmadia, *J. Chem. Soc. B*, 136 (1971).
8. S. Wolfe, *Accounts Chem. Res.*, **5,** 102 (1972).
9. E. Hirota, T. Tanaka, A. Sakakibara, Y. Ohashi, and Y. Morino, *J. Mol. Spectrosc.*, **34,** 222 (1970).
10. W. G. Penney and G. B. B. M. Sutherland, *Trans. Faraday Soc.*, **30,** 898 (1934); *J.Chem. Phys.*, **2,** 492 (1934).
11. R. H. Hunt, R. A. Leacock, C. W. Peters, and K. T. Hecht, *J. Chem. Phys.*, **42,** 1931 (1965).
12. R. H. Hunt and R. A. Leacock, *J. Chem. Phys.*, **45,** 3141 (1966).
13. T. Kasuya, *Sci. Papers Inst. Phys. Chem. Res.* (Tokyo), **56** (1), 1 (1962); *Chem. Abstr.*, **58,** 6206b (1963).
14. R. L. Redington, W. B. Olson, and P. C. Cross, *J. Chem. Phys.*, **36,** 1311 (1962).
15. Cf. A. Streitwieser, Jr., *Molecular Orbital Theory for Organic Chemists*, Wiley, New York, 1961, Sec. 1-3.
16. G. J. Karabatsos and S. S. Lande, *Tetrahedron*, **24,** 3907 (1968).
17. P. B. Woller and E. W. Garbisch, Jr., *J. Org. Chem.*, **37,** 4281 (1972).
18. H. Sachse, *Ber.*, **23,** 1363 (1890); *Z. Phys. Chem.*, **10,** 203 (1892).
19. O. Hassel, *Tidsskr. Kjemi, Bergv. Met.*, **3,** 32 (1943).
20. K. S. Pitzer, *Science*, **101,** 672 (1945).
21. M. Davis and O. Hassel, *Acta Chem. Scand.*, **17,** 1181 (1963).
22. P. Hazebroek and L. J. Oosterhoff, *Discuss. Faraday Soc.*, **10,** 87 (1951).
23. J. B. Hendrickson, *J. Amer. Chem. Soc.*, **83,** 4537 (1961); **85,** 4059 (1963).
24. J. A. Hirsch, *Topics Stereochem.*, **1,** 199 (1967).
25. F. R. Jensen, C. H. Bushweller, and B. H. Beck, *J. Amer. Chem. Soc.*, **91,** 344 (1969).
26. T. N. Sarachman, *J. Chem. Phys.*, **39,** 469 (1963).
27. A. Almenningen, O. Bastiansen, and P. N. Skancke, *Acta Chem. Scand.*, **15,** 711 (1961).
28. W. G. Rothschild and B. P. Dailey, *J. Chem. Phys.*, **36,** 2931 (1962).
29. H. Kim and W. D. Gwinn, *J. Chem. Phys.*, **44,** 865 (1966).
30. S. Kaarsemaker and J. Coops, *Rec. Trav. Chim. Pays-Bas*, **71,** 261 (1952).
31. J. Coops, H. Van Kamp, W. A. Lambregts, B. J. Visser, and H. Dekker, *Rec. Trav. Chim. Pays-Bas*, **79,** 1226 (1960).
32. R. U. Lemieux, *Advan. Carbohydr. Chem.*, **9,** 1 (1954): *Pure Appl. Chem.*, **25,** 527 (1971).
33. J. T. Edward, *Chem. Ind.* (London), 1102 (1955).
34. E. L. Eliel and C. A. Giza, *J. Org. Chem.*, **33,** 3754 (1968).

35. F. H. Westheimer, in *Steric Effects in Organic Chemistry*, M. S. Newman, Ed., Wiley, New York, 1956, Chap. 12.
36. J. E. Williams, P. J. Stang, and P. v. R. Schleyer, *Ann. Rev. Phys. Chem.*, **19,** 531 (1968); E. M. Engler, J. D. Andose, and P. v. R. Schleyer, *J. Amer. Chem. Soc.*, **95,** 8005 (1973).
37. N. L. Allinger and L. A. Freiberg, *J. Amer. Chem. Soc.*, **82,** 2393 (1960).
38. W. S. Johnson, V. J. Bauer, J. L. Margrave, M. A. Frisch, L. H. Dreger, and W. N. Hubbard, *J. Amer. Chem. Soc.*, **83,** 606 (1961).
39. J. B. Hendrickson, *J. Amer. Chem. Soc.*, **84,** 3355 (1962).
40. N. L. Allinger, M. T. Tribble, M. A. Miller, and D. H. Wertz, *J. Amer. Chem. Soc.*, **93,** 1637 (1971).
41. N. L. Allinger, J. A. Hirsch, M. A. Miller, and I. J. Tyminski, *J. Amer. Chem. Soc.*, **91,** 337 (1969).
42. A. I. Kitaigorodsky and V. G. Dashevsky, *Tetrahedron*, **24**, 5917 (1968).
43. H. A. Scheraga, *Advan. Phys. Org. Chem.*, **6,** 103 (1968).
44. C. Boelhouwer, G. A. M. Diepen, J. van Elk, P. T. van Raaij, and H. I. Waterman, *Brennst.-Chem.*, **39,** 299 (1958).
45. E. L. Eliel, H. Haubenstock, and R. V. Acharya, *J. Amer. Chem. Soc.*, **83,** 2351 (1961).
46. D. S. Seigler and J. J. Bloomfield, *J. Org. Chem.*, **38,** 1375 (1973).
47. G. M. Lampman, G. D. Hager, and G. L. Couchman, *J. Org. Chem.*, **35,** 2398 (1970).
48. K. B. Wiberg and G. M. Lampman, *J. Amer. Chem. Soc.*, **88,** 4429 (1966).
49. A. C. Cope, P. T. Moore, and W. R. Moore, *J. Amer. Chem. Soc.*, **81,** 3153 (1959); **82,** 1744 (1960).
50. R. B. Turner and W. R. Meador, *J. Amer. Chem. Soc.*, **79,** 4133 (1957).
51. W. H. Puterbaugh and M. S. Newman, *J. Amer. Chem. Soc.*, **81,** 1611 (1959).
52. R. B. Turner, D. E. Nettleton, Jr., and M. Perelman, *J. Amer. Chem. Soc.*, **80,** 1430 (1958).
53. H. Akimoto, J. L. Sprung, and J. N. Pitts, Jr., *J. Amer. Chem. Soc.*, **94,** 4850 (1972).
54. R. Maurel, M. Guisnet, and L. Bove, *Bull. Soc. Chim. Fr.*, 1975 (1969).
55. K. W. Egger and T. L. James, *J. Chem. Soc. B*, 348 (1971).
56. M. Procházka, J. Zelinka, A. Vilím, and J. V. Černý, *Collect. Czech. Chem. Commun.*, **35,** 1224 (1970).
57. M. B. Hocking, *Can. J. Chem.*, **50,** 1224 (1972).
58. S. J. Rhoads, J. K. Chattopadhyay, and E. E. Waali, *J. Org. Chem.*, **35,** 3352 (1970).
59. J. N. Butler and R. D. McAlpine, *Can. J. Chem.*, **41,** 2487 (1963).
60. G. Hesse and V. Jäger, *Justus Liebigs Ann. Chem.*, **740,** 79 (1970).

61. P. I. Abell and P. K. Adolf, *J. Chem. Thermodyn.*, **1,** 333 (1969).
62. D. E. O'Connor and W. I. Lyness, *J. Amer. Chem. Soc.*, **86,** 3840 (1964).
63. K. W. Egger, *J. Amer. Chem. Soc.*, **89,** 504 (1967).
64. G. Fischer, K. A. Muszkat, and E. Fischer, *J. Chem. Soc. B*, 1156 (1968).
65. H. Kloosterziel and J. A. A. van Drunen, *Rec. Trav. Chim. Pays-Bas*, **89,** 32 (1970).
66. H.-G. Viehe, *Chem. Ber.*, **97,** 598 (1964).
67. N. C. Craig, L. G. Piper, and V. L. Wheeler, *J. Phys. Chem.*, **75,** 1453 (1971).
68. H.-G. Viehe, J. Dale, and E. Franchimont, *Chem. Ber.*, **97,** 244 (1964).
69. H.-G. Viehe and E. Franchimont, *Chem. Ber.*, **96,** 3153 (1963).
70. A. Ohno, T. Saito, A. Kudo, and G. Tsuchihashi, *Bull. Chem. Soc. Japan*, **44,** 1901 (1971).
71. D. D. Perrin, *Dissociation Constants of Organic Bases in Aqueous Solution*, Butterworths, London, 1965.

5

Brønsted Acidity and Basicity—Solvent Effects[1-6]

5-1. DEFINITIONS

We shall follow the suggestion of Swain and Scott and use the terms *acidity* and *basicity* to refer to *equilibria* and the terms *electrophilicity* and *nucleophilicity* to refer to reaction rates.[7] To indicate whether we are discussing Brønsted basicity or Lewis basicity (and what kind of Lewis basicity) we may add suitable prefixes to these terms.[8] Thus the hydrogen basicity (or proton basicity) of a base is its Brønsted basicity and is measured by equilibrium constants for reactions in which the base removes protons from various acids (i.e., forms a bond to hydrogen). Similarly, the carbon basicity of a base is measured by equilibrium constants for reactions in which the base forms a bond to carbon. Carbon basicity is a general term that includes methyl basicity, ethyl basicity, acetyl basicity, phenyl basicity, etc., as more specific subdivisions. Unless it is otherwise made clear, the unprefixed terms acidity, basicity, nucleophilicity, electrophilicity, acid, acidic, etc., will refer to Brønsted acidity (or Brønsted electrophilicity).

Although this system of definitions is fairly widely used,[9a] it is incompatible with another fairly widely used system in which the term "base" refers to Brønsted basicity and "nucleophilic reagent" to Lewis basicity.[9b] Nevertheless, some chemists use both systems of definitions without appearing to notice the conflict between them.[9]

The strength of a Brønsted acid in the solvent S may be measured by the equilibrium constant for the reaction

$$HA^n + S \rightleftharpoons SH^+ + A^{n-1} \qquad (5\text{-}1)$$

The equilibrium constant for such a process is usually written in a form in which the constant concentration of solvent has been absorbed by K_a.

$$K_a = \frac{[SH^+][A^{n-1}]}{[HA^n]} \qquad (5\text{-}2)$$

For an electrically neutral acid ($n = 0$), the acidity constant may also be called an ionization constant. This is not such a reasonable term for a monocationic acid ($n = 1$), where there are no more ionic products in Eq. 5-1 than there were reactants.

5-2. SOLVENT ACIDITY AND BASICITY

The "ionization" of a Brønsted acid in solution is actually the donation of a proton from the acid to the solvent, and the "ionization" of a base is the abstraction of a proton from the solvent. It is not surprising then that the magnitudes of ionization constants are highly dependent on the acidity and basicity of the solvent. Thus, although 2,4-dinitroaniline is such a weak base in aqueous solution

$$2,4\text{-}(O_2N)_2C_6H_3NH_2 + H_2O \rightleftharpoons 2,4\text{-}(O_2N)_2C_6H_3NH_3^+ + OH^-$$

that it is difficult to detect any 2,4-dinitroanilinium ions, in the solvent sulfuric acid

$$2,4\text{-}(O_2N)_2C_6H_3NH_2 + H_2SO_4 \rightleftharpoons 2,4\text{-}(O_2N)_2C_6H_3NH_3^+ + HSO_4^-$$

it is so strong a base that it is difficult to detect any unprotonated 2,4-dinitroaniline. Similarly, acetic acid is largely un-ionized at moderate concentrations in aqueous solution; it is essentially completely ionized in liquid ammonia solution.

If two acids are too strongly acidic toward a given solvent, their ionization may be so complete as to make it impractical to compare their strengths in that solvent. For example, if the ionization constants of HX and HY are 10^4 and 10^2, the hydrogen ion concentrations of their 0.1 M solutions will be 0.099999 M and 0.0999 M, respectively. Thus equal concentrations of the two acids would yield equal concentrations of hydrogen ions, within the experimental error. On the other hand, if the ionization constants had been 10^{-5} and 10^{-7} (differing by 100-fold, just as in the previous case) the hydrogen ion concentrations of 0.1 *M* solutions would be 10^{-3} *M* and 10^{-4} *M*, respectively, and the difference in the strengths of the acids would be easily measurable. This tendency to make all acids (or bases) whose strengths are at a certain level or higher to appear to be of equal strength is known as the *leveling effect* of the solvent.

Because of the leveling effect, it may not be practical to compare the strengths of two acids or bases in a given solvent. If the strengths of two acids are indistinguishable because both acids are so strong that they are essentially completely deprotonated in a given solvent, we may compare their strengths by going to a more weakly basic solvent. However, we cannot guarantee that the relative strengths of the two acids in the first solvent will be the same as in the second solvent. There are several factors that can cause the relative strengths of acids to vary from one solvent to another. These other solvation factors are discussed in the next two sections.

5-3. SOLVENT EFFECTS ON ACIDS OF DIFFERENT CHARGE TYPE

When two acids are of different electrical charge type, their relative strength is particularly likely to vary from solvent to solvent. For example, in water, acetic acid (whose acidity constant is 1.8×10^{-5} *M*) is almost three times as strong an acid as the pyridinium ion, whose acidity constant is 6.2×10^{-6} *M*. In methanol, where the acidity constants are 2.2×10^{-10} *M*[10] and 2.8×10^{-6} *M*,[11] respectively, the pyridinium ion is more strongly acidic by more than 12,000-fold. The equilibrium constant for the reaction

$$C_5H_5NH^+ + S \overset{S}{\rightleftharpoons} SH^+ + C_5H_5N$$

decreases only slightly on going from the solvent water to methanol (perhaps because methanol is slightly less basic than water*). The equilibrium constant for the reaction

$$AcOH + S \overset{S}{\rightleftharpoons} SH^+ + AcO^-$$

decreases by almost 10^5-fold on going from water to methanol, probably largely because the reaction involves the net formation of two ions, and water is a much better ion solvator (for most ions, at least) than methanol is. Much larger changes in relative acidity may be observed when the ion-solvating power of the medium is varied more widely.

*Goldschmidt and co-workers described evidence that water is a somewhat stronger base than methanol in aqueous methanol solution.[12–14]

5-4. SPECIFIC SOLVATION EFFECTS

Even when two acids are of the same electrical charge type, there may be rather large differences in their relative acidities in two different solvents. Consider the ionization of two electrically neutral acids, HA and HB, in two solvents, S_1 and S_2. Let the acidity constants of HA in the two solvents be denoted $K_1{}^A$ and $K_2{}^A$ and those of HB denoted $K_1{}^B$ and $K_2{}^B$. For each of the two acids, we may write a thermodynamic acidity constant K_t, which is expressed in terms of activities and is therefore independent of the nature of the solvent.

$$K_t = \frac{a_{SH^+} a_{A^-}}{a_{HA}} = \frac{[SH^+][A^-]}{[HA]} \cdot \frac{\gamma_{SH^+} \gamma_{A^-}}{\gamma_{HA}} \tag{5-3}$$

where the a's are activities and the γ's activity coefficients. If activities are defined to approach concentrations in dilute solution in S_1 (i.e., activity coefficients are defined to approach unity in dilute S_1), then

$$K_t{}^A = K_1{}^A \quad \text{and} \quad K_t{}^B = K_1{}^B \tag{5-4}$$

It follows that

$$K_1{}^A = K_t{}^A = K_2{}^A \frac{\gamma_{SH^+} \gamma_{A^-}}{\gamma_{HA}} \tag{5-5}$$

where the γ's are activity coefficients in S_2 relative to S_1. Analogously,

$$K_1{}^B = K_2{}^B \frac{\gamma_{SH^+} \gamma_{B^-}}{\gamma_{HB}} \tag{5-6}$$

Dividing Eq. 5-5 by Eq. 5-6,

$$\frac{K_1{}^A}{K_1{}^B} = \frac{K_2{}^A}{K_2{}^B} \cdot \frac{\gamma_{A^-} \gamma_{HB}}{\gamma_{B^-} \gamma_{HA}} \tag{5-7}$$

The activity coefficients of the electrically neutral acids may be obtained from solubility or vapor pressure data, or in other ways. It is not possible to determine the activity coefficient of a single ion by thermodynamic measurements,[15] but one can determine an activity coefficient ratio, such as $\gamma_{A^-}/\gamma_{B^-}$, in various ways. If the activity coefficient ratio $(\gamma_{A^-} \gamma_{HB})/(\gamma_{B^-} \gamma_{HA})$ is not unity, the relative strength of HA and HB in solvent S_1 will differ from what it is in S_2. To the extent that log γ may be expressed exactly by the rule of additivity of atomic contributions, this activity coefficient ratio will always be unity. Experimentally, however, it is often far from unity. For example, chloroacetic acid is more than 15 times as strong an acid as 2,4-dinitrophenol (pK_a's 2.86 and 4.10) in water, but in dimethylformamide solution it is only one-thousandth as strong (pK_a's 9.0 and 6.0).[16] It is believed that the difference in hydrogen-bonding ability of the two solvents is an important factor in producing this 20,000-fold reversal in relative acid strengths. The negative charge in the chloroacetate anion is concentrated largely on the two oxygen atoms. These relatively highly charged atoms interact rather strongly with water, which forms hydrogen bonds to them. The negative charge in the 2,4-dinitrophenoxide anion is distributed over five oxygen atoms and three carbon atoms, none of which are sufficiently more highly charged than they were in the phenol to accept much stronger hydrogen bonds from water. Thus, on going from water to dimethylformamide, the chloroacetate ion loses much more solvation than the dinitrophenoxide ion does, and the tendency to form chloroacetate ions decreases much more than the tendency to form dinitrophenoxide ions does.

Even larger changes in relative acidities of acids of the same electrical charge type may be observed when an acid whose conjugate base has its charge concentrated largely on one small atom is compared with an acid whose conjugate base has a highly diffused charge. For example, in methanol solution, fluoradene[17] (p$K \sim 17$) is only slightly more acidic than methanol (pK 18):[18]

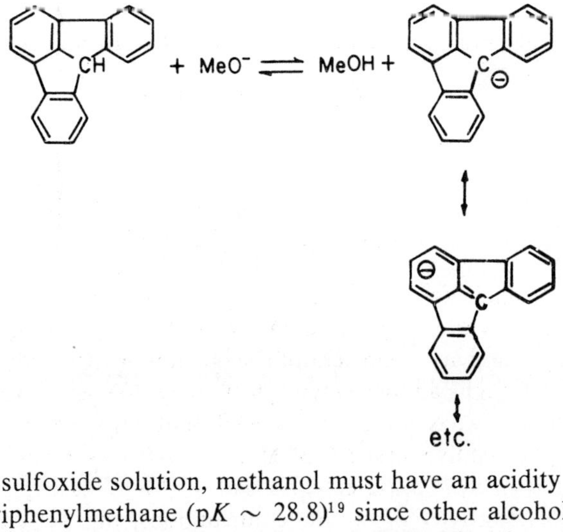

CH + MeO$^-$ ⇌ MeOH + C$^\ominus$

↕

C

↕

etc.

In dimethyl sulfoxide solution, methanol must have an acidity comparable to that of triphenylmethane (p$K \sim 28.8$)[19] since other alcohols, including ethanol, 1-propanol, and *t*-butyl alcohol do.[19,20] Therefore, in dimethyl sulfoxide, fluoradene (pK 10.5)[18] is about 10^{18} times as acidic as methanol. This corresponds to a change of more than 10^{15}-fold in the relative strengths of these two acids on going from methanol to dimethyl sulfoxide. It seems clear that a major factor in producing this enormous change in relative acidities is the strong solvation via hydrogen bonding that occurs when an alkoxide ion is dissolved in a hydroxylic solvent. An anion with a highly diffuse charge, such as the conjugate base of fluoradene, in which the negative charge is distributed over 19 carbon atoms, has much less tendency to be hydrogen bonded by a hydroxylic solvent. Hence such a carbanion loses much less stabilization due to solvation on going from a hydroxylic to a nonhydroxylic solvent than an alkoxide ion does. Even a nonhydroxylic solvent must interact strongly with any ion with a highly concentrated exposed* charge, especially if the solvent contains an exposed dipole. The stabilization of an alkoxide ion in dimethyl sulfoxide by charge-permanent dipole and charge-induced dipole interactions must be large enough to make the acidity of methanol relative to that of fluoradene much greater than it would be in the absence of such interactions. In this connection it is interesting that in the gas phase water (whose acidity in hydroxylic solvents is in within a factor of 10 of that of methanol) is a weaker acid than toluene[21] (whose acidity in dimethyl sulfoxide is much too small to determine by equilibrium measurements).

The tendency of anions with fairly concentrated exposed charges to accept hydrogen bonds is so great that in solvents that have no hydrogen atoms capable of forming strong hydrogen bonds such anions often form hydrogen-bonded complexes with other species that are present. Thus, for example, at moderate concentrations in acetonitrile solution, the dissociation of nitric acid leads largely to the nitric acid-conjugated nitrate ion

$$2HNO_3 + MeCN \rightleftharpoons MeCNH^+ + NO_3^- \cdot HNO_3$$

instead of to nitrate ions that are solvated only by the acetonitrile.[22] Similar behavior is observed for sulfuric, hydrobromic, and hydrochloric acids. In the case of hydrofluoric acid such behavior is observed even in aqueous solution, where the reaction

$$2HF + H_2O \rightleftharpoons H_3O^+ + HF_2^-$$

has long been known to occur. The tendency for such hydrogen bonding to be observable in this case is especially strong, at least partly because the fluoride ion has a particularly concentrated charge and because hydrogen fluoride is highly polar and much more acidic than water. Ordinarily

*Although the charge on a quaternary ammonium ion is concentrated rather largely on the nitrogen atom, it is not exposed. That is, because of the intervening alkyl groups the solvent may not get immediately adjacent to the positively charged nitrogen atom. Therefore the interaction between solvent and charge will be much smaller than it would otherwise have been.

hydroxylic solvents hydrogen bond to anions so effectively that potential hydrogen bonding species present at low concentrations in the solution cannot compete significantly. The tendency of anions to be hydrogen bonded by other solutes such as their conjugate acids is thus decreased when the solvent has hydrogen atoms capable of forming strong hydrogen bonds. It is also decreased by the ability of the solvent to accept hydrogen bonds. This may be seen in a comparison of acetonitrile, which neither forms nor accepts strong hydrogen bonds, with dimethyl sulfoxide, which does not form very strong hydrogen bonds but which does act as a strong acceptor of hydrogen bonds. For several phenols and carboxylic acids, equilibrium constants for association with their conjugate bases

$$HA + A^- \rightleftharpoons A^- \cdot HA$$

have been found to be 10–1000 times as large in acetonitrile as in dimethyl sulfoxide.[23] When this equilibrium is written in the more complete form

$$S \cdot HA + A^- \rightleftharpoons A^- \cdot HA + S$$

where S is the solvent, the smaller equilibrium constants obtained in dimethyl sulfoxide are seen to be explained in terms of the stronger hydrogen bonds formed by a given HA to this solvent.

In solvents that do not hydrogen bond to them strongly, anions tend to form complexes not only with solutes that are capable of hydrogen bonding but also with cations, especially those having concentrated exposed positive charges. The complexes formed with cations are ion pairs, which have a strong tendency to be formed in poor ion-solvating media of low dielectric constant, such as hydrocarbons, halocarbons, monoamines, ethers, etc. They are formed to a lesser but still significant extent in solvents often referred to as dipolar aprotic solvents, such as acetonitrile, dimethyl sulfoxide, nitromethane, acetone, *N,N*-dimethylformamide, hexamethylphosphoramide, sulfolane, etc., which have medium to high dielectric constants but contain no hydrogen atoms capable of forming strong hydrogen bonds. Ion pairs are also formed to major extents in hydrogen-bonding solvents, such as alcohols, amides, and carboxylic acids. Solutions of potassium *t*-butoxide in *t*-butyl alcohol are more basic, as measured by their effects on indicators, than solutions of sodium *t*-butoxide of the same concentration because of the greater tendency of the smaller sodium ions to coordinate with the *t*-butoxide ions.[24]

Although solvent effects on acidity and basicity are thus very strongly influenced by solvation of highly concentrated exposed charges, there are also other important factors present. Differences in solvation of electrically neutral species may be significant. For example, the solubility of *p*-nitrobenzoic acid increases by 125-fold on going from water to methanol whereas that of picric acid increases by only 11-fold.[25] Thus the term γ_{HB}/γ_{HA} in Eq. 5–7 has the value (1/11)/(1/125) or 11.4 when solvent 1 is water, solvent 2 is methanol, HA is *p*-nitrobenzoic acid, and HB is picric acid. Hence the 49-fold change in relative acidities that occurs on going from water (where picric acid is 400 times as acidic as *p*-nitrobenzoic acid) to methanol (where picric acid is 19,500 times as acidic as *p*-nitrobenzoic acid) is due to a 11.4-fold effect arising from differences in solvation of the neutral acids and a 4.3-fold effect arising from differences in solvation of the anions.[25]

Evidence for the importance of London forces, that is, electron-correlation energies, in the solvation of acids and bases has been described by Grunwald and Price.[26] The magnitude of London forces (cf. Section 2-2c) increases with increasing polarizability of the interacting species and decreases with increasing distance between the species. In any reaction in solution electron-correlation interactions with the solvent are fairly large, but in many cases they roughly cancel each other. From correlations of molecular refraction with molecular structure,[27] it may be seen that the polarizabilities of molecules may often be calculated rather reliably from contributions from the various constituent atoms, bonds, or groups. A comparison of the acidities of acids HA and HB may be stated in terms of the equilibrium constant for the reaction

$$HA + B^- \rightleftharpoons A^- + HB$$

The atoms in the reactants are exactly the same as those in the products, and the bonds are usually almost all, if not all, the same. Therefore it is not surprising that the sum of the polarizabilities of the reactants is often very nearly equal to the sum of the polarizabilities of the products. It has been noticed that species containing highly delocalized electrons are more polarizable than would be expected on the basis of the contributions due to the constituent atoms or bonds. It was therefore expected that reactions that are accompanied by major changes in the degree of delocalization of electrons could be affected by changes in the polarizability of the solvent. The effective polarizability of the solvent water is particularly low. Not only is single-bonded oxygen less polarizable than carbon, nitrogen, sulfur, or double-bonded oxygen, the hydrogen-bonded structure of water orients the water molecules in such a way that there are "holes" in the solvent structure. These holes are completely nonpolarizable. Furthermore, the triatomic character of water molecules tends to minimize the number of atoms of solvent molecules that can get close to a solute molecule. When a molecule of solute is surrounded by water molecules, each of the three atoms of each surrounding water molecule is very close to some part of the solute molecule. The molecules of water in the second layer of solvent are farther away from the solute by the van der Waals distance between two water molecules. When a molecule of solute is surrounded by a solvent with large molecules, the first layer of atoms around it is probably no closer to it than when water is the solvent, but the second layer of atoms consists largely of atoms in the same molecules as the first layer of atoms. Therefore the atoms in the second layer are farther away from the solute by a covalent bond distance, which is much smaller than a van der Waals distance.

Grunwald and Price measured equilibrium constants for the reaction

$$\mathrm{HA} + \mathrm{BH^+{\cdot}OAc^-} \xrightleftharpoons{\mathrm{HOAc}} \mathrm{BH^+{\cdot}A^-} + \mathrm{HOAc}$$

where $BH^+{\cdot}OAc^-$ and $BH^+{\cdot}A^-$ are ion pairs held together by hydrogen bonding and electrostatic attraction, in the cases where B is ammonia and trimethylamine and HA is picric acid and trichloroacetic acid. The equilibrium constant for the formation of trimethylammonium picrate is almost 20 times as large as that for the formation of ammonium picrate. In contrast, the equilibrium constant for the formation of trimethylammonium trichloroacetate is only about 25% larger than that for the formation of ammonium trichloroacetate. The extent of electron delocalization that occurs when picric acid ionizes is far greater than that which occurs when trichloroacetic acid ionizes. The highly polarizable picrate ion is stabilized more by interaction with a trimethylammonium ion than by interaction with an ammonium ion, in which fewer polarizable atoms are held near the picrate ion. Calculations concerning this case show that electron correlation energies should be of an order of magnitude large enough to explain the observed data.[26]

5-5. ACIDITY FUNCTIONS

Early work on acidity functions, which were originated by Hammett,[28] was reviewed by Paul and Long,[29] and the entire field has been covered by Rochester.[30] Acidity functions in strongly basic media were discussed by Bowden[31] and acidity functions in mixtures of water and strong acids by Arnett and Mach.[32,33]

5-5a. A Method for Determining Acidity Constants in Water-Sulfuric Acid Mixtures. For the conjugate acid of an electrically neutral base (A),

$$\mathrm{AH^+} \rightleftharpoons \mathrm{H^+} + \mathrm{A}$$

the thermodynamic acidity constant K_A may be written

$$K_A = \frac{a_{H^+}a_A}{a_{AH^+}} = \frac{[H^+][A]}{[AH^+]} \cdot \frac{\gamma_{H^+}\gamma_A}{\gamma_{AH^+}} \quad (5\text{-}8)$$

and for pK_A we may write

$$pK_A = -\log \frac{a_{H^+}a_A}{a_{AH^+}} = \log \frac{a_{AH^+}}{a_A} - \log a_{H^+} \quad (5\text{-}9)$$

If we define activities to approach concentrations in dilute aqueous solution, then for such solutions it follows that

$$pK_A = \log \frac{[AH^+]}{[A]} + pH \quad (5\text{-}10)$$

If A is a strong enough base to become protonated to a measurable extent in dilute aqueous solution, pK_A may be determined by measuring the ratio $[AH^+]/[A]$ in a solution of known pH. In many cases this may be done by spectrophotometric measurements on A or AH^+. If A is too weak a base, however, it may not be possible to transform a reliably measurable amount of it to AH^+ in dilute aqueous solution. It may be possible to obtain measurable concentrations of AH^+ in rather strongly acidic solutions, but it would be unreasonable to assume that activities are equal to concentrations in a rather strongly acidic solution. The problem of determining pK_A under such conditions was attacked by Hammett and Deyrup, who used the following approach.[34]

Let A and B be bases whose conjugate acids have thermodynamic pK values of pK_A and pK_B, respectively. For B, the following equation, analogous to Eq. 5-9, may be written.

$$pK_B = \log \frac{a_{BH^+}}{a_B} - \log a_{H^+} \quad (5\text{-}11)$$

Subtracting Eq. 5-11 from Eq. 5-9

$$pK_A - pK_B = \log \frac{a_{AH^+}}{a_A} - \log \frac{a_{BH^+}}{a_B} \quad (5\text{-}12)$$

or, in terms of concentrations and activity coefficients,

$$pK_A - pK_B = \log \frac{[AH^+]}{[A]} - \log \frac{[BH^+]}{[B]} + \log \frac{\gamma_{AH^+}\gamma_B}{\gamma_A\gamma_{BH^+}} \quad (5\text{-}13)$$

The bases chosen for study were mostly aromatic amines containing enough electron-withdrawing substituents to make complete protonation impossible in dilute aqueous solution. For most of the bases, in fact, protonation was not even detectable in dilute aqueous solution (strong acid concentrations less than 0.1 *M*). For each base the extent of protonation was measured by spectral means over as wide a range of mixtures of sulfuric acid and water as possible, that is, from solutions containing too little acid to protonate a reliably measurable fraction of the base to those too acidic for a reliably measurable fraction of the base to remain unprotonated. (In most cases these measurements contain some uncertainty resulting from the fact that the spectral changes observed on changing the concentration of strong acid arise not only from changes in the relative concentrations of the two species present—AH^+ and A, for example—but also from changes in the individual spectra of one or both of the two species; the corrections made for such solvent effects on spectra are somewhat arbitrary.) Then various pairs of bases whose extents of protonation had been measured in several of the same water-sulfuric acid mixtures were compared. For each water-sulfuric acid mixture in which extents of protonation were measured, it was possible to calculate the value of $[AH^+]/[A]$ and $[BH^+]/[B]$ for the two bases A and B that were being compared. For each pair of bases that Hammett and Deyrup compared it was found that the value of the function $\log([AH^+]/[A]) - \log([BH^+]/[B])$ was constant, within the experimental uncertainty, over the range of water-sulfuric acid mixtures in which it was measured. (Unfortunately, this range was usually a fairly narrow one, e.g., from 39 to 51% sulfuric acid.) Since the values of pK_A and pK_B are solvent independent, it follows from Eq. 5-13 that over any range of solvent composition in which $\log([AH^+]/[A]) - \log([BH^+]/[B])$ is constant, the term $\log[(\gamma_{AH^+}\gamma_B)/(\gamma_A\gamma_{BH^+})]$ must also be constant. The ranges of solvent composition over which $\log[(\gamma_{AH^+}\gamma_B)/(\gamma_A\gamma_{BH^+})]$ was found to be constant for one pair of bases or another covered most of the possible range from 0 to 100% sulfuric acid. It therefore seemed plausible that for each of the compounds studied $\log[(\gamma_{AH^+}\gamma_B)/(\gamma_A\gamma_{BH^+})]$ is a constant throughout the range 0–100% sulfuric acid. This range includes pure water and dilute aqueous solutions, where acidity coefficients are defined as unity and where $\log[(\gamma_{AH^+}\gamma_B)/(\gamma_A\gamma_{BH^+})]$ must therefore have the value zero. If this function is to be a constant it must then have the value zero in any water-sulfuric acid mixture.

For a pair of bases for which the activity coefficient term in Eq. 5-13 is zero, the equation takes the form

$$pK_A - pK_B = \log\frac{[AH^+]}{[A]} - \log\frac{[BH^+]}{[B]} \tag{5-14}$$

The series of bases studied by Hammett and Deyrup included one, *p*-nitroaniline, that was strong enough to be protonated to a measurable extent in dilute aqueous sulfuric acid but weak enough to require more than 20% sulfuric acid for essentially complete protonation. The pK value for a compound like this can be determined in dilute aqueous solution. Somewhat weaker bases, such as *o*-nitroaniline and 4-chloro-2-nitroaniline, cannot be protonated to a significant extent in dilute aqueous solution but can be in aqueous solutions containing less than 20% sulfuric acid. The values of $[AH^+]/[A]$ for *p*-nitroaniline and $[BH^+]/[B]$ for some such weaker base, and the value of pK_A determined for *p*-nitroaniline in dilute aqueous solution were substituted into Eq. 5-14 in order to calculate pK_B for the weaker base. Since this weaker base may require more than 50% sulfuric acid for essentially complete protonation, its extent of protonation may be measured in solutions that are sufficiently strongly acidic to bring about the measurable protonation of some still weaker base such as 2,4-dichloro-6-nitroaniline. From the resultant measurements, Eq. 5-14 can be used again to obtain the pK value for the still weaker base. By continuation of this stepwise procedure it was possible to calculate pK values for 2,4,6-trinitroaniline, a base so weak that almost 100% sulfuric acid is required for its complete protonation.

Values for pK's determined by the method described are, of course, no more reliable than the assumption that the activity coefficient term in Eq. 5-13 vanishes. It was originally hoped that this term would be negligibly small for any two electrically neutral bases A and B in any water-sulfuric acid mixture. However, subsequent investigations showed that if A and B are not rather closely related chemically the activity coefficient term in Eq. 5-13 may not necessarily be neglected. For example, Deno, Groves, and Saines found that if A is one of the nitroaniline bases studied by Hammett and Deyrup and B is a 1,1-diarylethylene derivative, which also acts as a base,

$$Ar_2C{=}CH_2 + H^+ \rightleftharpoons Ar_2\overset{+}{C}CH_3$$

the value of $\log([AH^+]/[A]) - \log([BH^+]/[B])$ changes considerably with changing composition of the water-sulfuric acid mixtures in which the measurements were made.[35] It follows that the values of $\log[(\gamma_{AH^+}\gamma_B)/(\gamma_{BH^+}\gamma_A)]$ must change by the same amount over the same range of solvent composition. Analogous observations have been made for a number of other classes of electrically neutral weak organic bases. Jorgenson and Hartter used modern experimental techniques to update the work of Hammett and Deyrup. Using only ring-substituted derivatives of nitroanilines (including a number of compounds in addition to those studied by Hammett and Deyrup), these workers showed that the activity coefficient term of Eq. 5-13 was negligible for all the pairs of bases for which comparisons were made (in solvent mixtures with compositions spanning the entire range from 0 to 100% sulfuric acid).[36] Furthermore, for several of the compounds essentially the same pK value was obtained from measurements in aqueous sulfuric acid, aqueous hydrochloric acid, and aqueous phosphoric acid.[37] It therefore seems that the activity coefficient term for all the pairs of ring-substituted nitroanilines studied is zero, within the experimental uncertainty, in all mixtures of water and sulfuric acid. Unfortunately, the experimental uncertainty in the application of Eq. 5-13 is large enough that undetected systematic trends in activity coefficient ratios could produce errors in the acidity constants that could easily reach 10-fold for the conjugate acids of some of the weakest bases studied. However, this error is small compared to the difference in strength between such a base and the weakest bases that can be studied in dilute aqueous solution.

5-5b. The Hammett Acidity Function H$_0$. As Hammett and Deyrup pointed out,[34] for any series of compounds for which the activity coefficient term in Eq. 5-13 is zero,

$$\log \frac{\gamma_{AH^+}\gamma_B}{\gamma_A\gamma_{BH^+}} = 0 \tag{5-15}$$

it follows that

$$\log \frac{\gamma_{AH^+}}{\gamma_A} = \log \frac{\gamma_{BH^+}}{\gamma_B} \tag{5-16}$$

That is, the ratio γ_{AH^+}/γ_A, in any given mixture of water and sulfuric acid, has the same numerical value for every compound in the series. Since a_{H^+} in a water-sulfuric acid mixture has a value that depends only on the composition of the mixture and its temperature, the function $(a_{H^+}\gamma_A)/\gamma_{AH^+}$ has a value in a given water-sulfuric acid mixture at a given temperature that depends only on what series of compounds the base A belongs to. This function is called the Hammett acidity function, h_0; the subscript zero refers to the electrical charge on the base. It is most commonly dealt with in the form of its negative logarithm, the Hammett acidity function, H_0.

$$H_0 = -\log h_0 = \log \frac{\gamma_{AH^+}}{\gamma_A a_{H^+}} \tag{5-17}$$

Although H_0 does not have the same value for all electrically neutral bases, as had originally been hoped, it is still a very useful function. Its evaluation for any series of compounds may be illustrated by the following specific example for ring-substituted nitroanilines. The evidence that Eq. 5-15 is a good approximation and the determination of pK values for a number of compounds of this type by Jorgenson and Hartter have been described.

If a value of pK_A so determined for the base A is substituted into the following equation, which follows readily from Eq. 5-8 or 5-9,

$$pK_A = \log \frac{[AH^+]}{[A]} - \log \frac{a_{H^+}\gamma_A}{\gamma_{AH^+}} \qquad (5\text{-}18)$$

along with an experimentally determined value of $[AH^+]/[A]$, the only unknown is $-\log [(a_{H^+}\gamma_A)/\gamma_{AH^+}]$, that is, H_0. Values of H_0 determined in this way are listed in Table 5-1. The values at up to 60% sulfuric acid are those of Paul and Long.[29] Those at more than 60% sulfuric acid are due to Jorgenson and Hartter[36] and are larger in absolute magnitude than the values previously used, which had been determined using bases (not all ring-substituted nitroanilines) to which Eq. 5-15 had not been shown to be applicable.

From its definition, it may be seen that H_0 becomes identical to pH in dilute aqueous solution. From this fact and the data in Table 5-1, it may be seen that this measure of the tendency of the solvent to protonate bases of the type in question changes from 7.0 in pure water to about -12.5 in pure sulfuric acid. Thus, if the following slightly more convenient form of Eq. 5-18 is used (note the similarity to Eq. 5-10)

$$pK_A = \log \frac{[AH^+]}{[A]} + H_0 \qquad (5\text{-}19)$$

we may calculate that for the bases studied the ratio $[AH^+]/[A]$ is $10^{19.5}$

Table 5-1. The Acidity Function H_0 for Ring-substituted Nitroanilines in Water-Sulfuric Acid Mixtures at about 25° C[29,36]

% H_2SO_4[a]	H_0	% H_2SO_4	H_0	% H_2SO_4	H_0
5	0.24	55	−3.91	92	−9.29
10	−0.31	60	−4.46	94	−9.68
15	−0.66	64	−4.95	96	−10.03
20	−1.01	68	−5.50	97	−10.21
25	−1.37	72	−6.10	98	−10.41
30	−1.72	76	−6.71	98.82	−10.62
35	−2.06	80	−7.34	99.27	−10.92
40	−2.41	84	−7.97	99.44	−11.12
45	−2.85	88	−8.61	100	−12.5[b]
50	−3.38	90	−8.92		

[a] By weight.

[b] Extrapolated from the data of Ref. 36, using the data of Ref. 29 as a guide.

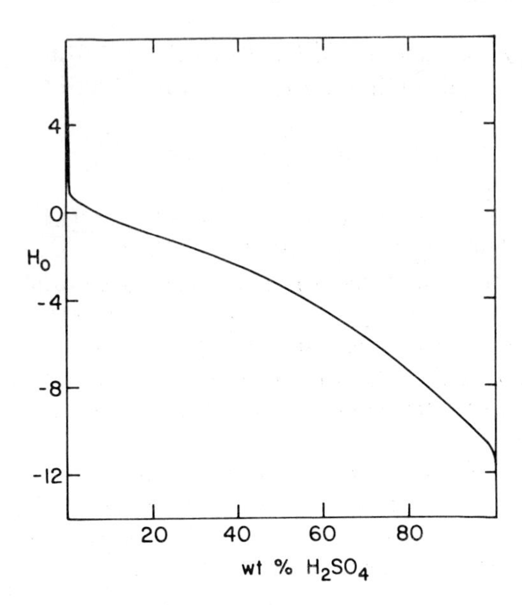

Figure 5-1. Plot of H_0 for ring-substituted nitroanilines versus percent sulfuric acid in water-sulfuric acid mixture.

times as large in pure sulfuric acid as in pure water. Most of this increase in the acidic strength of the solution takes place when the first sulfuric acid is added (between 0 and 5% sulfuric acid) and when the last of the water is replaced (between 90 and 100% sulfuric acid). This may be seen from Table 5-1, or more clearly, from Figure 5-1, a plot of H_0 versus the percent sulfuric acid. It is not surprising that this plot resembles a titration curve, since sulfuric acid is a strong acid in aqueous solution and water is a strong base in sulfuric acid solution.

Values of pK for the bases used to determine the H_0 values in Table 5-1 are listed in Table 5-2. The compounds are referred to as Hammett bases, although this term is sometimes used to include all the bases studied by Hammett and Deyrup, some of which do not appear to display the same activity-coefficient behavior as the compounds listed in Table 5-2.

5-5c. Other Acidity Functions. Acidity-function measurements have also been made on Hammett bases in mixtures of water with perchloric, phosphoric, nitric, hydrochloric, hydrobromic, hydrofluoric, trifluoroacetic, trichloroacetic, and other acids.[29,34,37] In the cases in which enough measurements were made to tell, the Hammett activity coefficient postulate (that Eq. 5-15 is a good approximation) was found to hold. However, fairly

Table 5-2. pK Values for Hammett Base Used in Determining the H_0 Values in Table 5-1.

Base	pK
p-Nitroaniline	0.99
o-Nitroaniline	−0.25
4-Chloro-2-nitroaniline	−0.97
2,5-Dichloro-4-nitroaniline	−1.78
2-Chloro-6-nitroaniline	−2.43
2,6-Dichloro-4-nitroaniline	−3.27
2,4-Dichloro-6-nitroaniline	−3.34
2,4-Dinitroaniline	−4.50
2,6-Dinitroaniline	−5.54
4-Chloro-2,6-dinitroaniline	−6.14
2-Bromo-4,6-dinitroaniline	−6.68
3-Methyl-2,4,6-trinitroaniline	−8.22
3-Bromo-2,4,6-trinitroaniline	−9.46
3-Chloro-2,4,6-trinitroaniline	−9.71
2,4,6-Trinitroaniline	−10.10

small changes in molecular structure are sometimes enough to cause the Hammett activity coefficient postulate to fail. Arnett and Mach, for example, found that when such Hammett bases as those listed in Table 5-2 were compared with similar tertiary amines (including *N,N*-dialkyl and *N*-aryl derivatives of some of the compounds in Table 5-2) in water-strong acid mixtures, Eq. 5-14 (and hence Eq. 5-15) was not obeyed.[33,37] Thus the activity coefficient behavior of *N,N*-disubstituted nitroaniline derivatives is different from that of the corresponding primary amines. The activity coefficient behavior of the tertiary amines is internally consistent, however, and for this reason one may define an acidity function based on *N,N*-disubstituted nitroaniline derivatives. Values of this acidity function, which Arnett and Mach denote H_0''', and the acidity function for Hammett bases, which they denote H_0', in aqueous hydrochloric acid solutions are listed in Table 5-3. Also listed are values of H_R, an acidity function for triarylcarbinols, which act as bases by the following type of reaction:

$$Ar_3COH + H^+ \rightleftharpoons Ar_3C^+ + H_2O \qquad (5\text{-}20)$$

Acidity functions for reactions like this, in which water is split out from the base and attacking proton, have also been denoted J_0, C_0, and in other ways.[29–33] One reason why H_R increases in absolute value with in-

Table 5-3. Values of Acidity Functions in Aqueous Hydrochloric Acid at About 25° C[37]

% HCl	H_0'[a]	H_0'''[b]	H_R[c]	$H_R - \log a_{H_2O}$
0.1	1.55	1.57	1.70	1.70
0.4	0.94	0.91	1.08	1.08
1	0.52	0.49	0.56	0.56
2	0.16	0.09	0.11	0.12
6	−0.55	−0.83	−0.97	−0.94
10	−1.01	−1.51	−1.74	−1.67
14	−1.46	−2.12	−2.50	−2.39
18	−1.90	−2.68	−3.38	−3.20
22	−2.34	−3.27	−4.27	−4.01
26	−2.81	−3.94	−5.26	−4.90
30	−3.33	−4.63	−6.23	−5.75
34	−3.85	−5.30	−7.21	−6.58
38	−4.37	−5.96		

[a] For ring-substituted nitroanilines.

[b] For *N,N*-disubstituted nitroanilines.

[c] For triarylcarbinols.

creasing hydrochloric acid concentration more rapidly than H_0' and H_0''' do is that the decreasing activity of water that accompanies increasing hydrochloric acid concentration must drive reaction 5-20, in which water is a by-product, to the right. The value of H_R may be corrected for this complication by subtracting $\log a_{H_2O}$ from it. Although this correction does move H_R nearer H_0' and H_0''', as shown in Table 5-3, by no means does it bridge the entire gap. The difference between H_0' and H_0''' may also be partly due to changes in the activity of water. The conjugate acids of the primary amines are probably fairly strongly hydrogen bonded to at least three water molecules via the three acidic protons on the positively charged nitrogen atom. The conjugate acids of the tertiary amines are probably less highly hydrogen bonded. With reasonable corrections for hydration and the formation of water as a by-product, the three acidity functions in Table 5-3 may be brought considerably closer together. However, only by varying the magnitudes of the water-activity corrections rather arbitrarily with the hydrochloric acid concentration can the three acidity functions be brought within the experimental uncertainty of each other.

Acidity functions have also been determined, most commonly in aqueous sulfuric acid, for other classes of electrically neutral organic bases, including benzophenone derivatives,[38] indole derivatives,[39] and amides and pyridine *N*-oxides,[40,41] among others. It is still not clear just how much structural variation is permissible within a given class of bases, but in general, structural changes made near the site of protonation are more likely to cause deviant activity-coefficient behavior.

A useful alternative to the procedure of defining a new acidity function for each new class of compounds studied in strong aqueous acid solution has been devised by Bunnett and Olsen.[42] The base (B) being investigated is studied experimentally in the usual way, with the determination of $[BH^+]/[B]$ over as wide a range of acid concentrations as possible. Then $\log([BH^+]/[B]) + H_0$ is plotted against $H_0 + \log[H^+]$, where $[H^+]$ is taken as equal to the concentration of strong acid and H_0 is the acidity function for ring-substituted nitroaniline derivatives (Table 5-1). The intercept of the plot is pK_B, and the slope is referred to as ϕ. It may be shown that this treatment is equivalent to the assumption of the following linear free energy relationship involving activity coefficients:

$$\log \frac{\gamma_{AH^+}\gamma_B}{\gamma_{BH^+}\gamma_A} = \phi \log \frac{\gamma_{AH^+}}{\gamma_A \gamma_{H^+}}$$

where A is an ideal Hammett base, that is, a ring-substituted nitroaniline derivative to which the H_0 values in Table 5-1 are applicable. Values of pK_B calculated by the procedure of Bunnett and Olsen agree quite well with those determined using alternative acidity functions up to about 80% sulfuric acid. The procedure is not recommended for use at sulfuric acid concentrations above 85%, and it appears to be less reliable in perchloric than in sulfuric acid solutions.

The preceding discussion of acidity functions concerns only electrically neutral bases. Nevertheless, acidity functions have been applied to both positively and negatively charged (including some polycharged) bases. The most important applications have been to bases with a unit negative charge, that is, to bases whose conjugate acids are electrically neutral, under strongly basic conditions. For such an acid-base pair, the thermodynamic acidity constant may be written

$$K_A = \frac{a_{H^+}a_{A^-}}{a_{HA}} = \frac{[H^+][A^-]}{[HA]} \cdot \frac{\gamma_{H^+}\gamma_{A^-}}{\gamma_{HA}} \tag{5-21}$$

For the comparison of the two electrically neutral acids HA and HB, the following equation, analogous to 5-13, may be derived.

$$pK_A - pK_B = \log \frac{[HA]}{[A^-]} - \log \frac{[HB]}{[B^-]} + \log \frac{\gamma_{HA}\gamma_{B^-}}{\gamma_{A^-}\gamma_{HB}} \tag{5-22}$$

For any series of acids for which the function log ([HA]/[A$^-$]) − log ([HB]/[B$^-$]) is independent of the composition of the solvent for any pair of acids for which the function can be measured, it follows that the activity coefficient term vanishes. That is,

$$\log \frac{\gamma_{HA}}{\gamma_{A^-}} = \log \frac{\gamma_{HB}}{\gamma_{B^-}}$$

for any two of the acids. The acidity function H_- for that series of acids is then defined as

$$H_- = -\log \frac{a_{H^+}\gamma_{A^-}}{\gamma_{HA}} \tag{5-23}$$

This acidity function is related to the thermodynamic acidity constant of any acid in the series upon which it is based by the following equation (analogous to Eq. 5-19).

$$pK_A = \log \frac{[HA]}{[A^-]} + H_- \tag{5-24}$$

Values of H_- have been reported for mixtures of water with such electrically neutral bases as hydrazine and ethylenediamine, but none of these mixtures is as strongly basic as certain solutions containing anionic bases such as alkoxide or hydroxide ions. Schwarzenbach and Sulzberger determined values of H_- using indigo derivatives as indicators in concentrated aqueous solutions of sodium, potassium, and lithium hydroxide.[43] These three bases were studied up to concentrations of 18, 12, and 5 *M*, respectively, where their solutions gave H_- values of 18.68, 18.12, and 14.31. Solvent mixtures containing water and/or alcoholic and alkali-metal hydroxides and/or alkoxides have also been studied. In not all cases is it clear that there is a class of bases for which the activity coefficient term in Eq. 5-22 vanishes. In the case of 0.011 *M* tetramethylammonium hydroxide solutions in aqueous dimethyl sulfoxide Dolman and Stewart found that the activity coefficient term is too small to detect for a series of compounds consisting of aniline and diphenylamine derivatives (which act as acids in these media).[44] Their values of H_-, listed in Table 5-4, show that the replacement of the solvent water by dimethyl sulfoxide increases the effective basicity of hydroxide ions (relative to that of the anions derived from aniline and diphenylamine derivatives) by more than 10^{14}-fold. This type of effect of a dipolar aprotic solvent was discussed in Section 5-4. It is noteworthy that the value of H_- rises most rapidly as the last of the water is replaced by dimethyl sulfoxide.

If the acidity function assumption, that is, Eq. 5-15, is exact for a series

Table 5-4. Values of H_- in 0.011 M Tetramethylammonium Hydroxide in Aqueous Dimethyl Sulfoxide[a]

Mole % Me_2SO	H_-	Mole % Me_2SO	H_-
0	12.0	76.12	20.14
10.32	13.17	80.78	20.68
15.20	13.88	85.46	21.27
20.18	14.49	90.07	21.98
26.95	15.22	92.47	22.45
33.42	15.87	94.74	23.01
39.86	16.48	96.21	23.48
46.54	17.12	97.89	24.25
52.55	17.73	98.71	24.84
58.56	18.34	99.14	25.30
64.20	18.92	99.59	26.19
71.35	19.65		

[a] For aniline and diphenylamine derivatives.[44]

of acids or bases throughout all mixtures of water and a second solvent, a plot of the pK values in any mixture versus those in water must give a straight line of slope 1.00. Kreevoy and Baughman have noted that the slopes are considerably larger than 1.00 for phenols and carboxylic acids in certain mixtures of water and dimethyl sulfoxide (including pure dimethyl sulfoxide).[45] They pointed out that pK values referred to aqueous solution but determined by acidity function measurements in mixtures rich in dimethyl sulfoxide will therefore be greatly in error in some cases. It is noteworthy that some of the most widely used and widely accepted acidity functions may have the same flaw, but it is not experimentally feasible to test them. Any nitroaniline basic enough to be studied quantitatively in dilute aqueous solution is much too basic for its pK to be determined in 100% sulfuric acid. Hence we cannot plot a set of pK values in sulfuric acid versus the values in water for Hammett bases. Consequently we cannot test the Hammett H_0 function in this way and for similar reasons we cannot test the H_- function listed in Table 5-4. It would be a remarkable coincidence if either of the appropriate plots had a slope of 1.00, but since we do not know what the slopes would be we cannot tell how well these acidity functions meet this test of their reliability.

5-6. ACIDITY AND BASICITY IN THE GAS PHASE[46,47]

One fruitful method of studying solvent effects is to extend the study of the reactions in question to the gas phase, where there is no solvent. Gas phase acidities and basicities are commonly discussed in terms of proton affinities (*PA*). The proton affinity of the base B is the enthalpy of reaction 5-25, that is, the amount of heat given off when the base combines with a proton.

$$BH^+ \rightarrow B + H^+ \tag{5-25}$$

For most simple bases ΔS^{chem} for these reactions has about the same value, so that ΔG^{chem} varies with structure in about the same way the proton affinity does. Changes in proton affinities with temperature tend to be relatively small. Table 5-5 is a list of basicities of binary hydrides, as measured by their proton affinities, and acidities, as measured by the proton affinities of their conjugate bases.[47]

The data illustrate the tremendous acidity of the bare proton in the gas phase. Even a saturated compound like methane is protonated (to give CH_5^+) highly exothermically. (The "noble" gases have proton affinities ranging from 42 kcal/mole for helium to 124 kcal/mole for xenon.[47]) Also illustrated is the much greater importance of electrical charge type in the gas phase than in solution. Iodide ion, the weakest of the anionic bases listed, is far more basic (a factor of 10^{78} at 25° C) than ammonia, the strongest of the electrically neutral bases. Ammonia is a stronger base

Table 5-5. Acidities and Basicities of Binary Hydrides in the Gas Phase[47]

Hydride	$PA(XH_n)$ (kcal/mole)	$PA(XH_{n-1}^-)$ (kcal/mole)
CH_4	126	≥405
SiH_4	≥146	369
NH_3	207	405
PH_3	185	368
AsH_3	175	356
H_2O	164	390
H_2S	170	350
H_2Se	170	339
HF	131	370
HCl	140	333
HBr	141	323
HI	145	313

than iodide ion in any solvent; it is about 10^{20} times as strong in aqueous solution.

In Table 5-6 are gas phase proton affinities for a number of hydrocarbons and organic halides.[47–49] The marked increase in basicity observed in the series ethylene, propene, isobutene reflects the series of carbonium-ion stabilities: primary $<$ secondary $<$ tertiary. The fact that the alkyl halides are all more basic than the corresponding hydrogen halides might be expected in view of the electron-donating inductive effects of alkyl groups relative to hydrogen. However, it is likely that much of the observed substituent effect of the alkyl groups is due to their polarizability. Polarizability effects are much more important for gas phase reactions than for reactions in solution. Strong evidence for polarizability effects has been obtained in studying the acidity and basicity of organic oxygen compounds. A number of entries in Table 5-7, which consists of data on the gas phase basicities of organic oxygen compounds,[50] show that the replacement of hydrogen by alkyl substituents increases the proton affinity. Thus ethers are more basic than alcohols, which are more basic than water. Furthermore, large alkyl groups increase the basicity more than small ones do. If these increases in the basicity of oxygen arise solely from electron-donating inductive effects by alkyl substituents, with larger alkyl groups having larger inductive effects, then alcohols should be weaker acids than water and alcohols with large alkyl groups should be weaker acids than alcohols with small alkyl groups in the gas phase. This is essentially the opposite of what is found experimentally. Brauman and Blair observed the following order of gas phase acidities: PhOH $>$ Me_3CCH_2OH $>$ *t*-BuOH $\sim$ *n*-$C_5H_{11}OH$ $\sim$ *n*-BuOH $>$ *i*-PrOH $>$ EtOH $>$ PhMe $>$ MeOH $>$ H_2O.[51] As far as the alcohols are concerned this series is roughly that which would be expected if polarizability were the only significant effect. A polarizable group will interact in a stabilizing manner with a nearby charge whether that charge is positive or negative. All alkyl groups are more polarizable than hydrogen, and their polarizability increases with their size. Hence both the gas phase acidity and basicity of alcohols should increase with increasing size of the alkyl group and with the increasing fraction of the alkyl group that is near the oxygen atom. Thus although neopentyl alcohol has no larger an alkyl group than *n*-pentyl alcohol does, a larger fraction of that alkyl group is probably nearer the oxygen atom so that there are larger polarizability interactions when the oxygen atom is charged. The fact that *t*-butyl alcohol is not significantly more acidic than *n*-butyl alcohol, even though most of its alkyl group is nearer the oxygen atom, suggests that, in addition to the polarizability effect, there is an effect operating that tends to make primary alcohols more acidic than tertiary alcohols.

Table 5-6. Gas Phase Basicities of Hydrocarbons and Organic Halides[a]

Base	*PA* (kcal/mole)	Base	*PA* (kcal/mole)
Methane	126	*m*-Xylene	188[b]
Ethane	147[c]	*p*-Xylene	188[b]
Ethylene	159[c]	MeF	151
Propene	179	EtF	163
cis-2-Butene	181	MeCl	160
Isobutene	195	EtCl	167
Acetylene	152	MeBr	163
Cyclopropane	179[b]	EtBr	170
Benzene	178[b]	MeI	170
Toluene	187[b]	EtI	175

[a] From Ref. 47 unless otherwise noted.

[b] From Ref. 48.

[c] From Ref. 49.

Table 5-7. Gas Phase Basicity of Organic Oxygen Compounds[50]

Base	*PA* (kcal/mole)	Base	*PA* (kcal/mole)
MeOH	182	HCO_2H	175
EtOH	187	$MeCO_2H$	188
n-PrOH	189	CF_3CO_2H	167
i-PrOH	193	HCO_2Me	188
t-BuOH	198	HCO_2Et	198
Me_2O	190	$MeCO_2Me$	202
Et_2O	205	$MeCO_2Et$	205
CH_2O	168	$MeCO_2Pr$-*n*	207
MeCHO	185	$EtCO_2Me$	205
EtCHO	187	$CH_2{=}CO$	201
Me_2CO	202		

The gas phase basicity of amines gives particularly valuable information about solvent effects because there are so many reliable data available on amine basicity in solution. Taft and co-workers showed that the effects of 4-substituents on the gas phase basicity of pyridine, as measured by proton affinities, are almost linearly related to the effects on the basicity in aqueous solution, as measured by pK values or ΔG values for ionization.[52] Figure

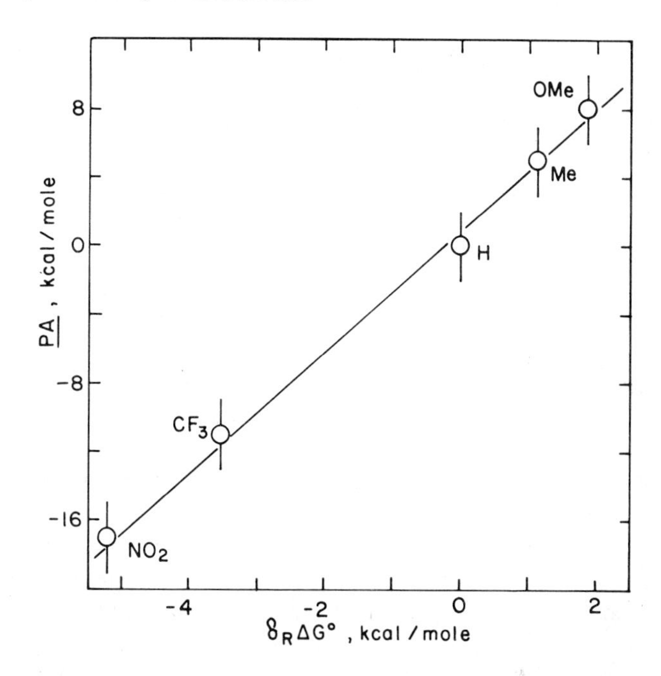

Figure 5-2. Plot of effects of 4-substituents on proton affinity of pyridine in the gas phase versus their effects on the free energy of deprotonation of the corresponding pyridinium ions in aqueous solution.

5-2 is a plot of the substituent effects on the proton affinity ($\delta_R PA$) versus effects on the free energy of deprotonation of the pyridinium ions in aqueous solution ($\delta_R \Delta G°$). The slope of the best line through the points (3.5) shows that the gas phase basicity is about 3.5 times as sensitive to the effect of a substituent as the basicity in aqueous solution is. There are probably two major reasons for this. First, polar substituent effects should be transmitted more efficiently through empty space than through water. Second, in the gas phase there is a larger difference between the extent of protonation of a pyridine and a pyridinium ion than there is in aqueous solution. In aqueous solution the pyridine nitrogen atom is hydrogen bonded by the solvent and therefore is already slightly protonated. The acidic hydrogen in a pyridinium ion is hydrogen bonded to the solvent in aqueous solution and therefore is not as fully attached to the ring nitrogen atom as it would be in the gas phase. The fact that the unsubstituted compound is not the weakest base shows that polarizability effects are not dominant (since hydrogen is certainly less polarizable than any of the larger substituents).

The deviation of the point for hydrogen from the best straight line in Figure 5-2 is smaller than the experimental uncertainty in the proton affinity, but it is interesting to note that it is in the direction that would be expected from a polarizability effect that is larger in the gas phase than in solution.

As in the case of oxygen acids and bases polarizability effects on the gas phase acidity and basicity of amines can be quite large. The order of acidities, $Et_2NH > Me_3CCH_2NH_2 \geq t\text{-}BuNH_2 \geq Me_2NH \geq i\text{-}PrNH_2 > n\text{-}PrNH_2 > EtNH_2 > MeNH_2 > NH_3$, observed by Brauman and Blair in the gas phase, shows that the acidity of ammonia is increased by any alkyl substituent and that it is increased most by those that are largest and that have the largest fraction of themselves nearest the nitrogen atom.[53] The same kind of substituent also increases the gas phase *basicity* of ammonia most efficiently. Table 5-8 lists the free energies of reaction of various protonated amines with ammonia in the gas phase ($\delta_R\Delta G_g°$) and in aqueous solution ($\delta_R\Delta G_w°$).[54] In addition to these measures of the basicities of the amines relative to ammonia, free energies of transfer of the amine [$\delta_R\Delta G_s°(B)$] and the substituted ammonium ion [$\delta_R\Delta G_s°(BH^+)$], relative to ammonia and ammonium ions, respectively, from the gas phase to aqueous solution are also listed. A thermochemical cycle may be used to show that Eq. 5-26 holds for any amine. In fact, the values of $\delta_R\Delta G_s°$

$$\delta_R\Delta G_s°(BH^+) - \delta_R\Delta G_s°(B) = \delta_R\Delta G_g° - \delta_R\Delta G_w° \qquad (5\text{-}26)$$

Table 5-8. Basicities of Amines Relative to Ammonia in the Gas Phase and in Water[a]

Amine	$\delta_R\Delta G_g°$	$\delta_R\Delta G_w°$	$\delta_R\Delta G_s°(B)$	$\delta_R\Delta G_s°(BH^+)$
$MeNH_2$	9.5	1.92	−0.27	7.3
Me_2NH	15.8	2.09	0.00	13.7
Me_3N	20.4	0.75	1.07	20.7
$EtNH_2$	12.0	1.96	−0.21	9.8
Et_2NH	20.6	2.42	0.22	18.4
Et_3N	27.1	2.01	1.26	26.4
i-$PrNH_2$	14.4	1.94	−0.06[b]	12.4[b]
t-$BuNH_2$	16.5	1.96	0.03[b]	14.5[b]
Azetidine	18.3	2.79	0.80[b]	16.3[b]
Pyrrolidine	20.1	2.81	1.20	18.5
Piperidine	21.4	2.56	1.56	20.4

[a] All values in kcal/mole.[54]
[b] Estimated from data on closely related compounds.

(BH^+) listed were calculated from such equations. The gas phase basicities of the amines are seen to increase with increases in the amount of polarizable hydrocarbon groups in the vicinity of the nitrogen atom. The same substituents that increase the gas phase basicity of the amines are seen to increase the free energy of transfer of the ammonium ions from the gas phase to aqueous solution by about the same amount. Three ethyl groups, for example, increase the basicity by 27.1 kcal/mole and make the free energy of transfer from the gaseous to the aqueous phase less favorable by 26.4 kcal/mole. Since $\delta_R \Delta G_s°(BH^+)$ and $\delta_R \Delta G_g°$ values tend to be of about the same magnitude and since the values of $\delta_R \Delta G_s°(B)$ are relatively small, it follows from Eq. 5-26 that the values of $\delta_R \Delta G_w°$ are relatively small; that is, that alkyl substituent effects on the basicity of ammonia tend to be much smaller in aqueous solution than in the gas phase. The polarizability effect of alkyl substituents in increasing the gas phase basicity of ammonia may be thought of as a type of internal solvation of the ammonium ion in the gas phase. The alkyl substituent will continue to exert a stabilizing interaction in aqueous solution, but the fact that it is present rather than a hydrogen atom means that at least one water molecule which could have been solvating the ammonium ion will be pushed further away from the positively charged nitrogen atom. If the substituent effects of alkyl groups were due entirely to their polarizability and if the polarizability of the alkyl groups stabilized the ammonium ions in aqueous solution by exactly the same amount that the water that has been displaced by the alkyl groups would have (by polarizability, charge-dipole interaction, and hydrogen bonding), then alkyl substituents would have no net effect on the basicity of ammonia in aqueous solution.

It is difficult to say to what extent the observed variations of ammonia and alkyl amine basicities in aqueous solution arise from imperfect compensation of the polarizability effect of the alkyl groups by solvation and to what extent they arise from sources other than polarizability and solvation. The inductive effect of the alkyl substituents (relative to that of hydrogen) must be operative, but we have no reliable measure of how large this effect is. There must also be steric effects, and a steric rationalization has been proposed for the observed order of amine basicities in aqueous solution: $NH_3 < RNH_2 < R_2NH > R_3N$.[55,56]

PROBLEMS

1. In each of the following cases explain your answer.
   a. In which solvent would the relative acidities of nitric and hydrochloric acid be more readily measurable, water or ethanol?

b. In which solvent would the relative basicities of methylamine and dimethylamine be more readily measurable, water or ammonia?

c. In which solvent would the relative acidities of aniline and diphenylamine be more readily measurable, water or ammonia?

2. The acidity constants of propionic acid and of *N*-methylanilinium ions are almost identical in water at 25° C. Which has the larger acidity constant in ethanol at 25° C? Explain.

3. If 1.0 mmole of 2,4-dinitroaniline is dissolved in 100 ml of 50% (by weight) water-sulfuric acid at 25 °C, what is the concentration of 2,4-dinitroanilinium ions in the resultant solution?

4. For proton transfer from the trimethylammonium ion to 1,3-diaminopropane in the gas phase, $\Delta H°$ is -13.0 kcal/mole and $\Delta S°$ is -20.6 eu.[57] Noting that the gas phase proton affinity of trimethylamine is significantly larger than that of any of the primary amines listed in Table 5-8 and that $\Delta S°$ for proton transfers from an ammonium ion to an amine in the gas phase is usually near ΔS_σ, explain why this reaction is so exothermic and why the entropy change is so unfavorable.

REFERENCES

1. J. F. Coetzee and C. D. Ritchie, Eds., *Solute-Solvent Interactions*, Dekker, New York, 1969.
2. R. P. Bell, *The Proton in Chemistry*, 2nd ed., Cornell University Press, Ithaca, N.Y., 1973, Chaps. 4–6.
3. J. E. Leffler and E. Grunwald, *Rates and Equilibria of Organic Reactions*, Wiley, New York, 1963, Chap. 8.
4. D. J. Cram, *Fundamentals of Carbanion Chemistry*, Academic, New York, 1965.
5. L. P. Hammett, *Physical Organic Chemistry*, 2nd ed., McGraw-Hill, New York, 1970, Chap. 9.
6. A. J. Parker, *Pure Appl. Chem.*, **25,** 345 (1971).
7. C. G. Swain and C. B. Scott, *J. Amer. Chem. Soc.*, **75,** 141 (1953).
8. Cf. A. J. Parker, *Proc. Chem. Soc.*, 371 (1961).
9. J. Hine, *Physical Organic Chemistry*, 2nd ed., McGraw-Hill, New York, 1962. (a) p. 77. (b) p. 44.
10. I. M. Kolthoff and L. S. Guss, *J. Amer. Chem. Soc.*, **61,** 330 (1939).
11. H. Goldschmidt and E. Mathiesen, *Z. Phys. Chem.*, **119,** 439 (1926).
12. H. Goldschmidt and A. Thuesen, *Z. Phys. Chem.*, **81,** 30 (1913).
13. H. Goldschmidt and P. Dahll, *Z. Phys. Chem.*, **108,** 121 (1924).
14. Cf. L. Thomas and E. Marum, *Z. Phys. Chem.*, **143,** 191 (1929).
15. E. A. Guggenheim, *J. Phys. Chem.*, **33,** 842 (1929); **34,** 1540 (1930).

16. B. W. Clare, D. Cook, E. C. F. Ko, Y. C. Mac, and A. J. Parker, *J. Amer. Chem. Soc.*, **88,** 1911 (1966).
17. H. Rapoport and G. Smolinsky, *J. Amer. Chem. Soc.*, **82,** 934 (1960).
18. C. D. Ritchie and R. E. Uschold, *J. Amer. Chem. Soc.*, **90,** 2821 (1968).
19. C. D. Ritchie and R. E. Uschold, *J. Amer. Chem. Soc.*, **89,** 2960 (1967).
20. E. C. Steiner and J. M. Gilbert, *J. Amer. Chem. Soc.*, **85,** 3054 (1963).
21. D. K. Bohme, E. Lee-Ruff, and L. B. Young, *J. Amer. Chem. Soc.*, **94,** 5153 (1972).
22. I. M. Kolthoff, S. Bruckenstein, and M. K. Chantooni, Jr., *J. Amer. Chem. Soc.*, **83,** 3927 (1961).
23. I. M. Kolthoff, M. K. Chantooni, Jr., and S. Bhowmik, *J. Amer. Chem. Soc.*, **90,** 23 (1968).
24. D. Bethell and A. F. Cockerill, *J. Chem. Soc., B*, 913 (1966).
25. I. M. Kolthoff, J. J. Lingane, and W. D. Larson, *J. Amer. Chem. Soc.*, **60,** 2512 (1938).
26. E. Grunwald and E. Price, *J. Amer. Chem. Soc.*, **86,** 4517 (1964).
27. R. J. W. Le Fèvre, *Advan. Phys. Org. Chem.*, **3,** 1 (1965).
28. L. P. Hammett, *Chem. Rev.*, **16,** 67 (1935).
29. M. A. Paul and F. A. Long, *Chem. Rev.*, **57,** 1 (1957).
30. C. H. Rochester, *Acidity Functions*, Academic Press, New York, 1970.
31. K. Bowden, *Chem. Rev.*, **66,** 119 (1966).
32. E. M. Arnett, *Progr. Phys. Org. Chem.*, **1,** 223 (1963).
33. E. M. Arnett and G. W. Mach, *J. Amer. Chem. Soc.*, **86,** 2671 (1964).
34. L. P. Hammett and A. J. Deyrup, *J. Amer. Chem. Soc.*, **54,** 2721 (1932).
35. N. C. Deno, P. T. Groves, and G. Saines, *J. Amer. Chem. Soc.*, **81,** 5790 (1959).
36. M. J. Jorgenson and D. R. Hartter, *J. Amer. Chem. Soc.*, **85,** 878 (1963).
37. E. M. Arnett and G. W. Mach, *J. Amer. Chem. Soc.*, **88,** 1177 (1966).
38. T. G. Bonner and J. Phillips, *J. Chem. Soc., B*, 650 (1966).
39. R. L. Hinman and J. Lang, *J. Amer. Chem. Soc.*, **86,** 3796 (1964).
40. K. Yates, J. B. Stevens, and A. R. Katritzky, *Can. J. Chem.*, **42,** 1957 (1964).
41. C. D. Johnson, A. R. Katritzky, and N. Shakir, *J. Chem. Soc., B*, 1235 (1967).
42. J. F. Bunnett and F. P. Olsen, *Can. J. Chem.*, **44,** 1899 (1966).
43. G. Schwarzenbach and R. Sulzberger, *Helv. Chim. Acta*, **27,** 348 (1944).
44. D. Dolman and R. Stewart, *Can. J. Chem.*, **45,** 911 (1967).
45. M. M. Kreevoy and E. H. Baughman, *J. Amer. Chem. Soc.*, **95,** 8178 (1973).
46. E. M. Arnett, *Accounts Chem. Res.*, **6,** 404 (1973).
47. J. L. Beauchamp, *Ann. Rev. Phys. Chem.*, **22,** 527 (1971).
48. S.-L. Chong and J. L. Franklin, *J. Amer. Chem. Soc.*, **94,** 6347, 6630 (1972).

49. D. K. Bohme, P. Fennelly, R. S. Hemsworth, and H. I. Schiff, *J. Amer. Chem. Soc.*, **95**, 7512 (1973).
50. J. Long and B. Munson, *J. Amer. Chem. Soc.*, **95**, 2427 (1973).
51. J. I. Brauman and L. K. Blair, *J. Amer. Chem. Soc.*, **90**, 6561 (1968); **92**, 5986 (1970).
52. M. Taagepera, W. G. Henderson, R. T. C. Brownlee, J. L. Beauchamp, D. Holtz, and R. W. Taft, *J. Amer. Chem.* Soc., **94**, 1369 (1972).
53. J. I. Brauman and L. K. Blair, *J. Amer. Chem. Soc.*, **91**, 2126 (1969); **93**, 3911 (1971).
54. E. M. Arnett, F. M. Jones, III, M. Taagepera, W. G. Henderson, J. L. Beauchamp, D. Holtz, and R. W. Taft, *J. Amer. Chem. Soc.*, **94**, 4724 (1972).
55. H. C. Brown, H. Bartholomay, Jr., and M. D. Taylor, *J. Amer. Chem. Soc.*, **66**, 435 (1944).
56. H. C. Brown, D. H. McDaniel, and O. Häfliger, in *Determination of Organic Structures by Physical Methods*, E. A. Braude and F. C. Nachod, Eds., Academic Press, New York, 1955, pp. 613, 614, 621, 622.
57. R. Yamdagni and P. Kebarle, *J. Amer. Chem. Soc.*, **95**, 3504 (1973).

6

Brønsted Acidity and Basicity—Structural Effects

Structural effects on Brønsted acidity and basicity have been discussed in more detail by several other authors. Albert and Serjeant have given particular attention to the experimental methods used.[1] The theoretical interpretation of the data has been emphasized by Bell[2] and by Brown, McDaniel, and Häfliger.[3] Cram has treated the acidity of hydrogen atoms attached to carbon.[4] Körtum, Vogel, and Andrussow[5] and Perrin[6,7] have compiled the extensive data on the strengths of acids and bases in aqueous solution. A collection of stability constants for the complexing of metal ions with bases also gives data on complexing of protons with bases.[8,9] Collumeau listed data on several hundred acids and bases, most of which are too weakly acidic or basic to be studied in purely aqueous solution.[10] Arnett included a similar collection in a critical review of the experimental techniques and theoretical interpretation of the data.[11] Fischer and Rewicki discussed the acidity of hydrocarbons.[12]

6-1. EFFECT OF THE IDENTITY OF THE ATOM TO WHICH THE PROTON IS ATTACHED

There is a tendency for the acidity of compounds to increase with increasing electronegativity of the atom to which the acidic proton is attached, provided that there are no changes in the number of inner electronic shells.

As evidence for this generalization, the following order of increasing acidities may be cited: $CH_4 < NH_3 < H_2O < HF$. It could be argued that the order of acidities in aqueous solution $(NC)_3CH$ (pK −5.0)[10] > C_6H_5-CONHCN (pK 2.7)[13] > $BrCH_2CO_2H$ (pK 2.9)[5] > HF (pK 3.2)[7] supports a diametrically opposite generalization. To the objection that the first three acids in the latter series have their acidities increased by electron-withdrawing substituents, it may be replied that in the former series methane, ammonia, and water have electron-donating hydrogen substituents that decrease their acidities. Even when such substituent effects are minimized as much as possible the correlation of acidity with electronegativity still seems applicable. Thus, if we wish to compare the acid HF with an HOX such that the X substituent will neither donate nor withdraw electrons, we might select OH as the substituent. The resultant compound HOOH (pK 11.6 in water)[8] is a much weaker acid than HF. Although the OH substituent can bring about neither net electron donation nor net electron withdrawal from the OH group to which it is attached in H_2O_2, it probably does withdraw electrons from the O^- atom to which it is attached in the HOO^- anion. It is impossible, of course, to have a substituent X that would act as neither an electron donor nor withdrawer in HOX or OX^-. However, there is a species in which there is no substituent X. The

$$\cdot OH \rightleftharpoons H^+ + O^- \cdot$$

hydroxy radical has a pK of 11.8[14] in water at 23° C.

Another useful generalization is that Brønsted acidity usually increases with an increase in the number of inner electronic shells in the atom to which the acidic proton is attached. Thus in aqueous solution the pK of HF is 3.2 whereas those of HCl, HBr, and HI have been estimated to be −7, −9, and −10, respectively.[2] Analogously, the pK_1 values for H_2O, H_2S, H_2Se, and H_2Te in aqueous solution are 15.7, 7.0, 3.9, and 2.6.[8] The acidities in the gas phase (Table 5-5) stand in the same order, but the differences in strength are much larger because the solvation energies of small ions such as fluoride and hydroxide are so much larger than those of the bigger ions in aqueous solution.

The state of hybridization of an atom also has a profound influence on the acidity of hydrogen atoms attached to it. In the case of hydrocarbons, for example, the fact that such reactions as

$$C_5H_{11}Na + C_6H_6 \rightarrow C_5H_{12} + C_6H_5Na$$

$$C_6H_5Na + CH{\equiv}CH \rightarrow C_6H_6 + CH{\equiv}CNa$$

proceed in the direction indicated, shows that hydrogen attached to *sp* carbon is more acidic than that attached to sp^2 carbon, which in turn is more acidic than that attached to sp^3 carbon. Similarly, trialkylammonium ions have pK values around 10; the pyridinium ion, in which the acidic proton is attached to sp^2 nitrogen, has a pK of 5.2[6]; and the conjugate acid of acetonitrile, in which the acidic proton is attached to sp nitrogen, is so strongly acidic that acetonitrile is only about half protonated in 100% sulfuric acid.[15]

In applying the generalizations given in this section, it should be noted that solvent effects on relative acidities are particularly likely when the different acids being compared have their acidic protons attached to atoms of different elements, atoms of different electrical charge type, or atoms in different states of hybridization (cf. Sections 5-2 and 5-6).

6-2. POLAR SUBSTITUENT EFFECTS ON ACIDITY AND BASICITY

In general, electron-withdrawing substituents increase the strength of acids and decrease that of bases to an extent that decreases with increasing distance between the substituent and the acidic proton or the basic atom. A number of Hammett reaction constants for various types of acids[16–25] are listed in Table 6-1. The substituent effects decrease, that is, ρ decreases, with every additional methylene group (or other divalent atom or group) placed between the substituents and the reaction center. However, as pointed out in Section 3-6b, this decrease is smaller than the 2.8-fold decrease in Taft substituent constants for common small substituents found when an added methylene group was inserted. The table contains several illustrations of the increases in substituent effects that accompany decreases in the dielectric constant of the solvent. Comparison of the ρ constants for β-arylpropionic acids, *trans*-cinnamic acids, and arylpropiolic acids (complicated by the different solvent used for the latter) suggests that substituent effects are transmitted more efficiently through unsaturated than through saturated systems. However, as noted in Section 2-2b, no clear difference of this kind could be seen in comparing substituent effects on the strengths of bicyclo[2.2.2]octane-1-carboxylic acids, bicyclo[2.2.2]-2-octene-1-carboxylic acids, and dibenzobicyclo[2.2.2]-2,5-octadiene-1-carboxylic acids. Perhaps it is significant that the unsaturation introduced into the aryl substituted acids is conjugated with the reaction center and with most of the substituents, or that the β-arylpropionic acids are relatively flexible.

Data on Taft equation correlations of the acidities of several types of aliphatic acids[26–29] are shown in Table 6-2. For all the acids except the RCH_2OH's, data on compounds where R is aryl or α,β-unsaturated (e.g., $CH_3CH{=}CH{-}$) deviated from the Taft equation, apparently because of

Table 6-1. Values of Hammett's ρ for Various Acidity Constants[a]

Acid	Solvent[b]	n[c]	ρ	$-\log K_0$ [d]	Ref.
$ArCO_2H$	H_2O		1.00	4.20	16
$ArCO_2H$	40% EtOH	8	1.67	4.87	16
$ArCO_2H$	70% EtOH	9	1.74	6.17	16
$ArCO_2H$	EtOH	9	1.96	7.21	16
$ArCO_2H$	MeOH	5	1.54	6.51	16
o-$HOArCO_2H$[e]	H_2O	6	1.10	4.00	16
$ArCH_2CO_2H$	H_2O	14	0.49	4.30	16
$ArCH_2CH_2CO_2H$	H_2O	8	0.21	4.55	16
trans-$ArCH{=}CHCO_2H$	H_2O	9	0.47	4.45	16
$ArCO_2H$	50% butyl Cellosolve	7	1.42	5.63	16
p-$ArC_6H_4CO_2H$	50% butyl Cellosolve	9	0.48	5.64	16
$ArC{\equiv}CCO_2H$	35% dioxane	11	0.80	3.26[f]	17,18
$ArOCH_2CO_2H$	H_2O	16	0.30	3.17[f]	19
$ArSCH_2CO_2H$	H_2O	17	0.32	3.38[f]	19
$ArSeCH_2CO_2H$	H_2O	19	0.35	3.75[f]	19
$ArSOCH_2CO_2H$	H_2O	13	0.17	2.73[f]	20
$ArSO_2CH_2CO_2H$	H_2O	14	0.25	2.51[f]	20
$ArB(OH)_2$	25% EtOH	14	2.15	9.70	16
$ArPO(OH)_2$	H_2O	10	0.76	1.84	16
$ArPO_2OH^-$	H_2O	12	0.95	6.96	16
$ArAsO(OH)_2$[g]	H_2O	9	1.05	3.54	16
$ArSeO_2H$	H_2O	16	0.90	4.74	16
α-$ArCH{=}NOH$	H_2O	6	0.86	10.70	16
ArOH	H_2O	14	2.23	9.92	21
ArOH[h]	48.9% EtOH	14	2.54	11.56	16
ArSH[h]	48.9% EtOH	12	2.24	7.67	16
$ArNH_3^+$	H_2O	22	2.89	4.58	21
$ArCH_2NH_3^+$	H_2O	14	1.06	9.39	22
$ArNHNH_3^+$	H_2O	10	1.17	5.19	23
ArNHPh	H_2O-Me_2SO	9	4.07	22.4	24
$ArNH_2$[i]	NH_3	6	5.3		25

[a] At 25° C unless otherwise indicated.
[b] "x% S" means x% solvent S, 100 − x% water.
[c] Number of compounds studied.
[d] Unless otherwise indicated, this is the intercept of the best straight line through the experimental points. It is not necessarily exactly equal to the experimental value of $-\log K$ for the unsubstituted compound.
[e] 4- and 5-substituted.
[f] $-\log K$ for the unsubstituted compound.
[g] At 18–25° C.
[h] At 20–22° C.
[i] At 31° C.

resonance interactions between the substituent and the reaction center. The data on mercaptans and ammonium ions deviated in the cases where R is hydrogen. In the case of the mercaptans, the compound in which R is hydrogen, namely, hydrogen sulfide, is about 30 times as strong an acid as would be expected from the Taft equation correlation. A similar deviation would probably be found in a correlation of the acidity of phosphonium ions. The p*K*'s of *n*-butylphosphonium ions, di-*n*-butylphosphonium ions, and tri-*n*-butylphosphonium ions are −0.03, 4.51, and 8.43, respectively.[11] The difference between these values is larger than would be calculated from a Taft equation correlation if the ρ^* value is near those found for mercaptans and ammonium ions. These deviations suggest that, relative to hydrogen, alkyl groups are stronger electron donors to sulfur and phosphorus than would be expected from observations on compounds in which the hydrogen and alkyl groups are attached to sp^2-hybridized carbon (as they are in the compounds from which Taft σ^* constants were determined). This generalization recalls certain other nonparallels between measures of the polar character of substituents. For example, since the Pauling electronegativity of chlorine is the same as that of nitrogen, and the electronegativities of iodine and sulfur are the same as that of carbon, we might expect σ^* for $ClCH_2$ to be equal to that for H_2NCH_2 and σ^*'s for ICH_2, CH_3SCH_2, and CH_3CH_2 all to be equal. Instead, the substituents containing the atoms of higher atomic weight have larger σ^* values. As Taft pointed out in a similar connection, "In the case of C—X bonds, the electron-withdrawing powers of I, Br, and Cl relative to F are much greater than expected on the basis of the relative electronegativities."[30] Taft attributes this extra electron-withdrawing power to the ability of the higher

Table 6-2. Values of Taft's ρ^* for Various Acidity Constants[a]

Acid	Solvent[b]	n^c	ρ^*	$-\log K_0{}^d$	Ref.
RCO_2H	H_2O	16	1.72	4.65	26
RCH_2OH	H_2O	8	1.42	15.9	27
RSH	H_2O	8	3.50	10.22	28
RNH_3^+	H_2O	25	3.14	10.16	29
$RR'NH_2^{+e}$	H_2O	21	3.23	11.15	29
$RR'R''NH^{+e}$	H_2O	31	3.30	9.61	29

[a] At 25° C.
[b,c,d] See the corresponding footnotes of Table 6-1.
[e] In these cases $\sigma_R^* + \sigma_{R'}^*$ or $\sigma_R^* + \sigma_{R'}^* + \sigma_{R''}^*$ was used.

halogens to use their *d* orbitals to accept electrons from the other bonds of the carbon atom. The anomalously strong electron-donating power of alkyl groups, relative to hydrogen, toward sulfur and phosphorus that produces the deviations from the Taft equation referred to may be explained in the same way.

Polarizability and solvation effects probably also help make alkyl substituents (relative to hydrogen) anomalously strongly base-strengthening substituents on phosphines in aqueous solution. We have already described how the base-strengthening polarizability effect of an alkyl group in an amine is roughly canceled by the extra solvation that occurs in aqueous solution when a hydrogen atom is present instead of the alkyl group (Section 5-6). Much of this solvation of the ammonium ion arises from hydrogen bonding of its acidic hydrogen atoms to the solvent. Hydrogen atoms attached to sulfur, phosphorus, and other high molecular weight elements give much weaker hydrogen bonds than hydrogen atoms attached to first-row atoms of the same group, such as oxygen and nitrogen.[31,32] Therefore the solvation energy does a much less complete job of canceling the polarizability effect of an alkyl group.

6-3. RESONANCE EFFECTS ON ACIDITY AND BASICITY

The fact that carboxylic acids (α-oxoalcohols) are more strongly acidic than most alcohols must be due in part to the polar substituent effect of the strongly electron-withdrawing substituent located so near the acidic proton. From the Taft equation correlation of the acidity of alcohols (Table 6-2) and the substituent constant of the acetyl group (Table 3-6), the pK of the oxygen-bound proton in hydroxyacetone would be estimated to be 13.56 [lower than pK_0 (15.9) by 1.65 (1.42) or 2.34]. If we use the generalization that removing one methylene group between the substituent and reaction center will increase the substituent effect by 2.8-fold, we would calculate a pK of 15.9 − 2.8(2.34) or 9.34 for acetic acid. Actually, acetic acid is much stronger than this, having a pK^{chem} of 4.46 (pK^{exptl} is 4.76; for the reference compound, ethanol, pK^{chem} and pK^{exptl} are identical). The use of the Taft equation and the factor 2.8 would not be expected to give a perfect calculation of the magnitude of the polar substituent effect, but it seems unlikely that an error in pK as large as 4.88 would result. It is much more likely that resonance stabilization of the acetate ion is producing this effect.

$$\left[CH_3-C(=\bar{O})-O-H \leftrightarrow CH_3-C(-\bar{O}^{\ominus})=\overset{\oplus}{O}-H \right] + H_2O \rightleftharpoons H_3O^+ + \left[CH_3-C(=\bar{O})-\bar{O}^{\ominus} \leftrightarrow CH_3-C(-\bar{O}^{\ominus})=\bar{O} \right]$$

The ΔpK of 4.88 corresponds to 6.66 kcal/mole more resonance stabilization of the acetate ion than of acetic acid, for which two equivalent contributing structures cannot be written.

In the preceding case, of a carboxylic acid, the polar effect of the substituent acted in the same direction as its resonance effect. It was therefore not possible to see the need for the resonance effect in explaining the data until the polar effect had been evaluated quantitatively and found (estimated) to be too small to explain the observed substituent effect. In view of the uncertainties that can accompany the quantitative evaluation of polar effects, it is instructive to consider a case in which the polar and resonance effects would be expected to operate in opposite directions. Therefore let us consider the basicity of amidines, that is, the acidity of amidinium ions, for example,

$$\left[CH_3-C(=\overset{\oplus}{N}H_2)-\bar{N}H_2 \leftrightarrow CH_3-C(-\bar{N}H_2)=\overset{\oplus}{N}H_2 \right] + H_2O \rightleftharpoons H_3O^+ + \left[CH_3-C(=\bar{N}-H)-\bar{N}H_2 \leftrightarrow CH_3-C(-\bar{N}^{\ominus}-H)=\overset{\oplus}{N}H_2 \right]$$

Acetamidine can be considered to be 1-aminoethylidenimine. An iminium ion of the type $RCH{=}NHR^+$ has been found to have a pK_a of 6.9.[33] The ethylideniminium ion, with one less R group on its positively charged nitrogen atom, should be somewhat more acidic. Its inductive substituent constant (Table 3-7) shows that the amino group has an electron-withdrawing polar effect. Hence if the polar substituent effect of the amino group were its only substituent effect, the pK_a of acetamidinium ions would be less than 7. Actually this pK_a value is 12.7.[6] Thus the resonance effect has increased pK by more than 5.7. This means that the resonance stabilization of the acetamidinium ion exceeds that of acetamidine by more than 7.8 kcal/mole. Such resonance stabilization is even greater in the case of derivatives of guanidine, which are the strongest known electrically neutral bases. In this case three equivalent contributing structures may be written.

$$H_2N{-}\overset{\overset{NH}{\|}}{C}{-}NH_2 + H_2O \rightleftharpoons \left[H_2N{-}\overset{\overset{\overset{+}{N}H_2}{\|}}{C}{-}NH_2 \leftrightarrow H_2N{-}\overset{\overset{NH_2}{|}}{C}{=}NH_2^+ \leftrightarrow H_2\overset{+}{N}{=}\overset{\overset{NH_2}{|}}{C}{-}NH_2 \right] + OH^-$$

Resonance effects also contribute to the substituent effects correlated by the Hammett equation, such as those covered by Table 6-1. The ρ values for phenols and anilinium ions recorded there must be used with σ^- rather than ordinary σ values to obtain reliable estimates of pK_a for species with para substituents capable of resonance electron withdrawal. Evidence for such extra resonance effects of para substituents is not restricted to acids in which the acidic proton is attached directly to an atom that is conjugated with the para substituent. Wepster and co-workers calculated ρ values for the ionization of phenoxyacetic acids and anilinoacetic acids using data on the meta-substituted compounds only.[34] These ρ values were combined with the observed pK_a values for para-substituted compounds to get "effective" substituent constants. The effective substituent constants calculated from phenoxyacetic acid pK_a data for the *p*-acetyl, cyano, methylsulfonyl, and nitro substituents were larger than ordinary σ_p values, and the ones calculated from anilinoacetic acid data were as large as or larger than σ^- values. Therefore, direct resonance interaction with para substituents and the reaction center (taking the $NHCH_2CO_2H$ and $NHCH_2CO_2^-$ groups as the reaction center) is responsible for as large a fraction (or larger) of the total substituent effect in the case of anilinoacetic acids as in the case of anilinium ions. The extent of such extra resonance effects on the acidity of acids of the type $ArXCH_2CO_2H$ appears to increase linearly with the expected resonance electron-donating ability of $-XCH_2CO_2^-$ or $-XCH_2CO_2H$, as measured by $\Delta\sigma_R^+$ for $-XCH_3$.[34,35]

Methods for separating meta- and para-substituent effects into their polar and resonance components are discussed in Section 3-3. A rather direct method is to assume that the effect of a 4-substituent in a reaction of bicyclo[2.2.2]octane derivatives (with the reaction center at the 1-position) is purely a polar effect and that this polar effect is of exactly the same magnitude as the polar effect of the corresponding para substituent in the corresponding reaction of aromatic compounds. The fact that the distance between substituent and reaction center is very nearly the same with 4-substituted bicyclo[2.2.2]octane derivatives as with the corresponding para-substituted benzene derivatives makes this approach attractive. The results obtained using derivatives of bicyclo[2.2.2]octane-1-carboxylic acid and benzoic acid are plausible, but the use of 4-substituted quinuclidines[36–38] (**1**) and 4-substituted pyridines (1-aza derivatives of the bicyclooctane and

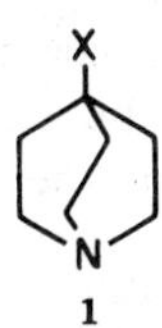

1

benzenoid compounds) provides a better test of the method since the observed substituent effects are so much larger. Taft and Grob pointed out that the acidities of 4-substituted pyridinium ions are fit well by Eq. 6-1 with ρ_I and ρ_R values of 5.15 and 2.69, respectively.[39] The acidities of

$$\log\left(\frac{K}{K_0}\right)_P = \rho_I\sigma_I + \rho_R\sigma_R^+ \qquad (6\text{-}1)$$

4-substituted quinuclidinium ions are well fit by Eq. 6-2 with ρ_I being within

$$\log\left(\frac{K}{K_0}\right)_Q = \rho_I\sigma_I \qquad (6\text{-}2)$$

the experimental uncertainty of that for pyridinium ions. Taking ρ_I to be equivalent in the two reactions, they subtracted Eq. 6-2 from 6-1 to obtain Eq. 6-3. A plot of the left side of this equation for 17 substituents versus

$$\log\left(\frac{K}{K_0}\right)_P - \log\left(\frac{K}{K_0}\right)_Q = \rho_R\sigma_R^+ \qquad (6\text{-}3)$$

the σ_R^+ values for the substituents gave satisfactory agreement with a straight line of slope 2.87. Deviations by resonance electron-withdrawing substituents suggest that the agreement would have been somewhat less impressive if the resonance effects had not been so large.

6-4. STERIC EFFECTS ON ACIDITY AND BASICITY

It is easy to imagine that the proton-accepting atom in some base might be so sterically hindered that there would be significant extra steric repulsion between the hindering groups and an added proton. It is not clear, however, that any case in which the strength of a base has been decreased for this reason has been observed. Brown and Kanner showed that 2,6-di-*t*-butylpyridine is less basic than would be expected in the absence of steric effects.[40] From the pK_a values for the conjugate acid of 2- and 6-alkylpyridines listed in Table 6-3, it may be seen that two methyl groups or two isopropyl groups on carbon atoms adjacent to the nitrogen atom increase the basicity about twice as much as one such group does. 2,6-Di-*t*-butylpyridine is not only 0.8 pK units less basic than pyridine, it is about 1.4 pK units less basic than it would be expected to be in the absence of steric effects. It is possible, however, that part or all of this decreased basicity is not due to repulsions between the *t*-butyl groups and the added proton but to steric hindrance of solvation of the 2,6-di-*t*-butylpyridinium ion via hydrogen bonding to water.

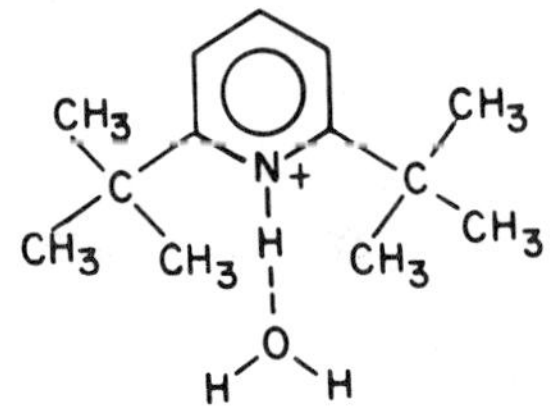

McDaniel and Özcan found that the relative basicities of pyridine, 2-*t*-butylpyridine, and 2,6-di-*t*-butylpyridine were about the same in aqueous methanol solutions containing 10, 20, 30, 40, 50, 60, and 70% (by volume) methanol as in pure water.[41] In aqueous isopropyl alcohol solutions the basicity of 2,6-di-*t*-butylpyridine, relative to that of the other two compounds, decreased with increasing isopropyl alcohol content and became too weak to measure in solutions containing 40% or more isopropyl alcohol. Intermediate behavior was observed in aqueous ethanol. It thus appears that there is steric hindrance to solvation of the 2,6-di-*t*-butylpyridinium ion by isopropyl alcohol or by ethanol. There may also be steric hindrance to the solvation of this cation by water, but this is less clear.

Steric hindrance of solvation may produce a decrease in the acidity of electrically neutral acids. Newman and Fukunaga measured the ionization

Table 6-3. pK_a Values for 2- and 6-Alkylpyridinium Ions in 50% Ethanol-Water at 27 ± 2° C[40]

Pyridine	pK_a
Pyridine	4.38
2-Picoline	5.05
2,6-Lutidine	5.77
2-Isopropylpyridine	4.82
2,6-Diisopropylpyridine	5.34
2-*t*-Butylpyridine	4.68
2,6-Di-*t*-butylpyridine	3.58

constants of highly hindered aliphatic carboxylic acids and found a clear tendency for the acid strength to decrease with increasing hindrance, as shown by the data in Table 6-4.[42]

Thus it is possible that steric hindrance sometimes interferes with the protonation of bases, and it seems clear that it sometimes interferes with solvation of the ions produced by protonations and deprotonations. Another way in which steric effects may operate is by modification of resonance effects. The acid-strengthening effect of an *o*-methyl group in benzoic acid has already been discussed (Section 3-4). A number of cases in which there is steric inhibition of resonance between groups para to each other on an aromatic ring are known. For example,[43] the introduction of two methyl groups meta to the hydroxy groups in *p*-cyanophenol increases the pK_a of the compound by 0.26, very nearly the effect observed in phenol, whose pK_a (9.99) is 0.19 lower than that of 3,5-dimethylphenol,

Table 6-4. pK_a Values in 50% Methanol-Water at 40° C[42]

Acid	pK
CH_3CO_2H	5.69
t-$BuCH_2CO_2H$	6.24
(*i*-Pr)$_2CHCO_2H$	6.48
(*t*-Bu)$_2CHCO_2H$	7.04
Et_3CCO_2H	6.65
(*i*-Pr)$_3CCO_2H$	7.36

whereas the introduction of two *m*-methyl groups into *p*-nitrophenol increases the pK_a by 1.09.

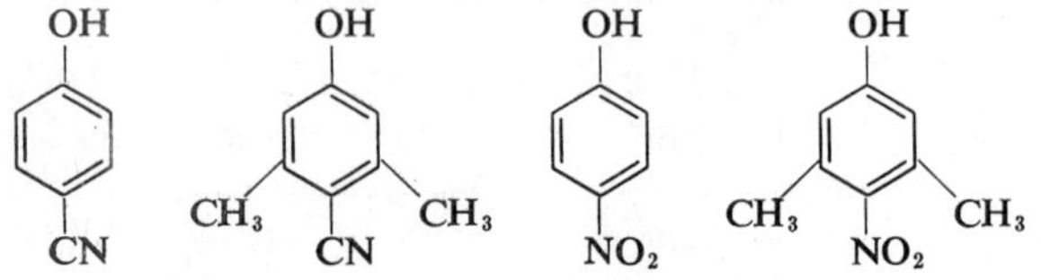

	OH, CN	OH, CH_3, CH_3, CN	OH, NO_2	OH, CH_3, CH_3, NO_2
pK_a	7.95	8.21	7.16	8.25

This difference in behavior is due to the necessity for a coplanar geometry of the entire ion (except for some of the hydrogen atoms in the methyl groups) if there is to be the maximum amount of resonance interaction between the O^- group and the nitro or cyano group, that is, if there is to be the maximum contribution of valence-bond structures such as the following:

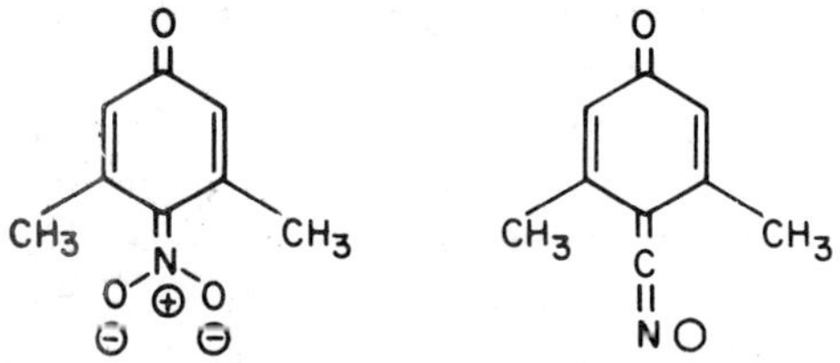

In the case of the 3,5-dimethyl-4-nitrophenoxide ion, such coplanarity can be achieved only at the expense of strong destabilizing steric interactions between the two oxygen atoms of the nitro group and the adjacent methyl groups. Because of the linear geometry of the cyano group there are no analogous steric repulsions in the 3,5-dimethyl-4-cyanophenoxide ion.

Several interesting results of steric inhibition of resonance are found with bicyclic compounds having a nitrogen atom at a bridgehead. Wepster found that benzoquinuclidine is a much stronger base than any other known electrically neutral aniline derivative.[44] The pK_a values of the conjugate acid of benzoquinuclidine and several related compounds are listed below.

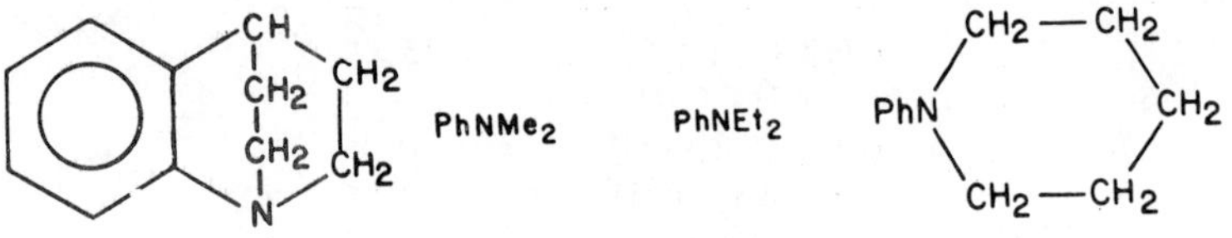

	Benzoquinuclidine	$PhNMe_2$	$PhNEt_2$	PhN(CH_2—CH_2)$_2CH_2$
pK_a	7.79	5.06	6.56	5.20

Most aniline derivatives are stabilized by resonance due to the contribution of structures having a double bond between the nitrogen atom and the ring.

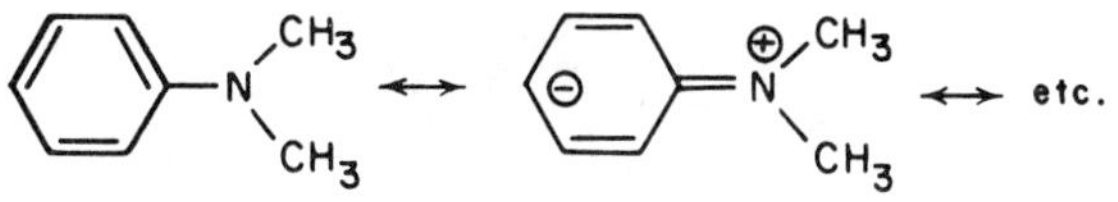

Such structures cannot contribute significantly unless the two saturated carbon atoms attached to nitrogen can get fairly near the plane of the ring, that is, unless the orbital containing the unshared electron pair on nitrogen is fairly nearly parallel to the *p* orbitals on the carbon atoms in the ring. In benzoquinuclidine the two saturated carbon atoms attached to nitrogen are about as far from the plane of the ring as they can get, and the orbital containing the unshared electron pair on nitrogen is in the plane of the ring, which is orthogonal to the *p* orbitals of the ring carbon atoms. It is not reasonable to attribute the strong basicity of benzoquinuclidine to the quinuclidine ring. Quinuclidine itself (whose conjugate acid has a pK_a of about 10.8)[6] is no stronger a base than many acyclic tertiary amines (e.g., triethylamine, pK_a 10.75).[6]

Pracejus and co-workers showed that the amide 2,2-dimethyl-1-azabicyclo[2.2.2]-6-octanone, also known as 2,2-dimethyl-6-quinuclidinone, is much more basic than typical amides.[45,46] Its conjugate acid has a pK_a of

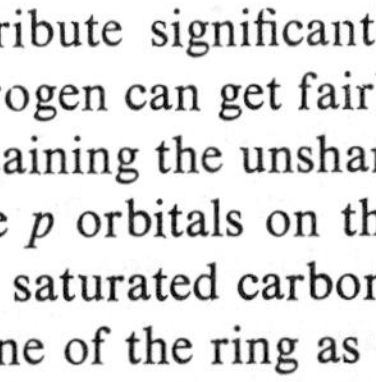

2,2-dimethyl-6-quinuclidinone

5.33, which is much higher than the pK_a values of 0.1, 0.4, 0.5, and 1.0 that have been reported for the conjugate acids of *N,N*-dimethylacetamide, *N*-acetylpiperidine, 2-azabicyclo[2.2.2]-3-octanone, and 2-methyl-2-azabicyclo[2.2.2]-3-octanone, respectively. When it is considered that the protonation of ordinary amides takes place very largely on the oxygen atom to yield the following resonance-stabilized cation,[47]

it may be seen that the basicity of the *nitrogen* atom of an amide is reduced by more than five powers of 10 by resonance interactions between the carbonyl group and the unshared electron pair on nitrogen. The appropriate contributing structure in the case of 2,2-dimethyl-6-quinuclidinone has a double bond at the bridgehead of a small bicyclic ring system and therefore violates Bredt's rule. Stated alternatively, the unshared electron pair on nitrogen is in an orbital orthogonal to the p orbitals of the carbonyl group. Although there is reason to believe that the carbonyl group exerts no resonance effect on the basicity of the nitrogen atom, there is a significant polar effect. The pK_a of quinuclidinium ions is 10.8 and that of 2,2,6-trimethylquinuclidinium ions is 11.3. It therefore appears that the polar effect of the oxo group in 2,2-dimethyl-6-quinuclidinone decreased the basicity by about 10^5 or 10^6. The change in structure between a quinuclidine and a quinuclidinone is analogous to replacing an ethyl group attached to nitrogen by an acetyl group. From the difference in σ^* values for these two groups [$1.65 - (-0.10) = 1.75$] and the ρ^* constant for the acidity of tertiary ammonium ions (Table 6-2), a change of $10^{5.8}$-fold in basicity would be expected.

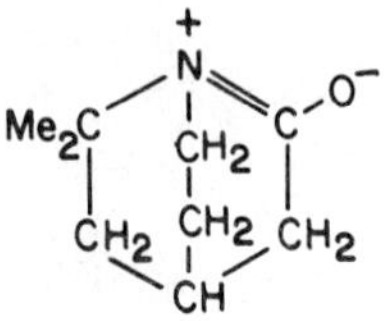

6-5. ACIDITY OF PROTONS ATTACHED TO CARBON[4]

6-5a. Acidity of Hydrocarbons.[12,48] Most hydrocarbons for which pK values have been determined yield highly resonance-stabilized carbanions. Because of the marked difference that usually exists between the spectrum of the hydrocarbon and that of the carbanion, the relative concentrations of the two species may usually be determined conveniently by spectrophotometric measurements. Some measurements of this type have been made in mixtures of dimethyl sulfoxide with methanol and potassium methoxide,[49] with ethanol and sodium ethoxide,[50,51] with water and tetramethylammonium hydroxide,[52] with acetic acid and sodium acetate,[51] etc. Such reaction mixtures become much more basic as the fraction of hydroxylic component present in the solvent is decreased (cf. Section 5-4). In addition, mixtures of dimethyl sulfoxide with tri-n-propylamine[51] and solutions of the potassium salt of dimethyl sulfoxide in pure dimethyl sulfoxide[49] were used. From some measurements, pK values referred to certain pure hydroxylic solvents were calculated by assuming that the H_- acidity function obtained from measurements of the acidities of aniline derivatives also holds for compounds in which the acidic proton is attached to carbon. In another case an attempt was made to establish an H_- scale based on the acidity of derivatives of fluorene.[51] In this and other cases one compound was chosen as a standard and arbitrarily assigned the pK value that had been obtained for it under another set of conditions, and the pK values for the other compounds were calculated by reference to the value for this standard. In addition, absolute pK values for several hydrocarbons in dimethyl sulfoxide have been determined potentiometrically.[53] All these measurements in dimethyl sulfoxide and its mixtures probably refer to equilibria between the hydrocarbons and their dissociated anionic conjugate bases. In contrast, the equilibrium constants for reactions such as

$$R^-M^+ + R'H \rightleftharpoons R'^-M^+ + RH$$

which have been determined in cyclohexylamine solution with the metal M being lithium or cesium, refer to carbanions that are present largely as ion pairs.[54–56]

Many of the p*K* values determined in the manners described in the preceding paragraph[48–58] are listed in Table 6-5, in which Binp and Biph are abbreviations for the following divalent radicals:

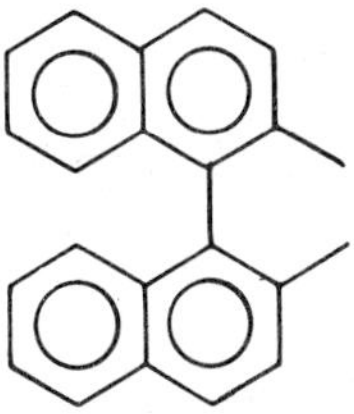

Binp

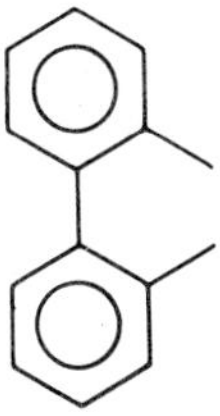

Biph

To facilitate a discussion of structural effects a numbered fluorene ring and the structural formulas of four of the hydrocarbons listed in Table 6-5 are given. In every case the acidic proton is on an sp^3-hybridized carbon atom that is attached to two or three sp^2-hybridized carbon atoms, and there are no highly polar substituents in the molecule. For this reason the relative acidities should be highly dependent on the differences in resonance energies between the carbanions and the respective hydrocarbons from which they are derived.[59] According to simple Hückel molecular orbital (HMO) theory this difference in resonance energies should be equal to $2\alpha + \Delta M\beta$, where α and β have the same value for all the hydrocarbons and only ΔM varies. The last column of Table 6-5 lists values of ΔM calculated by the HMO method, in which the carbon atoms in each π electron system are assumed to be coplanar, repulsions of electrons on different atoms are ignored, all carbon-carbon bond lengths in the π systems are taken to be the same, and steric repulsions are ignored. There is seen to be a rough tendency for p*K* to increase as ΔM decreases. It is also seen that compounds like 9-phenylfluorene, triphenylmethane, and others, whose conjugate bases would be greatly strained if coplanar, are weaker acids than would be expected from this rough tendency. Nevertheless, even if attention is restricted to those acids that can give carbanions that are probably coplanar or very nearly so, the correlation is still a rough one.[56] The deviations may arise from imperfections in the experimental data as well as imperfections in HMO theory.

8 9 1
7 2
6 3
5 4

Fluorene

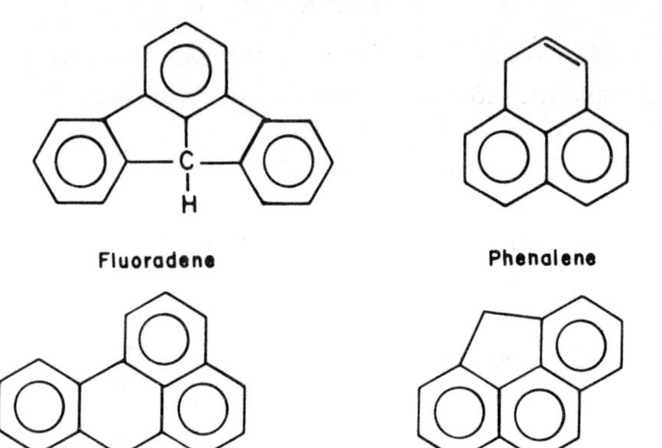

Fluoradene

Phenalene

Benzanthrene

4,5-Methylenephenanthrene

Table 6-5. Acidities of Aromatic Hydrocarbons

Hydrocarbon	Solvent		
	Me_2SO-X[a]	Cyclohexylamine[b]	ΔM[c]
$(BinpC{=}CH)_3CH$	5.9[d,e]		
$(BinpC{=}CH)_2CH_2$	8.2[d,e]		2.01
$(BiphC{=}CH)_3CH$	9.4[d]		2.50
$(BiphC{=}CH)_2C{=}CHCHBiph$	10.4[d]		2.34
$BiphC{=}CHCH_2CH{=}CBiph$	11.8[d]		1.93
$BiphC{=}CHCH{=}CHCHBiph$	12.2[d]		2.23
Fluoradene	13.9[d]		2.12
	10.5[f]		
$BiphC{=}CHCHBiph$	14.3[d]		2.23
Cyclopentadiene	~15[g]		2.00
3,4,5,6-Dibenzofluorene	16.8[d]		1.63
1,2,7,8-Dibenzofluorene	17.5[d]		1.69
9-Phenylfluorene	18.6[h]	18.5[h]	1.98
	16.4[f]		
Phenalene	19.5		1.70
3.4-Benzofluorene	19.6	19.5	1.58
Indene	18.8[i]	19.9	1.75
	18.5[f]		

Table 6-5.—Continued

Hydrocarbon	Solvent: Me_2SO-X[a]	Solvent: Cyclohexylamine[b]	ΔM[c]
1,2-Benzofluorene		20.1	1.61
Benzanthrene		21.2	1.50
9-Benzylfluorene	21.2		
9-Methylfluorene	21.8		
	19.7[f]		
4,5-Methylenephenanthrene	21.8	22.6	1.51
	20.0[f]		
Fluorene	22.1	22.7	1.52
	21.1[h,i]		
	20.5[f]		
9-Ethylfluorene	22.2		
9-Isopropylfluorene	22.7		
2,3-Benzofluorene		23.2	1.48
9-*t*-Butylfluorene	23.4		
$(p\text{-}PhC_6H_4)_3CH$	25.0[i]		
$Ph_2C{=}CHCHPh_2$		26.2[j]	
$Ph_2C{=}CHCH_2Ph$		26.6	
9,9-Dimethyl-10-phenyl-anthracene		29.0	
$p\text{-}PhC_6H_4CHPh_2$	27.5[i]	30.2	1.82
$p\text{-}PhC_6H_4CH_2Ph$	29.4[i]		
Ph_3CH	29.4[i]	31.5	1.80
	28.8[f]		
9-Methylanthracene	29.9[i]		
Ph_2CH_2	30.7[i]	33.1	1.30

[a] These pK values are referred to the solvent water assuming an H_- function and using 9-phenylfluorene, whose p*K* is taken to be 18.6, as a reference acid,[52] unless otherwise noted.
[b] Measurements on cesium salts using 9-phenylfluorene, whose p*K* is taken to be 18.5, as a reference acid,[56] unless otherwise noted.
[c] From Ref. 12, 52, or 55.
[d] Referred to the solvent water assuming an H_- function and using 9-cyanofluorene, whose p*K* is taken to be 10.4, as a reference acid.[12]
[e] This is a gross p*K* value; the acid may be a mixture of tautomers.
[f] Absolute p*K* value in pure dimethyl sulfoxide.[53]
[g] Unpublished work of K. Bowden and R. S. Cook.[57]
[h] This is a reference acid whose value is assigned as shown.
[i] Referred to the solvent methanol assuming an H_- function and using fluorene, whose p*K* is taken to be 21.1, as a reference acid.[49,58]
[j] Based on measurements on lithium salts.[55]

Breslow and co-workers worked out a method for the rough determination of the acidities of hydrocarbons that are weaker acids than those listed in Table 6-5. The critical step involved determination of the reduction potentials for transforming the appropriate carbonium ion to the radical and then the carbanion.

$$R^+ \overset{e}{\rightleftharpoons} R\cdot \overset{e}{\rightleftharpoons} R^-$$

This method, which can be applied only to cases in which the appropriate carbonium ion is stable enough to permit direct measurements to be made on it, was applied to 1,2,3-trialkylcyclopropenes, 1,2,3-triphenylcyclopropene, and 1,3,5-cycloheptatriene, as well as to the triarylmethanes that were studied as tests of the method. Unfortunately, the electrode processes studied were not fully reversible, but the actual pK values in dimethyl sulfoxide solution are probably not much above the values 62, 50, and 36 for 1,2,3-trimethylcyclopropene, 1,2,3-triphenylcyclopropene, and 1,3,5-cycloheptatriene, respectively.[60,61] The third result shows that cyclopentadiene is around 10^{21} times as strong an acid as cycloheptatriene. This difference in acidities is not explained by resonance theory in its simplest form, since the cycloheptatrienyl anion, for which seven equivalent valence-bond structures may be written, would be expected to be more stabilized by resonance than the cyclopentadienyl anion, for which only five equivalent valence-bond structures may be written. Neither is it explained by the simplest form of molecular orbital theory, which would envision each anion as composed of a ring of sp^2-carbon atoms with parallel p orbitals that overlap to form a racetrack for the delocalized electrons. By taking either theory, but probably more conveniently MO theory, to a more rigorous level, we arrive at the Hückel rule, according to which such a ring of overlapping p orbitals will lead to a particularly high degree of stabilization when the number of electrons in it is of the form $4n + 2$, that is, 2, 6, 10, etc. Thus the cyclopentadienyl anion, which has six π electrons, is particularly stable whereas the cycloheptatrienyl anion, which has eight π electrons, is not. The lesser acidity of cycloheptatriene is consistent with simple HMO theory, which gives a ΔM value of only 1.11 for transformation to the cycloheptatrienyl anion.[62]

Since ΔM for the transformation of cyclopropene to the cyclopropenyl anion (by loss of a methylene proton) is zero,[62] it is not surprising that 1,2,3-trimethylcyclopropene's ring hydrogen atom is very weakly acidic. It may be surprising, however, that the acidity is less than that which has been estimated for alkanes. The nonaromaticity of cyclopropenyl anions is also reflected in the ΔM value of 1.30 for 1,2,3-triphenylcyclopropene, which is smaller than the value for triphenylmethane, for example. The pK of about

50 for triphenylcyclopropene is much smaller than would be expected from the pK and ΔM values in Table 6-5, however. Part of the diminished acidity of cyclopropenes must result from angle strain. The sp^3 ring carbon atom in the acid, which would prefer a 109.5° C—C—C bond angle, is already strained by about 50°. Since this carbon atom has become sp^2-hybridized in the anion, with a preference for a 120° C—C—C angle, its strain has increased to 60°. Breslow and co-workers point out that the acidity of cyclopropene derivatives is probably also decreased by the "antiaromaticity" of cyclopropenyl anions.[60,61] That is, the presence of four electrons in the pi system of a three-membered ring does not just give *no* resonance stabilization. The pi electron energy is *higher* than that of a suitable reference species whose pi electrons are not cyclically delocalized. This effect shows up in approximate quantum chemical calculations that are more sophisticated than the simple HMO treatment.

6-5b. Hammett Equation Correlations of the Acidity of Carbon-Bound Protons. Several Hammett equation correlations of the acidity of carbon-bound protons have been made,[52,57,63–70] with the results shown in Table 6-6. The first six entries refer to compounds in which the acidic proton is separated from the aromatic ring by two intervening atoms. For these compounds the ρ values are as large as or larger than the ρ values for benzoic acids in the same solvent. The fact that ρ is larger for the keto forms of aroylacetones than for the enol forms would be expected, since ρ is proportional to the difference in substituent constants of the reaction center group in its reactant and product forms (Section 3-2). Because of its attachment via a carbonyl group, $COCH_2COMe$ would certainly be expected to be more electron withdrawing than C(OH)=CHCOMe. Since the product forms for these two groups are identical, the difference in substituent constants will be larger in the reaction of the keto form.

It is noteworthy that ρ for the series $ArCH_2CH(Me)NO_2$ is only 0.40, although in this case, too, the acidic proton is separated from the ring by two atoms. This small value of ρ must result in part from the location of most of the negative charge in $ArCH_2C(Me){=}NO_2^-$ anions on atoms further from the ring than the atom from which the proton was removed. Probably also relevant is the fact that in benzoic acids and in the first six classes of compounds listed in Table 6-6 there is an electron-withdrawing carbonyl group conjugated with the aromatic ring in the reactant, and its electron-withdrawing power is largely lost on ionization. There is no analogy for this in the ionization of compounds of the type $ArCH_2CH(Me)NO_2$.

The large ρ of 6.3 for the acidity of 2-substituted fluorenes shows the continuation of a trend whose beginnings could be seen in Table 6-1, where ρ for the acidity of anilines (5.3 in NH_3) is seen to be larger than ρ

Table 6-6. Hammett Equation Correlation of Acidity of Carbon-Bound Protons

Acids	Solvent[a]	Temperature (°C)	n	ρ[b]	$-\log K_0$[c]	Ref.
$ArCOCH_2PPh_3^+$	80% EtOH	25	3	2.4[d]	6.0	63
$ArCOCH_2CONHPh$	H_2O	25	9	0.79	9.4	64
$ArCOCH_2COMe$	H_2O	25	10	1.72	8.53	65
$ArC(OH){=}CHCOMe$	H_2O	25	10	1.10[e]	8.24	65
$ArCOCH_2\overset{+}{S}(Me)Ph$	H_2O	25	5	2.0[d]	7.32	66
$PhCOCH_2\overset{+}{S}(Me)Ar$	H_2O	25	5	1.4[d]	7.32	66
$ArCH_2NO_2$	H_2O	25	4	0.83[f]	6.88	69
$ArCH_2NO_2$	50% MeOH	25	4	1.22[f]	7.93	69
$ArCH(Me)NO_2$	H_2O	25	6	1.03[f]	7.39	69
$ArCH(Me)NO_2$	50% Dioxane	20	9	1.62	10.30	67
$ArCH(NO_2)_2$	H_2O	20	14	1.47	3.89	70
$ArCH_2CH(Me)NO_2$	50% MeOH	23	13	0.40	9.13	68
Fluorenes	H_2O-Me_2SO[g]	25	6	6.3[h]	22.1	52

[a] Percentages by volume; the unstated solvent component is water.
[b] Regular Hammett substituent constants (σ) used unless otherwise noted.
[c] Value observed for the unsubstituted compound.
[d] Based on para-substituted compounds only.
[e] Refers to acidity of oxygen-bound protons.
[f] Based on meta-substituted compounds only.
[g] Solvent composition varied.
[h] Based on 2-substituted fluorenes and meta-substituent constants.

for the acidity of phenols (2.2 in H_2O). An even greater fraction of the negative charge resulting from proton loss is distributed into the aromatic ring in the case of fluorene derivatives than with anilines or phenols. It might be mentioned that the data on fluorenes may be correlated better if, instead of a simple Hammett equation using σ_{meta} values, Eq. 6-4 is used. The values of ρ_1 (3.03) and ρ_2 (1.60, using σ^- values where they exist)

$$\log \frac{K}{K_0} = \rho_1\sigma_m + \rho_2\sigma_p \tag{6-4}$$

obtained show that the 2-substituents have both meta and para character.[52] This is because such substituents are conjugated with the reaction center through the other ring even though they are not conjugated with it directly.

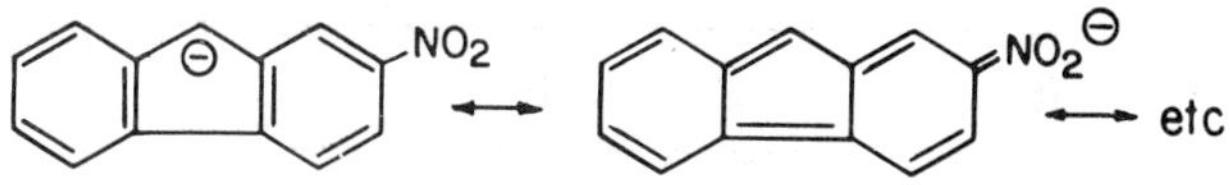

It is striking that the loss of benzyl protons in the cases of arylnitromethanes, 1-arylnitroethanes, and aryldinitromethanes gives so much smaller ρ values than in the case of fluorenes. This must be because the nitro groups in these carbanions take so much more of the negative charge upon themselves than the aromatic ring does. This is partly a result of the greater electron-withdrawing power of the nitro group and partly a result of steric hindrance. Kamlet and Glover pointed out that an aryldinitromethyl anion must not be coplanar and the largest deviation from coplanarity would be expected for the bulkier and less electron-withdrawing aryl group.[71] In agreement, an X-ray study showed that the aryl group in potassium *p*-chlorophenyldinitromethide is about 71° and the nitro groups only 7–10° out of the plane.[72]

It might be surprising that ρ is larger for the acidity of aryldinitromethanes than for arylnitromethanes. This may result from the greater electron-withdrawing power of the dinitromethyl than of the nitromethyl substituent. All the ortho substituents studied (Me, MeO, F, Cl, Br, and NO_2) increase the acidity of aryldinitromethanes relative to the unsubstituted compound and to the corresponding para-substituted compound. This suggests that the ortho-substituted compounds contain steric strains that are partly relieved in the anion. The acidity of the ortho-substituted compounds could be correlated fairly well by using steric as well as polar substituent constants.[70]

6-5c. Ionization Constants of Aliphatic 1,1-Dinitro Compounds. Several factors that influence acidity may be illustrated by data on aliphatic 1,1-dinitro compounds, which also provide examples of structural effects for which no convincing rationalization seems yet to have been proposed. The acidities of such compounds have been discussed by several groups of workers[73–77] and some of the more than 70 pK values that were determined are listed in Table 6-7. A ρ^* value of 3.60 was calculated from data on a set of 22 compounds[76] and a value of 3.29 from data on a set of 44 compounds.[77] These values of ρ^* for the series $RCH(NO_2)_2$ are of about the same magnitude as those for ammonium ions (Table 6-2), in which the R group is also attached directly to the atom bearing the acidic proton. Nevertheless, they are unexpectedly large in view of the considerably smaller ρ's obtained for $ArCH(NO_2)_2$'s and $ArCH_2CH(Me)NO_2$'s than for $ArNH_3^+$'s and $ArCH_2NH_3^+$'s, respectively (Tables 6-1 and 6-6).

Table 6-7. Ionization Constants for $RCH(NO_2)_2$ Compounds in Aqueous Solution[a]

R	pK_a	R	pK_a
H	3.63	F	8.0[b]
Me	5.30	Cl	3.53[c]
Et	5.61	Br	3.58
i-Pr	6.77	I	3.19
CH_2Pr-*i*	5.40	$CONH_2$	1.30
CH_2Bu-*t*	5.05	CO_2Me	0.98
$(CH_2)_2CN$	3.50	NO_2	0.14
CH_2CN	2.34	CN	−6.2

[a] Data from Ref. 77 at 20° C unless otherwise noted.

[b] From Ref. 73.

[c] From Ref. 75.

The acidity of *i*-$PrCH(NO_2)_2$ is less than one-tenth that which may be calculated from the Taft equation correlations, and other $RCH(NO_2)_2$ compounds in which R is secondary are also relatively weak. These observations are explained plausibly in terms of steric inhibition of resonance.[71,76] The secondary R group prevents the nitro groups from attaining the coplanarity required for maximum delocalization of the negative charge in the carbanion. This conclusion is supported by the extinction coefficients for $RC(NO_2)_2^-$ anions, which are smaller for the four secondary R compounds than for any of the 60 other compounds in which R has an sp^3-hybridized carbon at its point of attachment.[77] It is not surprising that when the branching is moved one atom further from the reaction center we find that the pK for *i*-$PrCH_2CH(NO_2)_2$ fits the correlations satisfactorily. It is noteworthy that an increase in the extent of this more distant branching, as in the case of *t*-$BuCH_2CH(NO_2)_2$, gives a compound that is too acidic to fit the Taft correlations satisfactorily. We may hypothesize that in the proper conformation the *t*-butyl group in *t*-$BuCH_2C(NO_2)_2^-$ has little effect on the coplanarity of the nitro groups, and there is a larger amount of steric destabilization in the acid, which is partly relieved in the carbanion. However, the situation has not received the detailed analysis that might make such an argument convincing. Why $(O_2N)_2CHCH_2CN$ is too weakly acidic to fit the correlations is even less clear.

The compounds with the electron-withdrawing R groups listed in the right-hand column of Table 6-7 are anomalously weakly acidic (in terms of the factors discussed so far, at least). It is true that the electron-withdrawing abilities of the carbamido, carbomethoxy, and nitro groups would be significantly decreased by steric inhibition of resonance. However, even if these groups were oriented perpendicular to the plane of the carbanion, where they would interfere little more than an alkyl group does with the coplanarity of the two nitro groups (cf. Table 2-2), they should still be able to exert their inductive substituent effect. Yet the substituent effect of the carbomethoxy group (relative to that of a methyl group), for example, is only 4.3 pK units compared with the 6.6 pK units that may be calculated from σ^* for carbomethoxy and the smaller of the two ρ^* values referred to. The effect of the cyano substituent (11.5 pK units) is in the vicinity of that which may be calculated (12.0 pK units) from a σ^* constant for the cyano group obtained by multiplying σ^* for the cyanomethyl group by 2.8. Yet this calculation based on the Taft equation takes no account of resonance, and the resonance effect of the cyano substituent cannot be reduced by rotation around the bond that joins the substituent to the reaction center.

The acid-strengthening substituent effects of the α-halogens fall even farther below those that would be calculated from the Taft equation. In fact, fluorine, the most electronegative of the elements, actually decreases the acidity markedly. Possible explanations for these facts are discussed in connection with further experimental evidence in the next section.

6-5d. Effect of α-Halogen and α-Alkyl Substituents on Acidity of Carbon-Bound Protons. The ionization constants of Adolph and Kamlet,[78] listed in Table 6-8, provide three new cases in which the α-fluorine substituent, relative to either hydrogen or chlorine, decreases the acidity of a nitro compound. It may also be seen that the α-chlorine substituent, which increases the acidity (relative to the hydrogen substituent) in each case, does not increase it to nearly the extent ($\sim$8 pK units) that would be expected from a ρ^* constant like that observed for dinitromethane derivatives and a σ^* constant for chlorine equal to 2.8 times the σ^* constant for the chloromethyl substituent. The report that a 6α-fluorine substituent increases the acidity of androst-4-ene-3,17-dione in *t*-butyl alcohol by only 50-fold[79] shows that the tendency of α-fluorine substituents to increase acidity by less than the amount expected from the usual inductive effect of fluorine is not limited to nitro compounds. Apparently the usual electron-withdrawing tendency of α-halogen substituents is opposed in these cases by some other factor(s). Although the larger halogens probably bring about some steric inhibition of resonance, a chlorine atom is about the size of a methyl group and a fluorine atom is considerably smaller (cf. Table 2-2). It seems likely that the α-halogen carbanions being considered are destabilized by repulsions between the unshared electrons on the halogen

Table 6-8. Ionization Constants of Compounds of the Type $XYCHNO_2$ in Aqueous Solution[a]

Y	X=H	X=Cl	X=F
CO_2Et	5.75	4.16	6.28
$CONH_2$	5.18	3.50	5.89
Cl	7.20	5.99	10.14

[a] p*K* values from Ref. 78 at 25° C.

and the electrons of the π system of the carbanion.[75,80–82] Such repulsions should be larger for fluorine than for any other halogen because the shorter carbon-halogen bond distance brings fluorine's unshared electrons particularly close to the π electrons and because fluorine's unshared electrons are in $2p$ orbitals, just like those of the π system.

It has also been suggested that the bond between an sp^2-hybridized carbon atom and a highly electronegative atom like fluorine tends to be weaker than the analogous bond involving sp^3-hybridized carbon (relative to the corresponding carbon-hydrogen bonds) if there is not too much stabilizing resonance interaction between the unshared electrons on fluorine and the π system of which the sp^2 carbon is a part.[83] This factor, which could be put in quantitative form by use of Eq. 1-14 if we had reliable electronegativities of sp^2- and sp^3-hybridized carbon, will tend to diminish the acid-strengthening effect of α-halogen (and alkoxy, etc.) substituents in the formation of sp^2-hybridized carbanions from sp^3-hybridized acids.[84]

Table 6-9 contains data on the effect of α-alkyl substituents on the acidity of carbon-bound protons.[85–90] The first three entries show increases in acidity when α-hydrogen is replaced by a methyl substituent. Another such case may be found in Table 6-5, where 9-methylfluorene is seen to be a stronger acid than fluorene. A 6α-methyl substituent increases the acidity of androst-4-ene-3,17-dione toward potassium *t*-butoxide by 12-fold.[79] The acidity of the α-methylene protons in methyl *n*-pentyl ketone is about the same as that of the methyl protons, and the α-methinyl protons of 2-methylcyclohexanone are about as acidic as the α-methylene protons (both cases referring to measurements on potassium salts in ethylene glycol dimethyl ether solution).[91] There are thus a number of cases in which a methyl substituent on an sp^3-hybridized carbon increases the acidity (or does not decrease the acidity) of a hydrogen atom attached to the same carbon atom. It may be relevant that in all these cases the carbanion formed is expected to be planar, and thus essentially sp^2-hybridized. (In this discussion we call a carbanion planar if the carbon from which the proton was removed is roughly in the same plane as the three atoms attached directly to it, regardless of the location of the rest of the atoms in the ion.) Carbanions stabilized by α-nitro and α-carbonyl substituents probably have the greatest tendency to be coplanar. The stability of an α-sulfone-stabilized carbanion at the bridgehead of a bicyclic ring system suggests that α-sulfone substituents do not impose any overwhelming tendency for coplanarity in carbanions.[92]

The tendency of alkyl substituents to prefer attachment to sp^2-hybridized carbon instead of sp^3-hybridized carbon (relative to the hydrogen substituent) may be illustrated by reactions 6-5 and 6-6, which are exothermic by 3 and 9 kcal/mole, respectively. The ionization constants in Table 6-9

$$\text{H—CH=CH}_2 + \text{Me—Et} \rightarrow \text{H—Et} + \text{Me—CH=CH}_2 \qquad (6\text{-}5)$$

$$\text{H—CH=O} + \text{Me—Et} \rightarrow \text{H—Et} + \text{Me—CH=O} \qquad (6\text{-}6)$$

show that the free energy change for reaction 6-7 is favorable by about

$$\text{H—CH=NO}_2^- + \text{Me—CH}_2\text{NO}_2 \rightarrow \text{H—CH}_2\text{NO}_2 + \text{Me—CH=NO}_2^- \qquad (6\text{-}7)$$

2.3 kcal/mole. This may result from a contribution from the same factor(s) that make(s) reactions 6-5 and 6-6 exothermic. It is true that there are a number of cases in which α-alkyl substituents decrease the ionization con-

Table 6-9. Effects of α-Alkyl Substituents on Acidity of Carbon-Bound Protons[a]

Acid	p*K*	Ref.
CH_3NO_2	10.2	85
$MeCH_2NO_2$	8.5	85,86
Me_2CHNO_2	7.7	86,87
$CH_2(COMe)_2$	9.0	85
$MeCH(COMe)_2$	11.0	85
$MeCOCH_2CO_2Et$	10.7	85
$MeCOCHEtCO_2Et$	12.7	85
$CH_2(CN)_2$	11.2	88
t-$BuCH(CN)_2$	13.1	88
$CH_2(SO_2Me)_2$	12.5	89
$MeCH(SO_2Et)_2$	14.6	90

[a] In water at 25° C.

stant in the formation of carbanions that are probably planar. However, the tendency, which increases in the order listed, of ethyl, isopropyl, and *t*-butyl groups to decrease the acidity of fluorene (Table 6-5), of α-methyl and α-ethyl groups to decrease the acidities of acetylacetone and ethyl acetoacetate (Table 6-9), and of α-methyl groups to decrease the acidity of dinitromethane (Table 6-7) may easily be rationalized in terms of steric inhibition of resonance. In all these cases more steric strain would be encountered in maximizing resonance stabilization than in the cases of the corresponding compounds for which α-methyl substituents have been found to increase the acidity. The last two entries in Table 6-9 show that if the ethylsulfonyl group is taken as equivalent to a methylsulfonyl group, then the α-methyl substituent decreases the acidity of a $CH_2(SO_2R)_2$. This fits the fact that α-sulfonyl carbanions have a decreased tendency to be coplanar.

It should be emphasized that substituent effects on acidity cannot be discussed solely in terms of effects on the stability of the carbanion. For example, although acinitro compounds yield the same carbanions that the corresponding nitro compounds do, the relative strengths of the acinitro compounds are $CH_2{=}NO_2H > MeCH{=}NO_2H > Me_2C{=}NO_2H$.[86] That is, methyl substituents do decrease the acidity when they are on sp^2-carbon in the acid as well as the carbanion. The preceding explanation for the acid-strengthening effect of α-alkyl substituents in certain cases where planar carbanions are formed is similar to that given by Streitwieser and co-workers.[93]

6-5e. Miscellaneous Structural Effects on the Acidity of Carbon-Bound Protons. A number of the irregularities that exist in structural effects on acidity may be seen in comparing the effects of replacing a hydrogen atom in water, ammonia, and methane with a given substituent. Table 6-10 gives results obtained with acetyl, cyano, methylsulfonyl, and nitro substituents as well as the results obtained by introducing one, two, and three substituents into methane.[77,85,89,94–106] The nitro and methylsulfonyl groups are seen to acidify water to the same extent, within the experimental error, but a nitro group acidifies ammonia somewhat more and methane far more than a methylsulfonyl group does. Steric inhibition of resonance explains why three nitro groups acidify methane no more than three methylsulfonyl groups or three cyano groups do, although one nitro group is much better than a methylsulfonyl or cyano group. It is interesting, however, that one or two acetyl groups acidify methane more than the same number of methylsulfonyl or cyano groups do even though the acetyl group is much poorer at acidifying water or ammonia.

Table 6-10. pK_a Values for Some Acetyl, Cyano, Methylsulfonyl, and Nitro Compounds in Water at 25° C[a]

X	HOX	H_2NX	CH_3X	CH_2X_2	CHX_3
NO_2	−1.4[b]	7.4[c]	10.2	3.6	0.1[d]
SO_2Me	−1.2[e]	10.8[f]	~28.5[g,h]	12.5[i]	~0
CN	<3.5[j,k]	10.3[l]	~25[m]	11.2	−5.1[h,n]
COMe	4.8[o]	~15.8[p]	~20[m,q]	9.0	5.9

[a] Data from Ref. 85 unless otherwise noted.

[b] From Ref. 94.

[c] From Ref. 95.

[d] From Ref. 77.

[e] From Ref. 96.

[f] From Ref. 97.

[g] From Ref. 98.

[h] Determined using an acidity function method.

[i] From Ref. 89.

[j] From Ref. 99.

[k] This is the value for the equilibrium mixture of HOCN and HNCO.[99] Since this mixture is composed largely of HNCO,[100,101] the pK_a of HOCN is considerably lower.

[l] From Ref. 102.

[m] Estimated from the rate of deprotonation.

[n] From Ref. 103.

[o] From Ref. 5.

[p] From the assumptions that the relative acidities of acetamide and *p*-nitrobenzamide are the same as those of acetic acid and *p*-nitrobenzoic acid,[5] that the relative acidities of *p*-nitrobenzamide and ethanol are the same in water as in ethanol,[104] and that the relative acidities of water and ethanol are the same in water as in isopropyl alcohol.[105]

[q] From Ref. 106.

It seems reasonable that the tendency of negative charge to reside on an atom in an anion would vary in the order O > N > C, other factors being equal. For this reason, there will be a greater tendency for the negative charge to be distributed by resonance into the substituent(s) in the case of a substituted methide ion than in the case of an amide ion, and even less tendency in the case of an oxy anion. Therefore resonance should increase in importance relative to inductive effects on going from oxygen to nitrogen to carbon-bound protons. Since σ_I is smaller for acetyl than for cyano or methylsulfonyl whereas the common measures of resonance electron-withdrawing ability, such as σ_R and σ_R^-, are larger for acetyl than for cyano or methylsulfonyl, this provides a qualitative explanation for acetyl being a stronger acidifying X group than cyano or methylsulfonyl in CH_3X but weaker in H_2NX and HOX. However, it does not appear to be possible to correlate the pK values very precisely in terms of ordinarily accepted σ_I and σ_R (or σ_R^-, etc.) values. By both measures, nitro is a stronger electron withdrawer than methylsulfonyl. This gives no way of explaining why methanesulfonic acid is as strong or stronger than nitric acid. There is no shortage of rationalizations for imperfections in linear free energy correlations when the substituents are as near the reaction center as they are in the present case. One that may be applied in the case of the acidity of HOX's is that the σ_I and σ_R values ordinarily used do not apply when the substituents are attached to oxygen.

No correlation of the acidities of the CH_3X's could be very satisfying in view of the poor quality of three of the experimental pK values listed in Table 6-10 and the fact that the acidity of methane, the most obvious reference compound in such a correlation, is even more poorly known. The rehybridization at carbon that probably takes place when CH_3X is transformed to its conjugate base could also produce deviations from simple linear free energy relationships. For example, the double-bond stabilization parameter D_X (Section 8-3) may relate in part to the strength of a bond between sp^2-carbon and X relative to that of a bond between sp^3-carbon and X. The fact that D_X is about 3 kcal/mole smaller for alkylsulfonyl than for cyano, nitro, and probably acetyl may explain why methylsulfonyl is the most weakly acidifying of the four substituents when it is attached to carbon but not when it is attached to oxygen or nitrogen.

No such change in hybridization at the atom bearing the substituent takes place in the transformation of *trans*-$(O_2N)_2CHCH{=}CHX$ to its conjugate base. This helps explain the success of Eq. 6-8 in correlating the acidities in methanol at 15° C of the compounds in which X is carbo-

$$\log K = 1.66\sigma_I + 7.95\sigma_R^- - 6.22 \qquad (6\text{-}8)$$

methoxy, methylsulfonyl, cyano, and nitro.[107] It seems unlikely that there are major changes in steric inhibition of resonance with changing X. It would be desirable to have data on more compounds to learn whether Eq. 6-8 continues to be applicable when the number of data greatly exceeds the number of disposable parameters.

6-6. VERY WEAK BASES

6-6a. Basicities from Acidity Function Studies in Aqueous Acid Solutions. Most of the numerical data that have been obtained on the strength of bases that are not significantly stronger than water have come from measurements in aqueous sulfuric acid using acidity functions. Most of the earlier measurements were based on the assumption that the Hammett acidity function H_0 is applicable to all electrically neutral bases. Many of the more than 500 values tabulated by Collumeau are in this category.[10] When this assumption is true the calculated pK value refers to dilute aqueous solution. In the more common cases, where it is not true, the calculated pK is merely the approximate value of H_0 for the water-sulfuric acid mixture in which the base is half protonated. Such pK values are still meaningful measures of base strength but they refer to a varying set of water-sulfuric acid mixtures rather than to any one fixed solvent. The illustrative set of pK_a values in Table 6-11 was calculated from acidity functions that the bases were found to follow at least fairly well or from a procedure like that of Bunnett and Olsen, which allows for deviations from H_0 (Section 5-5).[108–123] Even these procedures leave plenty of room for uncertainty, especially for the more negative pK_a values, some of which may easily be in error by 1.0 pK unit or more.

Many of the relative pK_a values shown are predictable. The electron-donating ability of *p*-methoxy substituents would be expected to make 1,1-di-*p*-anisylethylene a stronger base than 1,1-diphenylethylene. The electron-donating ability of methyl makes 3,4,4,5-tetramethyl-2,5-cyclohexadienone, for whose protonated form (**2**) tertiary carbonium-ion contributing

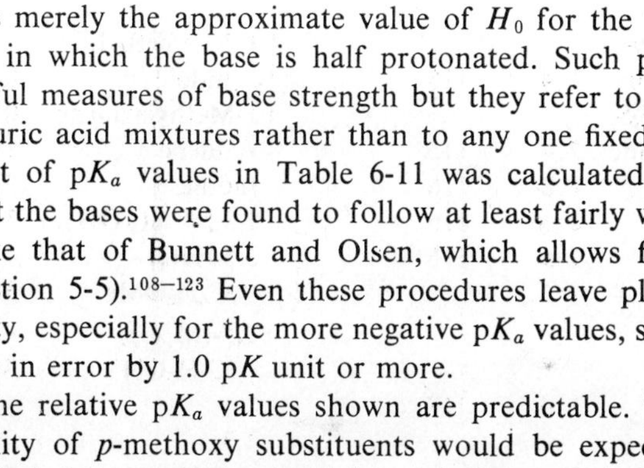

2

structures may be written, more basic than 4,4-dimethyl-2,5-cyclohexadienone, where the analogous carbonium-ion structures are secondary. A larger number of relative values are rationalizable. Thiobenzamide is a weaker base than thioacetamide because the greater electron-withdrawing inductive effect of a phenyl relative to a methyl group is more important than its greater resonance electron-donating effect.

The fact that not all the bases follow the same acidity function is another way of saying that the relative basicities vary with the composition of the

Table 6-11. pK_a Values for the Conjugate Acids of Weak Bases in Water[a]

Base	$-pK_a$[b]	Ref.	Base	$-pK_a$	Ref.
$(H_2N)_2CO$	−0.5	108	$(H_2N)_2CS$	1.5[c]	119
$MeCONH_2$	0.9[d]	109	$MeNHCSNH_2$	1.5[c]	119
$EtCONH_2$	1.0[d]	109	$MeCSNH_2$	2.6[c]	119
Pyrrole-2-carboxamide	1.1	110,111	$PhCSNH_2$	3.5[c]	119
$PhNHCONH_2$	1.3	112	*t*-BuSMe	6.7	120
$HCONH_2$	2.0[d]	113	*i*-PrSMe	6.8	120
$EtOCONH_2$	3.3[d]	114	EtSMe	6.9	120
2,6-Dimethyl-4-pyrone	−0.4	108	Me_2S	7.0	120
3,4,4,5-Tetramethyl-2,5-cyclohexadienone	1.4	115	1,2-Dimethylindole	−0.3[e]	111,121
			1-Methylazulene	0.8	122
4,4-Dimethyl-2,5-cyclohexadienone	2.4	115	Azulene	1.7	122
Me_2SO	1.8[f]	116	Indole	2.4[e]	111,121
MeSOPh	2.3[f]	116	$1,3,5\text{-}C_6H_3(OH)_3$	3.1[e]	123
Ph_2SO	2.5[f]	116	$3,5\text{-}(HO)_2\text{-}C_6H_3OMe$	3.5[e]	123
			$3,5\text{-}(MeO)_2\text{-}C_6H_3OH$	4.4[e]	123
$PhP(=O)(Me)CH_2Ph$	1.6	117	$1,3,5\text{-}C_6H_3(OMe)_3$	5.7[e]	123
$MeO\text{-}P(=O)(Me)Ph$	2.7	117	$1,3\text{-}(HO)_2C_6H_3Me$	5.8[e]	123
			$1,3\text{-}C_6H_4(OH)_2$	7.8[e]	123
H_2O	1.74[g]		$(p\text{-}MeOC_6H_4)_2C{=}CH_2$	4.4[e]	122
MeOH	2.0	118	$Ph_2C{=}CH_2$	9.4	122
Me_2O	2.5	118			
MeOEt	2.6	118			
MeOPr-*i*	2.6	118			

[a] At 25° C unless otherwise noted.

[b] All the values in this column refer to protonation on oxygen.

[c] Refers to protonation on sulfur.

[d] At 33.5° C.

[e] Refers to protonation on carbon.

[f] At "room temperature."

[g] From the definition of pK_a.

aqueous sulfuric acid solution used. The data on 1,3,5-trihydroxybenzene, 1,3,5-trimethoxybenzene, and the structurally intermediate species show that the hydroxy groups provide more basicity in water, the solvent to which the pK values listed refer. This is believed to result from stabilization of the protonated species by hydrogen bonding via the hydroxy substitu-

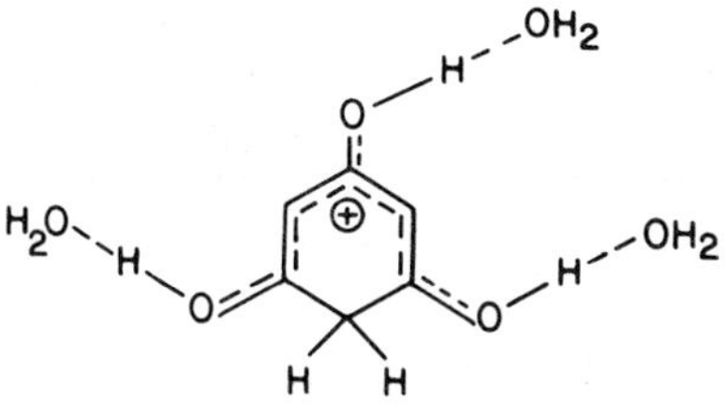

ents.[123] However, although increasing the sulfuric acid content of water-sulfuric acid mixtures increases the hydrogen ion activity, it decreases the concentration of water available for hydrogen bonding. For this reason the extent of protonation of 1,3,5-trihydroxybenzene increases with increasing sulfuric acid concentration more slowly than does the extent of protonation of 1,3,5-trimethoxybenzene. By the time 56 wt% sulfuric acid is reached the two compounds have become equally basic, and in 62% sulfuric acid, the most acidic solution in which measurements were made on 1,3,5-trimethoxybenzene, this compound is about twice as basic as 1,3,5-trihydroxybenzene.

A larger effect of the same type is found in comparing ethers and sulfides.[118,120] Although the values extrapolated to aqueous solution correspond to dimethyl ether being 30,000 times as basic as dimethyl sulfide, in 76 wt% sulfuric acid dimethyl sulfide is about 20 times as basic. It is significant that in the gas phase dimethyl sulfide is about 10^8 times as basic as dimethyl ether. (These measurements refer to the formation of free Me_2SH^+ and Me_2OH^+, not to the formation of such species as $Me_2{}^+OH$—OMe_2, for which gas phase equilibrium constants can also be determined.)

6-6b. Basicities in Strongly Acidic Solvents. A large number of qualitative observations on the behavior of very weak bases in media that are more acidic than aqueous sulfuric acid have been made. A much smaller number of equilibrium constants for protonation in such solvents have been measured. Kilpatrick and Luborsky[124] obtained results on the methylbenzenes in liquid hydrogen fluoride at 20° C that were recalculated by Perkampus[125] to give the basicity constants listed in Table 6-12. The facts that protonation puts positive charges on alternating ring atoms and that methyl groups on these positively charged atoms act as stabilizing substituents

Table 6-12. Basicity Constants of Methylbenzenes in Hydrogen Fluoride at 20° C[125]

Base	K_B	Base	K_B
C_6H_6	1.05×10^{-8}	1,2,4-$C_6H_3Me_3$	2.7×10^{-5}
C_6H_5Me	2.0×10^{-7}	1,3,5-$C_6H_3Me_3$	3.2×10^{-3}
o-$C_6H_4Me_2$	5.8×10^{-7}	1,2,3,4-$C_6H_2Me_4$	1.7×10^{-4}
m-$C_6H_4Me_2$	8.7×10^{-6}	1,2,3,5-$C_6H_2Me_4$	3.8×10^{-3}
p-$C_6H_4Me_2$	5.7×10^{-7}	C_6HMe_5	7.3×10^{-3}
1,2,3-$C_6H_3Me_3$	2.8×10^{-5}	C_6Me_6	3.7×10^{-3}

explain why the meta isomer is the most basic of the xylenes and why the 1,3,5 isomer is the most basic of the trimethylbenzenes.

Among the most useful quantitative data on the strengths of very weak bases are the "enthalpies of protonation" that were measured by Arnett and co-workers in highly acidic media.[126–129] These enthalpies of protonation are actually enthalpies of transfer from an inert solvent, usually carbon tetrachloride, to a strongly acid solvent (fluorosulfonic acid for all the enthalpies discussed here). For the compounds we are considering, which are strong enough bases to be almost completely protonated in fluorosulfonic acid solution, it is not possible to separate the process of dissolving in fluorosulfonic acid from the process of being protonated by it. Hence the enthalpies obtained when the bases dissolve will contain the desired enthalpies of protonation, but they will also contain the true enthalpies of solution, which will include the energies required to separate the molecules of the solute from each other as well as the energies of interaction of the unprotonated base with surrounding solvent. The energies of separation of the molecules from each other are simply enthalpies of vaporization, but reliable values are not available for enough of the desired compounds to be very useful in the present case. Instead, the enthalpies of solution of the various bases in carbon tetrachloride (or in 1,1,2,2-tetrachloroethane, if they were not soluble enough in carbon tetrachloride) were measured. Then these enthalpies of solution in an inert solvent were subtracted from the enthalpies of solution (and protonation) in fluorosulfonic acid to obtain the enthalpies of protonation listed in Table 6-13. These at least have the virtue of referring to transfer of all the bases from one common environment to a second common environment.

A plot of $\Delta H°$ versus pK_a values for the conjugate bases of more than 30 amines determined in water or calculated from acidity function measurements showed that the points deviated by an average of 1.2 kcal/mole

Table 6-13. Enthalpies of Transfer of Bases from an Inert Solvent to Fluorosulfonic Acid[a]

Base	$-\Delta H^\circ$ (kcal/mole)	Base	$-\Delta H^\circ$ (kcal/mole)
Et_3N	49.2	Et_2O	19.1
Et_2NH	47.7	EtOH	19.1
n-Bu_2NH	46.4	H_2O	16.4
n-Bu_3N	45.2	Ph_3PO	23.0
Pyridine	38.6	$PhNO_2$	6.6
$PhNMe_2$	37.7	Me_2CO	19.1
$PhNH_2$	34.0[b]	Cyclohexanone	18.2
p-$O_2NC_6H_4NH_2$	31.1[b]	Cyclobutanone	14.4
Ph_3N	19.1	$MeCH{=}CHCHO$	19.5
Ph_3P	28.7	$Me_2C{=}CHCOMe$	23.9
Me_3P	44.6[c]	PhCHO	16.1
PH_3	14.0[c]	PhCOMe	18.9
H_2S	5.3[c]	Ph_2CO	16.9
MeSH	19.2[c]	$PhCO_2Et$	14.5
Me_2S	18.1[c]	$MeCO_2Et$	17.4

[a] Measured at 25° C using carbon tetrachloride as the inert solvent[126,129] unless otherwise noted.

[b] The inert solvent was 1,1,2,2-tetrachloroethane.

[c] At −52° C.

from the line described by Eq. 6-9. This is an unusually wide-ranging

$$-\Delta H^\circ = (1.78\ pK_a + 28.1)\ \text{kcal/mole} \qquad (6\text{-}9)$$

correlation, covering K_a values differing by more than 10^{20}. There is no clear tendency for the various different types of nitrogen bases to describe different lines. In fact, it is significant that even when the bases are restricted to such a closely related group as the meta- and para-substituted anilines the points in a plot of ΔH° versus pK_a deviate by about 0.7 kcal/mole on the average. This suggests that there are differences in the way the solvation of the substituent groups changes on going from the inert solvent to fluorosulfonic acid.

Very few of the oxygen bases are strong enough for the pK_a values of their conjugate acids to be known well enough to tell whether they fit Eq. 6-9 or not. A few relatively strongly basic unsaturated ketones seem to fit satisfactorily, but water, for which pK_a is -1.74, deviates by 8.6 kcal/mole, probably because hydronium ions are much more weakly hydrogen bonded in fluorosulfonic acid than in aqueous solution.

6-7. ENTHALPIES AND ENTROPIES OF ACID-BASE EQUILIBRIA

6-7a. Enthalpies and Entropies of Ionization of Aliphatic Carboxylic Acids. Free energies of acid-base equilibria can be expressed in terms of the enthalpies and entropies as shown in Eq. 6-10. Values of $\Delta H°$ can be obtained directly by calorimetric measurements or indirectly, and usually

$$\Delta G° = \Delta H° - T\Delta S° \quad (6\text{-}10)$$

much less reliably, by determining how log K changes with changing temperature. Perhaps the form in which Eq. 6-10 is usually written has given some chemists the impression that $\Delta G°$ (or K) is a derived quantity, which may be expressed in terms of the more fundamental quantities $\Delta H°$ and $\Delta S°$. This is not a very meaningful impression; $\Delta H°$ could just as well be expressed as a function of $\Delta G°$ and $\Delta S°$. The question of whether $\Delta H°$ or $\Delta G°$ is the better measure of potential energy changes, which are the kind of energy changes implied by the most common types of interactions of substituents with reaction centers that physical organic chemists discuss, was treated in Section 1-5, but more in reference to gas phase than to solution phase reactions. In the gas phase, changes in substituents in reactions we think of as belonging to the same reaction series are usually accompanied by quite small changes in $\Delta S°$. In such cases the substituent changes $\Delta G°$ and $\Delta H°$ in the same way, and it makes no difference which we take as a measure of potential energy changes. The situation is different for reactions in solution, especially when the reactions involve ions.

Free energies, enthalpies, and entropies of ionization of several aliphatic carboxylic acids in water at 25° C are listed in Table 6-14.[130,131] The relative values of $\Delta G°$ certainly seem easier to explain in terms of the structures of the substituent groups than do the relative values of $\Delta H°$. The simple generalization that all saturated alkyl substituents should have about the same effect, which is plausible for a reaction governed largely by polar effects, fits the $\Delta G°$ values, whose range for such substituents is only 0.37 kcal/mole. The $\Delta H°$ values for the same substituents cover a range of 1.89 kcal/mole. The $\Delta H°$ values classify the isobutyl substituent as similar to chloromethyl and bromomethyl. They classify methyl and ethyl substituents as intermediate in character between hydroxymethyl and methoxymethyl substituents. The $\Delta G°$ values classify hydroxymethyl and methoxy-

Table 6-14. Thermodynamics of Ionization of Carboxylic Acids[a]

Acid	$\Delta G°$ (kcal/mole)	$\Delta H°$ (kcal/mole)	$\Delta S°$ (eu)
HCO_2H	5.12	−0.08	−17.4
$MeCO_2H$	6.49	−0.02	−21.9
$EtCO_2H$	6.65	−0.14	−22.8
i-$PrCO_2H$	6.62	−0.75	−24.7
t-$BuCO_2H$	6.86	−0.69	−25.3
n-$PrCO_2H$	6.58	−0.64	−24.2
n-$BuCO_2H$	6.61	−0.66	−24.4
n-$C_5H_{11}CO_2H$	6.66	−0.64	−24.5
n-$C_6H_{13}CO_2H$	6.68	−0.61	−24.4
n-$C_7H_{15}CO_2H$	6.68	−0.62	−24.5
i-$PrCH_2CO_2H$	6.52	−1.15	−25.7
Et_2CHCO_2H	6.48	−1.97	−28.4
FCH_2CO_2H[b]	3.54	−1.39	−17
$ClCH_2CO_2H$[b]	3.92	−1.12	−17
$BrCH_2CO_2H$[b]	3.96	−1.24	−17
ICH_2CO_2H[b]	4.34	−1.42	−19
$ClCH_2CH_2CO_2H$	5.45	−0.32	−19.3
$NCCH_2CO_2H$[b]	3.37	−0.90	−14
$HOCH_2CO_2H$	4.62	0.11	−17.2
$MeOCH_2CO_2H$[b]	4.87	−0.96	−19.6
$(CO_2H)_2$[c]	2.14	−1.02	−10.6
$^{-}O_2CCO_2H$	5.41	−1.50	−23.2
$CH_2(CO_2H)_2$[c]	4.27	0.29	−13.4
Fumaric[c]	4.63	0.11	−15.2
Maleic[c]	3.02	0.08	−9.9
$H_3\overset{+}{N}CH_2CO_2H$	3.21	0.98	−7.5

[a] From calorimetric measurements in water at 25° C unless otherwise noted.

[b] From pK_a measurements over a range of temperatures.

[c] $\Delta G°$ and $\Delta S°$ values for dicarboxylic acids have been calculated from K_1 values that have been divided by 2 and K_2 values that have been multiplied by 2 to put them on the same statistical basis as the K values for the monocarboxylic acids.

methyl substituents as similar to each other and considerably different from alkyl substituents. Free energy changes for ionic reactions in solution are rather generally believed to be better measures of potential energy changes than enthalpy or entropy changes are.[2,132]

In view of the imperfections in existing rationalizations of equilibrium constants, that is, free energies of reaction, it is not surprising that our ability to explain the effects of substituents on enthalpies of reaction is quite limited. Nevertheless, not only are the enthalpies of significance in their own right, but also they tell something about how equilibrium constants change with temperature. In fact, if the effect of a substituent on $\Delta H°$ differs too much from its effect on $\Delta G°$, there will be a qualitative reversal in the substituent effect at some experimentally accessible temperature.

In spite of the poor correlation between the values of $\Delta G°$ and $\Delta H°$ of ionization shown in Table 6-14, the changes in the relative values of $\Delta G°$ with changing temperature are not large enough to do any violence to the correlation of the $\Delta G°$ values by such relationships as the Taft equation. (This is partly because the Taft equation correlation is significantly imperfect at any temperature.) Consider the case of acetic acid and diethylacetic acid, whose $\Delta G°$ values are almost identical but whose $\Delta H°$ values differ by more than any other pair of saturated unsubstituted acids listed. If $\Delta H°$ were constant, diethylacetic acid would be stronger than acetic acid by 0.14 pK units at 0° C and weaker by 0.28 at 100° C. These differences are smaller than the differences between pK_a values for saturated unsubstituted carboxylic acids (excluding formic) that already exist at 25° C. Furthermore, the difference of 0.28 at 100° C is probably an overestimate because $\Delta H°$ values are not temperature-independent, and $\Delta C_p°$ values tend to be less negative for those alkanoic acids for which $\Delta H°$ is less negative.

A plot of all the $\Delta H°$ values in Table 6-14 versus the corresponding $\Delta G°$ values shows no hint of a correlation at first glance. However, if we omit the points for substituents such as hydroxy or ammonio groups (which might hydrogen bond with the reaction center), branched compounds, and compounds without an sp^3-hybridized carbon atom attached to the carboxy group, the remaining points give a crude correlation, as shown in Fig. 6-1.

6-7b. Correlations Among Enthalpies, Entropies, and Free Energies. It may be shown that if the Hammett or Taft equation with invariant substituent constants, but possibly a variable reaction constant, is to be precisely applicable to a given reaction series over a given temperature range, or, more generally, if a plot of log K values at one temperature versus log K values at any other temperature is to give a perfectly straight line,

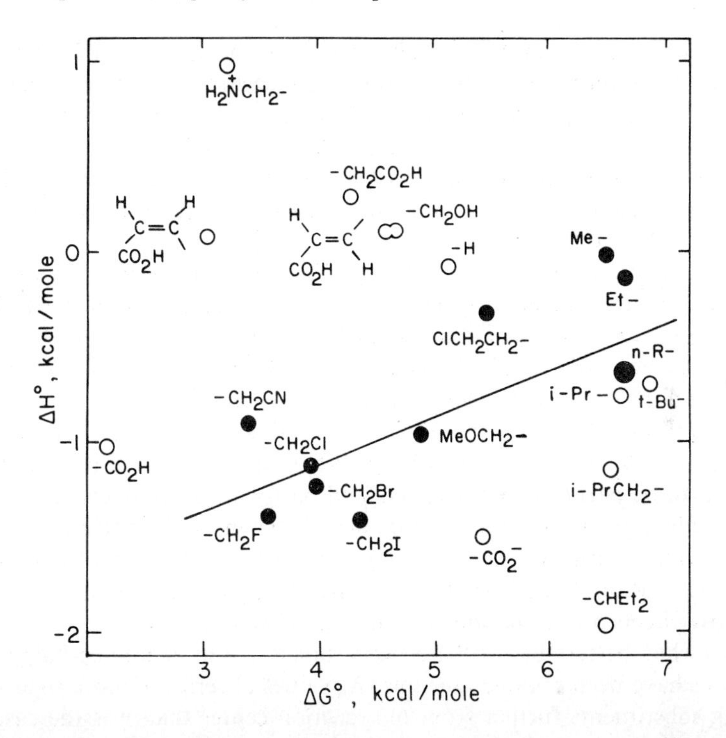

Figure 6-1. Plot of enthalpies of ionization versus free energies of ionization in water at 25° C for acids of the type RCO_2H. Each point labeled with the appropriate R. The point labeled *n*-R refers to the almost identical values for *n*-butanoic, *n*-pentanoic, *n*-hexanoic, *n*-heptanoic, and *n*-octanoic acids. The solid circles refer to electrically neutral, unbranched acids with sp^3-hybridized carbon attached to the carboxy group and without substituents thought capable of hydrogen bonding in the anion.

the effects of substituents on $\Delta G°$ must be exactly proportional to their effects on $\Delta H°$, $\Delta S°$, $\Delta C_p°$, etc. However, as seen in the comparison of $\Delta G°$ and $\Delta H°$ values for acetic acid and diethylacetic acid, fairly large imperfections in a correlation of $\Delta G°$ with $\Delta H°$ (or $\Delta S°$) may not lead to very large deviations from a linear free energy relationship over the experimentally attainable temperature range. A good correlation of $\Delta G°$ with $\Delta H°$ and $\Delta S°$ requires a correlation of $\Delta H°$ and $\Delta S°$ with each other. Correlations of $\Delta H°$ with $\Delta S°$ in a number of reaction series were observed and discussed by several early workers. Such correlations were brought to the attention of physical organic chemists by Leffler and subsequently have been discussed and reviewed by others.[133–140] They can be deceiving, especially when ΔH° and ΔS° are obtained from the variation of log K with temperature rather than from calorimetric measurements. For example, if the true values of K are the same for every reaction in a series at every temperature, random errors in the determination of K will tend to yield an excellent correlation of ΔH° with ΔS° if these functions are obtained from the observed K values. For such reasons correlations of this type require careful statistical analysis.

Linear relationships among ΔG°, ΔH°, and ΔS° values come in a variety of types. In some reaction series ΔS° does not change significantly. Thus ΔG° changes are essentially equal to ΔH° changes and the substituent effects are said to be enthalpy effects. Less commonly, the ΔH° changes are negligible and the substituent effects are said to be entropy effects. More commonly ΔH° and ΔS° change in the opposite direction, so that their changes both affect ΔG° in the same direction. When this occurs substituent effects are said to be partly enthalpy effects and partly entropy effects. To the extent to which the solid circles in Fig. 6-1 are taken as a positive correlation of ΔH° values with ΔG° values with a slope around 0.25, substituent effects on the ionization of acids in the series may be taken as roughly one-fourth enthalpy effects and three-fourths entropy effects.

Somewhat better, but still not very precise, free energy-enthalpy correlations have been obtained for the pK_a values of series of more rigid acids having substituents further from the reaction center than was the case for the acids in Table 6-14. For meta- and para-substituted benzoic acids and phenols Larson and Hepler's critical compilation of ΔG°, ΔH°, and ΔS° values was used.[141] With the s and I values given in Table 6-15, the data fit Eq. 6-11 with average deviations of 0.13 and 0.21 kcal/mole for the benzoic

$$\Delta H^\circ = s\Delta G^\circ + I \tag{6-11}$$

acids and phenols, respectively. Data on electrically charged substituents and on p-carbethoxyphenol, for which the data seemed incredible, were omitted. For meta- and para-substituted anilinium ions the calorimetrically measured ΔH° values of Liotta, Perdue, and Hopkins were used and an average deviation of 0.15 kcal/mole obtained.[142] Two almost simultaneously published determinations of the thermodynamic functions of the acidities of 3- and 4-substituted pyridinium ions[143,144] illustrate the difficulties of obtaining reliable values of ΔH° and ΔS° from determinations of pK_a at various temperatures. Although the ΔG° values for the 10 acids studied by both groups deviated by an average of 0.34 kcal/mole, the ΔH° values deviated by an average of 2.4 kcal/mole. The values of s and I in Table 6-15 are based on the measurements of Bellobono and Monetti, whose pK_a values seemed more reliable and whose ΔH° values deviated by an

Table 6-15. Free Energy-Enthalpy Correlations for 3- and 4-Substituted Aromatic Acids[a]

Acids	n[b]	s	I[c]
$ArCO_2H$	16	0.26	−1.2
ArOH	23	0.32	1.2
$ArNH_3^+$	21	0.72	3.0
Pyridinium ions	11	1.08	−2.6

[a] Correlation of data in water at 25° C using Eq. 6-11.

[b] Number of acids studied.

[c] In kcal/mole.

average of only 0.46 kcal/mole from calorimetrically measured values of Berthon and co-workers[145] for the five acids studied by both groups, and on the calorimetric results. Data on the formyl and acetyl substituents were omitted because these groups were probably hydrated to the 1,1-diols to different extents in the pyridine and the pyridinium ion. The average deviation, 0.24 kcal/mole, is a particularly small fraction of the range in $\Delta H°$ values (13 kcal/mole) in this case.

Since the entropies of gas phase deprotonation should be about the same for all the acids in Table 6-15 and *very* similar for those in a given series, the values of s for Eq. 6-11 would be near 1.0 for the gas phase reactions. Therefore it must be solvation effects that make three of the values in Table 6-15 significantly smaller than 1.0. We have already pointed out (Section 5-6) that substituent effects on the acidities of pyridinium ions in aqueous solution are only 29% as large as they are in the gas phase. In terms of the scheme shown in Eq. 6-12, substituent effects must change $\Delta G_s(\mathrm{HA}) - \Delta G_s(\mathrm{A})$ for pyridinium ions in the same direction as, and 71% as much as,

$$
\begin{array}{ccccc}
\mathrm{HA}(g) & \xrightarrow{\Delta G_g} & \mathrm{H}^+(g) & + & \mathrm{A}(g) \\
\downarrow{\scriptstyle \Delta G_s(\mathrm{HA})} & & \downarrow{\scriptstyle \Delta G_s(\mathrm{H}^+)} & & \downarrow{\scriptstyle \Delta G_s(\mathrm{A})} \\
\mathrm{HA}(w) & \xrightarrow{\Delta G_w} & \mathrm{H}^+(w) & + & \mathrm{A}(w)
\end{array}
\qquad (6\text{-}12)
$$

they change ΔG_g. It is certainly plausible that the change should be in the same direction. One of the major ways in which HA and A are solvated is by hydrogen bonding. A substituent that increases the acidity of HA and hence decreases ΔG_g will also decrease $\Delta G_s(\mathrm{HA})$, since the stronger acid will hydrogen bond to water more strongly, and increase $\Delta G_s(\mathrm{A})$, since A will be a weaker base that hydrogen bonds to water more weakly. In the case of phenols substituent effects in aqueous solution have been found to be only 21% as large as in the gas phase,[146] showing that $\Delta G_s(\text{HA}) - \Delta G_s(\text{A})$ changes 79% as much as ΔG_g does. A qualitatively similar situation would be expected for benzoic acids and anilinium ions. Thus for these families of similar acids there is a linear free energy relationship between gas phase acidities and free energies of solvation and therefore a linear free energy relationship between gas phase and solution phase acidities.

Table 6-15 reveals an important difference in the behavior of the four sets of acids. The value of s for pyridinium ions is probably within the experimental uncertainty of 1.00. This means that substituent effects on the values of $\Delta H_s(\text{HA}) - \Delta H_s(\text{A})$ are essentially equal to the effects on the corresponding values of $\Delta G_s(\text{HA}) - \Delta G_s(\text{A})$ and that substituents have a negligible effect on $\Delta S_s(\text{HA}) - \Delta S_s(\text{A})$ values. The s values for the other three series of acids, however, show that substituent effects on $\Delta G_s(\text{HA}) - \Delta G_s(\text{A})$ are smaller than on $\Delta H_s(\text{HA}) - \Delta H_s(\text{A})$; that is, the changes in enthalpies of solvation are partly compensated by changes in entropies of solvation. The latter type of behavior is easily rationalized qualitatively. When a water molecule is bound by a solute there is a decrease in entropy owing to the loss in translational freedom. If it is weakly bound, however, the water molecule may retain a significant amount of freedom. The same substituents that decrease the enthalpy of solvation by forming a stronger bond to the water molecules also decrease the entropy of solvation by further restricting their freedom.

The relative values of s in Table 6-15 may be at least partly rationalized in terms of an explanation that emphasizes changes in the solvation of the ions involved in the various equilibria. The strongest interactions with water should take place at the charged parts of the ions, and the strength of interaction should tend to increase with the charge density. It is probably significant that the two families of acids for which s is relatively small are carboxylic acids and phenols, both of which give ions with delocalized charges. In the two families for which s is relatively large the charge on the ion is concentrated largely on one atom. Hence the charged parts of the carboxylate and phenoxide ions are probably in contact with a relatively large number of molecules of water, but they do not bind these molecules very tightly. The charged parts of anilinium and pyridinium ions probably bind a smaller number of molecules of water but each molecule is bound more tightly. Acid strengthening (base weakening) substituents will decrease the strength with which solvating water molecules are held to carboxylate and phenoxide ions and increase the strength with which they are held to anilinium and pyridinium ions. Therefore, if changing the strength of binding has a large effect on the entropy of binding when the binding is weak but little or no effect when it is strong, we can explain the variation in s values. There are theoretical reasons why the entropy of solvation should be more sensitive to the strength of binding when it is weak than when it is strong and empirical evidence that it is. If we regard the solvated water molecule as being held by a bond, then when the bond is weak its vibrational energy levels will be closely spaced so that a number of the lower levels will be significantly populated at 25° C. A small increase in bond strength will increase the spacing so that more of the molecules will be in the zero-point vibrational level. This decrease in the number of states that can be attained with the available energy will produce a decrease in entropy. However, when the bond is so strong that practically all the molecules are in the zero level, further increases in bond strength will not significantly affect the entropy. Empirically, it is found in several series of closely related hydrogen bond association reactions between fairly strongly hydrogen bonding species that increases in strength of binding are not accompanied by systematic changes in $\Delta S°$. The most reliable measurements are probably those of Arnett and co-workers who found that substituent effects that caused $\Delta G°$ for hydrogen bonding of four different pyridine derivatives to *p*-fluorophenol to vary over a range of 2.0 kcal/mole produced $T\Delta S°$ values that were in a range of 0.5 kcal/mole and showed no tendency to decrease with decreasing $\Delta G°$.[147] Similarly, seven meta- and para-substituted phenols whose $\Delta G°$ values for association with diphenyl sulfoxide ranged over 2.0 kcal/mole gave a $T\Delta S°$ range of 0.4 kcal/mole and no clear trend.[148] Similar results were obtained with meta- and para-substituted phenols and *N,N*-dimethylacetamide.[149] The available evidence on relatively weak hydrogen bonds shows that $\Delta S°$ usually decreases with decreasing $\Delta G°$, but there are a number of exceptions.[147,150–153a] Unfortunately, the available $\Delta S°$ values are probably not very accurately known and no studies have been made on a series of closely related compounds covering a wide enough range of $\Delta G°$ values.

This rationalization for why s is smaller for carboxylic acids and phenols than for anilinium ions will also explain why the value for pyridinium ions is larger than that for anilinium ions if we can argue convincingly that the pyridinium ions bind their water molecules more tightly than the anilinium ions do. The gas phase proton affinity of pyridine is 9.7 kcal/mole greater than that of aniline[154] although $\Delta G°$ for ionization in aqueous solution is only 0.8 kcal/mole more negative. Most of this change in relative acidities is reasonably attributed to greater solvation of the anilinium ion than of the pyridinium ion. The sum of the strengths of the three hydrogen bonds formed by $PhNH_3^+$ to three water molecules is therefore presumably greater than the strength of the one hydrogen bond formed by the pyridinium ion. However, it does not follow that the individual hydrogen bonds to the anilinium ion are stronger than the one to the pyridinium ion. In fact, the observation that the basicity of pyridine increases relative to that of *N,N*-dimethylaniline[154] and *N,N*-diethylaniline[154] by 3 kcal/mole or more on going from the gas phase to aqueous solution suggests that the hydrogen bond from water to the pyridinium ion is stronger than that to an *N,N*-dialkylanilinium ion.

Although the preceding rationalization may explain why s is largest for pyridinium ions, it does not satisfyingly explain why s is around 1.00 or perhaps significantly larger. Changes in the acidity of the pyridinium ion might plausibly change the strength of the hydrogen bond between its acidic proton and the adjacent water molecule without changing the entropy of bonding to this water molecule. However, the acidity of this water molecule should increase with increasing acidity of the pyridinium ion bonded to it. This should change the strength of the relatively weak hydrogen bond between the first water molecule and those in the second solvation shell. Changing the strength of this weak hydrogen bonding should be accompanied by entropy changes, but the value of s of 1.08 gives no evidence of this. Perhaps the reader will find the alternative discussions of enthalpy and entropy effects in acid-base reactions that have been published[130–134,140–143] (largely before the work on pyridinium ions) to be more enlightening than the preceding one. Perhaps they will lead to an understanding of why s is so large for pyridinium ions.

6-8. EQUILIBRIUM IN HYDROGEN BONDING[153,155,156]

In hydrogen bonding the extent of transfer of the proton to the base is considerably smaller than it is in typical Brønsted acid-base reactions in solution and much smaller than it is in such reactions in the gas phase. For this reason substituent effects might be expected to operate in the same direction as they do in the more complete proton transfer reactions but to be smaller. There is certainly evidence that this is the case when the acids and bases involved are restricted to closely related species. There are a number of cases in which equilibrium constants have been determined for hydrogen-bond complex formation between a given base and a number of related acids or between a given acid and a number of related bases. Because of complications arising from hydrogen bonding with the solvent in most good ion-solvating media, most such equilibrium constants have been determined in poor ion-solvating media—most commonly carbon tetrachloride, whose lack of hydrogen atoms makes infrared and proton magnetic resonance measurements particularly convenient. Since the ionization constants of the acids and bases involved typically have not been determined in such media, a direct comparison of equilibrium constants for hydrogen bonding and Brønsted acidities and basicities under the same conditions is not possible. It has been shown, however, that the equilibrium constants are increased by electron-withdrawing substituents in the acids and electron-donating substituents in the bases. A number of studies have been made on the coordination of a series of meta- and para-substituted phenols with a given base. The results of some of these[148–150,156–166] are summarized in Table 6-16. The only solvent listed in which pK values for phenols have been determined is acetonitrile, in which ρ for the acidity of phenols is about 4.8.[166] The ρ value of 1.1 for the hydrogen bonding of phenols to chloride ions in this solvent therefore suggests that the extent of proton transfer in these reactions is about one-fourth. Since ρ for the ionization of phenols should be larger than 4.8 (but smaller than 10, the approximate value in the gas phase[146]) in the other solvents listed, it is not clear that the extent of proton transfer is significantly greater than this in any of the reactions covered. Since the extent of proton transfer should increase with increasing strength of binding, ρ would be expected to in-

Table 6-16. Correlations of Equilibrium Constants for Hydrogen Bonding of Meta- and Para-Substituted Phenols

Base	Solvent	Temperature (°C)	n^a	ρ^b	$K_{PhOH \cdot B}{}^c$	Ref.
$PhNH_2$	CCl_4	27	11	0.8	4.0	157,158
Ph_2CO	CCl_4	25	3	~0.7	7.7	159
AcOMe	*n*-Heptane	25	3	~0.8	11.5	160
Pyridine	CCl_4	27	11	1.1	46	157
Et_3N	CCl_4	30	12	1.4	47	159,161
Ph_2SO	C_2Cl_4	25	7	1.5	74	148
Me_3N	Cyclohexane	25	4	1.3	86	162
$AcNH_2$	CCl_4	27	7	1.1	71	163
$AcNMe_2$	CCl_4	25	17	1.7	127	149
$(Me_2N)_2CO$	CCl_4	30	6	1.8	121	164
Quinoline-1-oxide	CCl_4	25	5	1.4	150	165
Cl^-	MeCN	25	6	1.1	150	166

[a] Number of phenols studied.

[b] Calculated from data on monosubstituted phenols using σ^n or σ° values (but not necessarily calculated in the paper from which the experimental values were taken).

[c] In M^{-1}.

crease with increasing $K_{PhOH \cdot B}$. The data show a clear but imperfect trend in this direction. The trend is clearer when the different bases being compared are closely related to each other. This is seen in the results of Zeegers-Huyskens and co-workers, who studied complex formation by phenols with pyridines[167] and anilines[158] and varied the nature of 3- and 4-substituents in the acids and bases simultaneously. In both series the sensitivity of K to changes in substituents on the phenol increases with increasing basicity of the base.

If phenol acidities were perfectly correlated by the Hammett equation using the same substituent constants that were used to obtain the ρ values in Table 6-16, then the Hammett equation correlation in the table would be equivalent to a correlation with the acidities of the phenols. However, the equilibrium constants for hydrogen bonding are best correlated by use of σ^n values, whereas the ionization constants of phenols are best correlated by use of σ^- values (in the media in which measurements have been made, at least). That is, direct resonance between substituents and the reaction center seems to be of little importance in the hydrogen bonding of phenols with bases. In fact, many of the imperfections in correlations between acidity (or basicity) and hydrogen-bonding ability point to resonance effects being relatively less important and polar effects more important in hydrogen bonding than in Brønsted acidity and basicity. The data[5,7,27,28,151,152,157,159,168–176] in Table 6-17, for example, show that alcohols form stronger hydrogen bonds than phenols of the same acidity (in water). Hexafluoro-2-propanol has about the same acidity in water that *p*-chlorophenol does, but its equilibrium constant for hydrogen bonding to tetrahydrofuran is more than twice as large. The lack of correlation with acidity should be even more striking if pK_a values were available in carbon tetrachloride, where *p*-chlorophenol, whose anion has a delocalized charge, would be expected to have a significantly larger ionization constant than hexafluoro-2-propanol. The marked polarity of oxygen-hydrogen bonds leads to a stabilizing dipole-dipole interaction when a hydroxylic compound associates with a base. This explains the much weaker hydrogen-bonding ability of thiophenol and isocyanic and hydrazoic acids. However, it does not explain why 5-fluoroindole forms hydrogen bonds that are as strong as those formed from hydrazoic acid and stronger than those formed from methanol or ethanol. It is likely that 5-fluoroindole would have a larger ionization constant than methanol or ethanol in carbon tetrachloride solution but not likely that it would be as strong an acid as hydrazoic acid.

The largest and most reliable set of equilibrium constants for hydrogen bonding of a number of different bases to a given acid refer to the acid *p*-fluorophenol. The ^{19}F nmr measurements of Taft and co-workers, the

Table 6-17. Equilibrium Constants for Hydrogen Bonding[a]

		Tetrahydrofuran		Pyridine	
Acid	pK_a[b]	K	Ref.	K	Ref.
HNCO	3.5[c]	9.1	168	33.9	168
HN_3	4.7[d]	2.76	169	4.75	169
PhSH	6.5[e]			0.22	170
$(CF_3)_2CHOH$	9.3[f]	78	152		
p-ClC_6H_4OH	9.4[g]	29	172	120	157,173
PhOH	10.0[g]	11.1	174	42	159
CF_3CH_2OH	12.4[h]	10.0	152		
MeOH	15.5[h]			3.0	151
EtOH	15.9[h]			2.4	151
5-Fluoroindole	16.3[i]	2.6	176	5.6	176

[a] K in M^{-1} for reaction in carbon tetrachloride at 25°.
[b] In aqueous solution at 25° C.
[c] See Table 6-10.
[d] Ref. 7.
[e] Ref. 28.
[f] Ref. 171.
[g] Ref. 5.
[h] Ref. 27.
[i] Ref. 175.

calorimetric measurements of Arnett and co-workers, and the ir measurements of Schleyer and co-workers have been combined to obtain this set.[147,177,178] The equilibrium constants in carbon tetrachloride at 25° were called K_{HB}. A number of additional values of K_{HB} were estimated, largely from Gramstad's data on phenol. Some of the values of log K_{HB} for bases of various types are listed in Table 6-18. The observations that K_{HB} is smaller for $Cl_2CHCOMe$, p-$O_2NC_6H_4CHO$, and $CF_3CH_2NH_2$ than for Me_2CO, PhCHO, and n-$BuNH_2$, respectively, illustrate the decreases in hydrogen-bonding basicity brought about by electron-withdrawing substituents. Since the same substituents decrease Brønsted basicity this illustrates how hydrogen-bonding basicity and Brønsted basicity are correlated within families of closely related compounds. Outside of such families, however, there are major deviations from such correlations. The Brønsted basicity of 2,2,2-trifluoroethylamine, for example, is greater than that of any of the oxygen bases in the table (in aqueous solution, at least), and yet many of the oxygen bases are much stronger hydrogen bonders. In fact, it seems that the hydrogen-bonding basicity of an oxygen base is generally greater than that of a nitrogen base of comparable Brønsted basicity. This is but one bit of evidence that hydrogen bonding is usefully considered as a mixture of an acid-base reaction and an electrostatic interaction. Because of the greater electronegativity of oxygen, the basic atom of an oxygen base is ordinarily at the negative end of a larger dipole than is the basic atom of a nitrogen (or sulfur) base of comparable Brønsted basicity. Similarly, it is not surprising that bases with sp^2- and sp-hybridized nitrogen and sp^2-hybridized oxygen are stronger hydrogen bonders than are species that have comparable Brønsted basicity but sp^3-hybridized basic atoms. The especially great hydrogen-bonding basicity of such bases as triethylphosphine oxide, for which zwitterionic valence bond structures are probably particularly important, would also be expected on an electrostatic basis.

Table 6-18. Equilibrium Constants for Hydrogen Bonding to *p*-Fluorophenol[a]

Base	$\log K_{HB}$	Base	$\log K_{HB}$
PhCOF	−0.30	AcOMe	1.00
CF_3COMe	−0.19	MeCN	1.05
c-$C_6H_{11}F$	−0.05	Pyrimidine	1.05[b]
PhOMe	0.02	CF_3CONMe_2	1.18
Azulene	0.02	Me_2CO	1.18
Et_2S	0.11	Tetrahydrofuran	1.26
p-$O_2NC_6H_4CHO$	0.36	p-$Me_2NC_6H_4CHO$	1.55
$Cl_2CHCOMe$	0.48	$(n\text{-}Pr)_3N$	1.57
$CF_3CH_2NH_2$	0.59	$(PhO)_3PO$	1.73
EtCHO	0.71	$PhCH_2NH_2$	1.75
1,4-Dioxane	0.71[b]	Pyridine	1.88
$PhNO_2$	0.73	n-$PrNMe_2$	1.98
PhCN	0.79	$HCONMe_2$	2.06
PhCHO	0.83	n-$BuNH_2$	2.11
$PhCO_2Et$	0.88	$AcNMe_2$	2.38
Ph_2CO	0.97	Me_2SO	2.53
Et_2O	0.98	$(Me_2N)_3PO$	3.56
$(EtO)_2SO$	0.98	Et_3PO	3.64

[a] In carbon tetrachloride at 25° C. K_{HB} in M^{-1}. Some of the values were estimated by applying a linear free energy relationship to equilibrium constants obtained using other acids, usually phenol.[178]

[b] The experimental value was divided by 2 to put this K_{HB} value on a per oxygen or per nitrogen basis.

Table 6-19. Correlation of Equilibrium Constants for Hydrogen Bonding with K_{HB} Values[a]

Acid	Bases	m	c
EtOH	All	0.51	−0.53
MeOH	All	0.55	−0.3
α-Naphthol	All	0.92	0.18
CF_3CH_2OH	All	0.92	−0.17
$(CF_3)_2CHOH$	All	1.16	0.32
5-Fluoroindole[b]	Carbonyls, R_2SO, and X_3PO	0.70	−0.15
5-Fluoroindole[b]	Pyridines and guanidines	0.70	−0.50
5-Fluoroindole[b]	R_3N	~0.6	−0.6

[a] Correlation from Ref. 178 referring to data at 20° or 25° C unless otherwise noted.
[b] Correlations based on data in Ref. 179.

Taft and co-workers examined the possibility that log K for hydrogen bonding by other acids to a variety of bases is linearly related to log K_{HB} for hydrogen bonding of the same bases to *p*-fluorophenol.[178] Application of Eq. 6-13 to data on several acids gave the results summarized in Table

$$\log K = m \log K_{HB} + c \tag{6-13}$$

6-19. The correlations obtained with alcohols were good enough to leave open the possibility that Eq. 6-13 is rather general for hydroxy acids (although, almost undoubtedly, steric effects could give major deviations in some cases). However, not enough different hydroxy acids have been studied with enough different bases to be sure about this. In any event the applicability of Eq. 6-13 to data on hydroxy acids appears to be more general than its applicability to data on 5-fluoroindole, whose equilibrium constants for hydrogen bonding to 27 oxygen and nitrogen bases have been determined.[179] In this case the correlation is good only for families of related bases. One good linear plot was obtained from data on carbonyl bases, sulfoxides, and compounds of the type X_3PO. The points for the two ethers studied deviated from this line. A second line was described by the points for pyridine derivatives and a third, poorer, line by tertiary amines in which all three nitrogen-bound carbon atoms were sp^3-hybridized. Other NH acids such as isothiocyanic,[180] isocyanic,[168] and hydrazoic[169] give equilibrium constants that also tend to be correlated with those for *p*-fluorophenol, but the ranges of bases studied were not sufficient to tell how precise and how general the correlation is.

PROBLEMS

1. Use the ρ for anilinium ions in Table 6-1, use 2,4,6-trinitroaniline (see Table 5-2) as the reference compound, and assume that the first equation in Section 3-5 is applicable; then calculate the pK values of the conjugate acids of 3-methyl-2,4,6-trinitroaniline and 3-bromo-2,4,6-trinitroaniline. Explain the differences, if any, between the calculated and experimental values (Table 5-2).

2. For series of compounds in which the acidic OH is separated from the aromatic ring by one atom, the Hammett ρ is usually in the range 0.7–1.4 in a largely aqueous solvent (cf. Table 6-1). For compounds of the type $ArB(OH)_2$, however, ρ is 2.15. Suggest an explanation for this fact.

3. Using the data in Tables 3-1 and 6-1 select an Ar group such that the acidity of $ArNH_3^+$ would be expected to be essentially the same as that of $ArSeCH_2CO_2H$.

4. The pK_a values for the monoprotonated forms of 1,8-diaminonaphthalene, 1,8-bis(methylamino)naphthalene, and 1,8-bis(dimethylamino)naphthalene in water at 25° C are 4.61, 5.61, and 12.34, respectively.[181] Rationalize the relative strengths of these three bases and discuss the probable acidities of their diprotonated forms.

5. Suggest an explanation for why *N*,*N*-diethylaniline is so much more basic than *N*,*N*-dimethylaniline or *N*-phenylpiperidine (Section 6-4).

6. What do you expect for the relative values of r when the Yukawa-Tsuno equation (Section 3-3d) is applied to the basicities of ArCHO's, ArCOMe's, and $ArCONH_2$'s?[182] Give your reasons.

7. Assume that the dipole moments of 1.96 D for *t*-butyl fluoride[183] and 2.20 D for *t*-butyl iodide[184] arise solely from carbon-halogen bond dipoles consisting of a negative charge on the halogen atom and a positive charge on the carbon to which it is attached. Calculate the energy of interaction of such dipoles with a charge equal to the charge on an electron located 3.00 Å from the carbon atom on the extension of the halogen-carbon line in both the carbon-fluorine and carbon-iodine cases. Repeat the calculation for a distance of 15.00 Å. Do the results suggest anything about the way in which the relative polar substituent effects of the halogens may change as the distance from the reaction center is changed? Might the preceding have anything to do with the effects of halogen substituents on the acidities of acetic acid[3] and quinuclidinium ions?[38]

8. The ionization of acetic acid in water, for which $\Delta H°$ at 25° C is given in Table 6-14, has a ΔC_p of -32 cal/degree-mole.[131] What does this tell about the shape of a plot of log K versus T between 0° and 60° C?

9. In which medium—vapor, benzene, or cyclohexane—would you expect the equilibrium constant for the formation of a hydrogen-bonded dimer from trifluoroacetic acid[185] to be largest? In which would it be smallest? Give your reasons.

10. There are a number of species whose gas phase Brønsted acidities have been

compared with those of their monomethyl derivatives. Among the following pairs of species, the methyl derivative (the second species listed) has been found to be the stronger acid in six cases and the weaker acid in nine cases.

a. H_2O, CH_3OH[186]
b. CH_3OH, CH_3CH_2OH[186]
c. CH_3NH_2, $CH_3CH_2NH_2$[186]
d. $CH_2{=}CH_2$, $CH_3CH{=}CH_2$[187]
e. $HC{\equiv}CH$, $CH_3C{\equiv}CH$[186]
f. PhH, $PhCH_3$[187]
g. C_6H_5OH, *p*-$CH_3C_6H_4OH$[146]
h. Pyridinium ion, 4-methylpyridinium ion[192]
i. HCO_2H, CH_3CO_2H[188]
j. NH_4^+, $CH_3NH_3^+$[189]
k. $CH_3NH_3^+$, $CH_3CH_2NH_3^+$[190]
l. C_6H_5OH, *o*-$CH_3C_6H_4OH$[146]
m. $C_6H_5CO_2H$, *p*-$CH_3C_6H_4CO_2H$[191]
n. H_2, CH_3H[187]
o. $C_6H_5CH_3$, *p*-$C_6H_4(CH_3)_2$[186]

Assign each of the fifteen cases listed to one of the two categories and give your reasons.

REFERENCES

1. A. Albert and E. P. Serjeant, *Ionization Constants of Acids and Bases*, 2nd ed., Chapman and Hall, London, 1971.
2. R. P. Bell, *The Proton in Chemistry*, 2nd ed., Cornell University Press, Ithaca, N.Y., 1973.
3. H. C. Brown, D. H. McDaniel, and O. Häfliger, in *Determination of Organic Structures by Physical Methods*, E. A. Braude and F. C. Nachod, Eds., Academic Press, New York, 1955, Chap. 14.
4. D. J. Cram, *Fundamentals of Carbanion Chemistry*, Academic Press, New York, 1965, Chaps. 1 and 2.
5. G. Körtum, W. Vogel, and K. Andrussow, *Pure Appl. Chem.*, **1**, 189 (1961) (organic acids).
6. D. D. Perrin, *Dissociation Constants of Organic Bases in Aqueous Solution*, Butterworths, London, 1965; Supplement, 1972.
7. D. D. Perrin, *Pure Appl. Chem.*, **20**, 133 (1969) (inorganic acids and bases).
8. L. G. Sillén and A. E. Martell, *Stability Constants of Metal-Ion Complexes*, Special Publication No. 17, The Chemical Society, London, 1964.
9. L. G. Sillén and A. E. Martell, *Stability Constants of Metal-Ion Complexes, Supplement No. 1*, Special Publication No. 25, The Chemical Society, London, 1971.

10. A. Collumeau, *Bull. Soc. Chim. Fr.*, 5087 (1968).
11. E. M. Arnett, *Progr. Phys. Org. Chem.*, **1,** 223 (1963).
12. H. Fischer and D. Rewicki, *Progr. Org. Chem.*, **7,** 116 (1968).
13. J. C. Howard and F. E. Youngblood, *J. Org. Chem.*, **31,** 959 (1966).
14. J. L. Weeks and J. Rabani, *J. Phys. Chem.*, **70,** 2100 (1966).
15. N. C. Deno, R. W. Gaugler, and M. J. Wisotsky, *J. Org. Chem.*, **31,** 1967 (1966).
16. H. H. Jaffé, *Chem. Rev.*, **53,** 191 (1953).
17. M. S. Newman and S. H. Merrill, *J. Amer. Chem. Soc.*, **77,** 5552 (1955).
18. I. J. Solomon and R. Filler, *J. Amer. Chem. Soc.*, **85,** 3492 (1963).
19. L. D. Pettit, A. Royston, C. Sherrington, and R. J. Whewell, *J. Chem. Soc.*, *B*, 588 (1968).
20. D. J. Pasto, D. McMillan, and T. Murphy, *J. Org. Chem.*, **30**, 2688 (1965).
21. A. I. Biggs and R. A. Robinson, *J. Chem. Soc.*, 388 (1961).
22. L. F. Blackwell, A. Fischer, I. J. Miller, R. D. Topsom, and J. Vaughan, *J. Chem. Soc.*, 3588 (1964).
23. A. Fischer, D. A. R. Happer, and J. Vaughan, *J. Chem. Soc.*, 4060 (1964).
24. D. Dolman and R. Stewart, *Can. J. Chem.*, **45,** 911 (1967).
25. T. Birchall and W. L. Jolly, *J. Amer. Chem. Soc.*, **88,** 5439 (1966).
26. R. W. Taft, Jr., in *Steric Effects in Organic Chemistry*," M. S. Newman, Ed., Wiley, New York, 1956, p. 607.
27. P. Ballinger and F. A. Long, *J. Amer. Chem. Soc.*, **82,** 795 (1960).
28. M. M. Kreevoy, B. E. Eichinger, F. E. Stary, E. A. Katz. and J. H. Sellstedt, *J. Org. Chem.*, **29,** 1641 (1964).
29. H. K. Hall, Jr., *J. Amer. Chem. Soc.*, **79,** 5441 (1957).
30. Ref. 26, Chap. 13, Sec. VI-3.
31. G. C. Pimentel and A. L. McClellan, *The Hydrogen Bond*, Freeman, San Francisco, 1960, Appendix C.
32. J. M. Hopkins and L. I. Bone, *J. Chem. Phys.*, **58,** 1473 (1973).
33. J. Hine, J. C. Craig, Jr., J. G. Underwood II, and F. A. Via, *J. Amer. Chem. Soc.*, **92,** 5194 (1970).
34. A. J. Hoefnagel, J. C. Monshouwer, E. C. G. Snorn, and B. M. Wepster, *J. Amer Chem. Soc.*, **95,** 5350 (1973).
35. Cf. N. V. Hayes and G. E. K. Branch, *J. Amer. Chem. Soc.*, **65**, 1555 (1943).
36. J. Paleček and J. Hlavatý, *Z. Chem.*, **9,** 428 (1969).
37. W. Eckhardt, C. A. Grob, and W. D. Treffert, *Helv. Chim. Acta*, **55,** 2432 (1972).
38. E. Ceppi, W. Eckhardt, and C. A. Grob, *Tetrahedron Lett.*, 3627 (1973).
39. R. W. Taft and C. A. Grob. *J. Amer. Chem. Soc.*, **96,** 1236 (1974).

40. H. C. Brown and B. Kanner, *J. Amer. Chem. Soc.*, **75,** 3865 (1953); **88,** 986 (1966).
41. D. H. McDaniel and M. Ozcan, *J. Org. Chem.*, **33,** 1922 (1968).
42. M. S. Newman and T. Fukunaga, *J. Amer. Chem. Soc.*, **85,** 1176 (1963).
43. G. W. Wheland, R. M. Brownell, and E. C. Mayo, *J. Amer. Chem. Soc.*, **70,** 2492 (1948).
44. B. M. Wepster, *Rec. Trav. Chim. Pays-Bas*, **71,** 1171 (1952).
45. H. Pracejus, *Chem. Ber.*, **92,** 988 (1959).
46. H. Pracejus, M. Kehlen, H. Kehlen, and H. Matschiner, *Tetrahedron*, **21,** 2257 (1965).
47. G. Fraenkel and C. Niemann, *Proc. Natl. Acad. Sci. U.S.*, **44,** 688 (1958).
48. A. Streitwieser, Jr., and J. H. Hammons, *Progr. Phys. Org. Chem.*, **3,** 41 (1965).
49. E. C. Steiner and J. M. Gilbert, *J. Amer. Chem. Soc.*, **87,** 382 (1965).
50. K. Bowden and R. Stewart, *Tetrahedron*, **21,** 261 (1965).
51. R. Kuhn and D. Rewicki, *Justus Liebigs Ann. Chem.*, **704,** 9; **706,** 250 (1967).
52. K. Bowden and A. F. Cockerill, *J. Chem. Soc., B*, 173 (1970).
53. C. D. Ritchie and R. E. Uschold, *J. Amer. Chem. Soc.*, **89,** 1721, 2752, 2960 (1967); **90,** 2821 (1968).
54. A. Streitwieser, Jr., J. I. Brauman, J. H. Hammons, and A. H. Pudjaatmaka, *J. Amer. Chem. Soc.*, **87,** 384 (1965).
55. A. Streitwieser, Jr., J. H. Hammons, E. Ciuffarin, and J. I. Brauman, *J. Amer. Chem. Soc.*, **89,** 59 (1967).
56. A. Streitwieser, Jr., E. Ciuffarin, and J. H. Hammons, *J. Amer. Chem. Soc.*, **89,** 63 (1967).
57. K. Bowden, A. F. Cockerill, and J. R. Gilbert, *J. Chem. Soc., B*, 179 (1970).
58. E. C. Steiner and J. D. Starkey, *J. Amer. Chem. Soc.*, **89,** 2751 (1967).
59. G. W. Wheland, *J. Chem. Phys.*, **2,** 474 (1934).
60. R. Breslow and W. Chu, *J. Amer. Chem. Soc.*, **95,** 411 (1973).
61. R. Breslow, *Accounts Chem. Res.*, **6,** 393 (1973).
62. A. Streitwieser, Jr., *Molecular Orbital Theory for Organic Chemists*, Wiley, New York, 1961, Sec. 14–1.
63. S. Fliszár, R. F. Hudson, and G. Salvadori, *Helv. Chim. Acta*, **46,** 1580 (1963).
64. E. Pelizzetti and C. Verdi, *JCS Perkin II*, 808 (1973).
65. M. Bergon and J.-P. Calmon, *Bull. Soc. Chim. Fr.*, 1020 (1972).
66. C. Kissel, R. J. Holland, and M. C. Caserio, *J. Org. Chem.*, **37,** 2720 (1972).
67. M. Fukuyama, P. W. K. Flanagan, F. T. Williams, Jr., L. Frainier, S. A. Miller, and H. Shechter, *J. Amer. Chem. Soc.*, **92,** 4689 (1970).
68. F. G. Bordwell, W. J. Boyle, Jr., and K. C. Yee, *J. Amer. Chem. Soc.*, **92,** 5926 (1970).
69. F. G. Bordwell and W. J. Boyle, Jr., *J. Amer. Chem. Soc.*, **94,** 3907 (1972).

70. G. I. Kolesetskaya, I. V. Tselinskii, and L. I. Bagal, *Reakts. Sposobnost. Org. Soedin.*, **6**, 387 (1969).
71. M. J. Kamlet and D. J. Glover, *J. Org. Chem.*, **27**, 537 (1962).
72. B. Klewe and S. Ramsøy, *Acta Chem. Scand.*, **26**, 1058 (1972).
73. V. I. Slovetskii, L. V. Okhlobystina, A. A. Fainzel'berg, A. I. Ivanov, L. I. Biryukova, and S. S. Novikov, *Bull. Acad. Sci. USSR Div. Chem. Soc.*, 2032 (1965).
74. V. I. Slovetskii, S. A. Shevelev, V. I. Erashko, L. I. Biryukova, A. A. Fainzel'berg, and S. S. Novikov, *Bull. Acad. Sci. USSR Div. Chem. Sci.*, 621 (1966).
75. A. I. Ivanov, V. I. Slovetskii, S. A. Shevelev, V. I. Erashko, A. A. Fainzel'berg, and S. S. Novikov, *Russ. J. Phys. Chem.*, **40**, 1234 (1966).
76. M. E. Sitzmann, H. G. Adolph, and M. J. Kamlet, *J. Amer. Chem. Soc.*, **90**, 2815 (1968).
77. I. V. Tselinskii, A. S. Kosmynina, V. N. Dronov, and I. N. Shokhor, *Reakts. Sposobnost Org. Soedin.*, **7**, 50 (1970).
78. H. G. Adolph and M. J. Kamlet, *J. Amer. Chem. Soc.*, **88**, 4761 (1966).
79. G. Subrahmanyam, S. K. Malhotra, and H. J. Ringold, *J. Amer. Chem. Soc.*, **88**, 1332 (1966).
80. Cf. R. A. Abramovitch, G. M. Singer, and A. R. Vinutha, *Chem. Commun.*, 55 (1967).
81. D. Holtz, *Progr. Phys. Org. Chem.*, **8**, 1 (1971).
82. Cf. A. Streitwieser, Jr., and F. Mares, *J. Amer. Chem. Soc.*, **90**, 2444 (1968).
83. J. Hine and P. B. Langford, *J. Amer. Chem. Soc.*, **78**, 5002 (1956).
84. Cf. J. Hine and P. D. Dalsin, *J. Amer. Chem. Soc.*, **94**, 6998 (1972).
85. R. G. Pearson and R. L. Dillon, *J. Amer. Chem. Soc.*, **75**, 2439 (1953).
86. D. Turnbull and S. H. Maron, *J. Amer. Chem. Soc.*, **65**, 212 (1943).
87. V. M. Belikov, Yu. N. Belokon, Ts. B. Korchemnaya, and N. G. Faleev, *Bull. Acad. Sci. USSR Div. Chem. Sci.*, 317 (1968).
88. F. Hibbert, F. A. Long, and E. A. Walters, *J. Amer. Chem. Soc.*, **93**, 2829 (1971).
89. J. Hine, J. C. Philips, and J. I. Maxwell, *J. Org. Chem.*, **35**, 3943 (1970).
90. R. P. Bell and B. G. Cox, *J. Chem. Soc., B*, 652 (1971).
91. H. O. House and V. Kramar, *J. Org. Chem.*, **28**, 3362 (1963).
92. W. v. E. Doering and L. K. Levy, *J. Amer. Chem. Soc.*, **77**, 509 (1955).
93. A. Streitwieser, Jr., C. J. Chang, and D. M. E. Reuben, *J. Amer. Chem. Soc.*, **94**, 5730 (1972).
94. O. Redlich, R. W. Duerst, and A. Merbach, *J. Chem. Phys.*, **49**, 2986 (1968).
95. J. N. Brønsted and C. V. King, *J. Amer. Chem. Soc.*, **49**, 193 (1927).
96. A. K. Covington and T. H. Lilley, *Trans. Faraday Soc.*, **63**, 1749 (1967).
97. R. L. Hinman and B. E. Hoogenboom, *J. Org. Chem.*, **26**, 3461 (1961).

98. F. G. Bordwell, R. H. Imes, and E. C. Steiner, *J. Amer. Chem. Soc.*, **89,** 3905 (1967).
99. R. Caramazza, *Gazz. Chim. Ital.*, **88,** 308 (1958); *Chem. Abstr.*, **53,** 16659d (1959).
100. C. Reid, *J. Chem. Phys.*, **18,** 1544 (1950).
101. W. C. v. Dohlen and G. B. Carpenter, *Acta Cryst.*, **8,** 646 (1955).
102. N. Kameyama, *Chem. Abstr.*, **16,** 7 (1922).
103. R. H. Boyd, *J. Phys. Chem.*, **67,** 737 (1963).
104. R. S. Stearns and G. W. Wheland, *J. Amer. Chem. Soc.*, **69,** 2025 (1947).
105. J. Hine and M. Hine, *ibid.*, **74,** 5266 (1952).
106. R. P. Bell and P. W. Smith, *J. Chem. Soc.*, *B*, 241 (1966).
107. L. A. Kaplan, N. E. Burlinson, W. B. Moniz, and C. F. Poranski, *Chem. Commun.*, 140 (1970).
108. R. Schaal, *J. Chim. Phys.*, **52,** 719 (1955).
109. M. Liler, *J. Chem. Soc.*, *B*, 385 (1969).
110. K. Yates, J. B. Stevens, and A. R. Katritzky, *Can. J. Chem.*, **42,** 1957 (1964).
111. J. F. Bunnett and F. P. Olsen, *Can. J. Chem.*, **44,** 1899 (1966).
112. J. W. Barnett and C. J. O'Connor, *JCS Perkin II*, 1331 (1973).
113. M. Liler, *J. Chem. Soc.*, *B*, 334 (1971).
114. V. C. Armstrong and R. B. Moodie, *J. Chem. Soc.*, *B*, 275 (1968).
115. K. L. Cook and A. J. Waring, *JCS Perkin II*, 84 (1973).
116. D. Landini, G. Modena, G. Scorrano, and F. Taddei, *J. Amer. Chem. Soc.*, **91,** 6703 (1969).
117. R. Curci, A. Levi, V. Lucchini, and G. Scorrano, *JSC Perkin II*, 531 (1973).
118. P. Bonvicini, A. Levi, V. Lucchini, G. Modena, and G. Scorrano, *J. Amer. Chem. Soc.*, **95,** 5960 (1973).
119. C. Tissier and M. Tissier, *Bull. Soc. Chim. Fr.*, 2109 (1972).
120. P. Bonvicini, A. Levi, V. Lucchini, and G. Scorrano, *JCS Perkin II*, 2267 (1972).
121. R. L. Hinman and J. Lang, *J. Amer. Chem. Soc.*, **86,** 3796 (1964).
122. M. T. Reagan, *J. Amer. Chem. Soc.*, **91,** 5506 (1969).
123. A. J. Kresge, H. J. Chen, L. E. Hakka, and J. E. Kouba, *J. Amer. Chem. Soc.*, **93,** 6174 (1971).
124. M. Kilpatrick and F. E. Luborsky, *J. Amer. Chem. Soc.*, **75,** 577 (1953).
125. H.-H. Perkampus, *Advan. Phys. Org. Chem.*, **4,** 195 (1966).
126. E. M. Arnett, R. P. Quirk, and J. J. Burke, *J. Amer. Chem. Soc.*, **92,** 1260 (1970).
127. E. M. Arnett, R. P. Quirk, and J. W. Larsen, *J. Amer. Chem. Soc.*, **92,** 3977 (1970).
128. E. M. Arnett and J. F. Wolf, *J. Amer. Chem. Soc.*, **95,** 978 (1973).

129. E. M. Arnett, *Accounts Chem. Res.*, **6,** 404 (1973).
130. J. J. Christensen, R. M. Izatt, and L. D. Hansen, *J. Amer. Chem. Soc.*, **89,** 213 (1967).
131. J. J. Christensen, M. D. Slade, D. E. Smith, R. M. Izatt, and J. Tsang, *J. Amer. Chem. Soc.*, **92,** 4164 (1970).
132. M. G. Evans and M. Polanyi, *Trans. Faraday Soc.*, **32,** 1333 (1936).
133. J. E. Leffler, *J. Org. Chem.*, **20,** 1202 (1955).
134. J. E. Leffler and E. Grunwald, *Rates and Equilibria of Organic Reactions,* Wiley, New York 1963, Chap. 9.
135. J. E. Leffler, *J. Chem. Phys.*, **23,** 2199 (1955); **27,** 981 (1957).
136. R. F. Brown and H. C. Newsom, *J. Org. Chem.*, **27,** 3010 (1962).
137. R. F. Brown, *J. Org. Chem.*, **27,** 3015 (1962).
138. O. Exner, *Nature*. **201,** 488 (1964).
139. R. C. Petersen, *J. Org. Chem.*, **29,** 3133 (1964).
140. O. Exner, *Progr. Phys. Org. Chem.*, **10,** 411 (1973).
141. J. W. Larson and L. G. Hepler, in *Solute-Solvent Interactions*, J. F. Coetzee and C. D. Ritchie, Eds., Dekker, New York, 1969, Chap. 1.
142. C. L. Liotta, E. M. Perdue, and H. P. Hopkins, Jr., *J. Amer. Chem. Soc.*, **95,** 2439 (1973).
143. I. R. Bellobono and M. A. Monetti, *JCS Perkins II*, 790 (1973).
144. M. R. Chakrabarty, C. S. Handloser, and M. W. Mosher, *JCS Perkin II*, 938 (1973).
145. G. Berthon, O. Enea, and E. M'Foundou, *Bull. Soc. Chim. Fr.*, 2967 (1973).
146. R. T. McIver, Jr., and J. H. Silvers, *J. Amer. Chem. Soc.*, **95,** 8462 (1973).
147. E. M. Arnett, L. Joris, E. Mitchell, T. S. S. R. Murty, T. M. Gorrie, and P. v. R. Schleyer, *J. Amer. Chem. Soc.*, **92,** 2365 (1970).
148. S. Ghersettti and A. Lusa, *Spectrochim. Acta*, **21,** 1067 (1965).
149. B. Stymne, H. Stymne, and G. Wettermark, *J. Amer. Chem. Soc.*, **95,** 3490 (1973).
150. J. Rubin and G. S. Panson, *J. Phys. Chem.*, **69,** 3089 (1965).
151. E. D. Becker, *Spectrochim. Acta*, **17,** 436 (1961).
152. A. Kivinen, J. Murto, and L. Kilpi, *Suomen Kemistilehti. B*. **40,** 301 (1967); **42,** 19 (1969).
153. G. C. Pimentel and A. L. McClellan, *The Hydrogen Bond*, Freeman, San Francisco, 1960, (a) Appendix B.
154. R. Yamdagni and P. Kebarle, *J. Amer. Chem. Soc.*, **95,** 3504 (1973).
155. S. N. Vinogradov and R. H. Linnell, *Hydrogen Bonding*, Van Nostrand Reinhold, New York, 1971.
156. A. S. N. Murthy and C. N. R. Rao, *Appl. Spectrosc. Rev.*, **2,** 69 (1968).

157. A.-M. Dierckx, P. Huyskens, and T. Zeegers-Huyskens, *J. Chim. Phys.*, **62,** 336 (1965).
158. G. Lichtfus, F. Lemaire, and T. Zeegers-Huyskens, *Spectrochim. Acta*, A, **28,** 2069 (1972).
159. S. Singh, A. S. N. Murthy, and C. N. R. Rao, *Trans. Faraday Soc.*, **62,** 1056 (1966).
160. S. Nagakura, *J. Amer. Chem.. Soc.*, **76,** 3070 (1954).
161. D. Clotman, D. Van Lerberghe, and T. Zeegers-Huyskens, *Spectrochim. Acta*, *A*, **26,** 1621 (1970).
162. R. L. Denyer, A. Gilchrist, J. A. Pegg, J. Smith, T. E. Tomlinson, and L. E. Sutton, *J. Chem. Soc.*, 3889 (1955).
163. C. Dorval and T. Zeegers-Huyskens, *Tetrahedron Lett.*, 4457 (1972).
164. J. P. Muller, G. Vercruysse, and T. Zeegers-Huyskens, *J. Chim. Phys.*, **69,** 1439 (1972).
165. T. Kubota, *J. Pharm. Soc. Japan*, **75,** 1540 (1955).
166. I. M. Kolthoff and M. K. Chantooni, Jr., *J. Amer. Chem. Soc.*, **91,** 4621 (1969).
167. D. Clotman, J. P. Muller, and T. Zeegers- Huyskens, *Bull. Soc. Chim. Belges*, **79,** 689 (1970).
168. J. Nelson, *Spectrochim. Acta, A*, **26,** 109 (1970).
169. J. Nelson, *Spectrochim. Acta, A*, **26,** 235 (1970).
170. R. Mathur, E. D. Becker, R. B. Bradley, and N. C. Li, *J. Phys. Chem.*, **67,** 2190 (1963).
171. W. J. Middleton and R. V. Lindsey, Jr., *J. Amer. Chem. Soc.*, **86,** 4948 (1964).
172. I. M. Ginsburg, *Opt. Spektrosk.*, **17,** 28 (1964).
173. R. S. Drago and T. D. Epley, *J. Amer. Chem. Soc.*, **91,** 2883 (1969).
174. T. Gramstad, *Spectrochim. Acta*, **19,** 497 (1963).
175. G. Yagil, *Tetrahedron*, **23,** 2855 (1967).
176. J. Mitsky, L. Joris, and R. W. Taft, *J. Amer. Chem. Soc.*, **94,** 3442 (1972).
177. D. Gurka and R. W. Taft, *J. Amer. Chem. Soc.*, **91,** 4794 (1969).
178. R. W. Taft, D. Gurka, L. Joris, P. v. R. Schleyer, and J. W. Rakshys, *J. Amer. Chem. Soc.*, **91,** 4801 (1969).
179. J. Mitsky, L. Joris, and R. W. Taft, *J. Amer. Chem. Soc.*, **94,** 3442 (1972).
180. T. M. Barakat, J. Nelson, S. M. Nelson, and A. D. E. Pullin, *Trans. Faraday Soc.*, **62,** 2674 (1966); **65,** 41 (1969).
181. R. W. Alder, P. S. Bowman, W. R. S. Steele, and D. R. Winterman, *Chem. Commun.*, 723 (1968).
182. Y. Yukawa, Y. Tsuno, and M. Sawada, *Bull. Chem. Soc. Japan*, **39,** 2274 (1966).
183. D. R. Lide, Jr., and D. E. Mann, *J. Chem. Phys.*, **29,** 914 (1958).

184. A. Audsley and F. R. Goss, *J. Chem. Soc.*, 497 (1942).
185. S. D. Christian and T. L. Stevens, *J. Phys. Chem.*, **76,** 2039 (1972).
186. J. I. Brauman and L. K. Blair, *J. Amer. Chem. Soc.*, **90,** 6561 (1968); **91,** 2126 (1969); **92,** 5986 (1970); **93,** 3911, 4315 (1971).
187. D. K. Bohme, E. Lee-Ruff, and L. B. Young, *J. Amer. Chem. Soc.*, **94,** 5153 (1972).
188. R. Yamdagni and P. Kebarle, *J. Amer. Chem. Soc.*, **95,** 4050 (1973).
189. M. S. B. Munson, *J. Amer. Chem. Soc.*, **87,** 2332 (1965).
190. J. I. Brauman, J. M. Riveros, and L. K. Blair, *J. Amer. Chem. Soc.*, **93,** 3914 (1971).
191. R. Yamdagni, T. B. McMahon, and P. Kebarle, *J. Amer. Chem. Soc.*, **96,** 4035 (1974).
192. M. Taagepera, W. G. Henderson, R. T. C. Brownlee, J. L. Beauchamp, D. Holtz, and R. W. Taft, *J. Amer. Chem. Soc.*, **94,** 1369 (1972).

7

Lewis Acidity and Basicity

A Lewis base is ordinarily thought of as a species that uses a pair of its electrons to form a bond to the Lewis acid with which it reacts. There are a few special cases of reactions commonly thought of as acid-base reactions that do not clearly fit this definition. For example, in the reaction (observed in the gas phase) of a proton with methane to give CH_5^+ it is unlikely that the methane is using a pair of electrons just to hold the incoming proton, since in CH_5^+ there are not enough outer-shell electrons to use a pair for bonding each of the hydrogen atoms. In spite of these "exceptions," which could perhaps be regularized by appropriate modifications of definitions, the Lewis concept of acidity and basicity is extremely useful.

Brønsted basicity is a special case of Lewis basicity in which the base forms a bond to hydrogen. Therefore, any species that can act as a Brønsted base can also be thought of as acting as Lewis base. A number of factors that affect the strength of bases in the Brønsted sense were discussed in Chapters 5 and 6. These factors often have similar effects on the strength of bases in the Lewis sense. For example, an equilibrium involving an ionic base may be affected by a solvent change that changes the stability of the base; this change in the stability of the base will have the same effect on any reaction in which the base is involved whether it is a reaction in which the base acts in the Brønsted sense or one in which it acts in the Lewis sense. (This does not mean that a given change in solvent will have the same effect on the equilibrium constant for every reaction in which the base is involved; the equilibrium constant depends on the stability of *all* the species taking part in the reaction.) There are many cases in which a given modification in solvent or structure will change basicity toward one

Lewis acid in one direction and basicity toward another Lewis acid in the opposite direction. Considerable attention is given in this chapter to the way in which the strength of bases depends on the nature of the Lewis acid with which they react.

In Brønsted acid-base reactions the base always coordinates with a proton. A molecule or ion may act as a Lewis base by coordinating with any of millions of possible Lewis acids. It is convenient to subdivide Lewis acids into categories depending on the nature of the atom to which the base forms a bond. Thus we may call boron trifluoride a boron acid because bases react with it by forming a bond to boron. Analogously, a tertiary butyl cation in reacting with a chloride anion to give *t*-butyl chloride would be acting as a carbon acid. The nitronium ion (NO_2^+) acts as a nitrogen acid in most of its reactions. Following the suggestion of Parker we may refer to the strength of a base toward carbon acids as its carbon basicity.[1] Of course the carbon basicity of a given base would not be expected to be constant, even under a given set of reaction conditions; it should vary at least somewhat with the nature of the carbon acid with which the base was reacting. Thus carbon basicity can be subdivided into the more specific categories methyl basicity, phenyl basicity, acetyl basicity, etc.

The conjugate terms carbon acidity, oxygen acidity, etc., for referring to the strengths of various acids toward carbon bases (such as cyanide ions), oxygen bases, etc., do not seem to have come into usage.

Since carbon acids are particularly important, we discuss them first.

7-1. CARBON ACIDS[2]

7-1a. Carbonium-Ion Stability in the Gas Phase. Since carbonium ions commonly react as carbon acids, their stabilities are measures of the strengths of carbon acids. Many qualitative and quantitative observations have been made on equilibria involving carbonium ions. In using such observations to evaluate carbonium-ion stabilities, we should remember that to be quantitatively unambiguous we must specify the reactants and reaction conditions.

Suppose that the equilibrium constants for the formation of a series of carbonium ions from the corresponding olefins and acid could be measured,

$$\mathrm{R'{-}\underset{\displaystyle R}{\underset{|}{C}}{=}CHR''} + \mathrm{H^+} \rightleftharpoons \mathrm{R'{-}\overset{\oplus}{\underset{\displaystyle R}{\underset{|}{C}}}{-}CH_2R''}$$

and the equilibrium constants for the formation of the same carbonium ions by ionization of the corresponding chlorides could also be measured.

$$R'{-}\underset{\displaystyle R}{\overset{\displaystyle Cl}{C}}{-}CH_2R'' \rightleftharpoons R'{-}\underset{\displaystyle R}{\overset{\oplus}{C}}{-}CH_2R'' + Cl^-$$

The first set of equilibrium constants would not have to vary with the nature of the carbonium ion being formed in the same way that the second set did. Equilibrium constants in the first set should decrease with increasing ability of the group R″ to stabilize double bonds (other things being equal). The ability of R″ to stabilize double bonds should have no effect on equilibrium constants for formation of the carbonium ions from the corresponding chlorides.

Statements such as "electron-donating substituents stabilize carbonium ions," which ignore reference compounds, are often made and are useful. Such statements should be treated as abbreviations for longer but more reliable statements such as "equilibrium constants for formation of carbonium ions from common precursors are ordinarily increased by increasing the electron-donating power of substituents." The generality of the preceding statement arises from the fact that substituents ordinarily interact much more strongly with the positive charge of a carbonium ion than with the double bond, or the carbon-chlorine bond, or the carbon-oxygen bond, etc., of the carbonium-ion precursor.

One measure of the relative stabilities of carbonium ions has already been given in the proton affinities of several alkenes listed in Table 5-6. The conclusion that the stabilities of simple alkyl cations decrease in the order tertiary > secondary > primary agrees with that which is reached in examining the enthalpy of transformation of a number of compounds of the type RX into R^+ and X^- in the gas phase. Although such simple heterolysis is rarely the preferred reaction path in the gas phase, the enthalpy of heterolysis may be calculated by combining the enthalpy of homolysis (Eq. 7-1), which is often a preferred reaction path, with the ionization potential of R· ($\Delta H_2°$) and the electron affinity of X· ($\Delta H_3°$). The enthalpy of carbonium-ion formation ($\Delta H_c°$) is equal to $\Delta H_1° + \Delta H_2° - \Delta H_3°$. Egger and Cocks have compiled a large number of enthalpies of

$$\begin{array}{lll} RX \rightarrow R\cdot + X\cdot & \Delta H_1° & \quad (7\text{-}1) \\ R\cdot \rightarrow R^+ + e & \Delta H_2° & \\ X^- \rightarrow X\cdot + e & \Delta H_3° & \\ RX \rightarrow R^+ + X^- & \Delta H_c° & \end{array}$$

heterolysis (and also enthalpies of homolysis), some of which are estimates, for organic compounds, and have discussed the relative values in terms of the structures of the reactants.[3] Some of these values, for the formation of R^+ from RH, ROH, RCl, and RI, are listed in Table 7-1. The average experimental uncertainty is probably around 2 or 3 kcal/mole. Since the differences in entropies of heterolysis are probably fairly small, the *relative* values of $\Delta H_c°$ are probably nearly the same as the relative values of $\Delta G_c°$ would be. Comparison with the enthalpies of heterolysis of H_2, H_2O, HCl, and HI, which are 401, 390, 334, and 315 kcal/mole, respectively, shows that none of the carbonium ions listed is as strongly acidic as is the bare proton H^+ in the gas phase. It is also seen that the order of carbonium-ion stabilities *t*-Bu > *m*-$MeC_6H_4CH_2$ > $PhCH_2$ > *i*-Pr > $CH_2{=}CHCH_2$ > *n*-Pr > Et > Me holds regardless of which of the four types of reactant is the carbonium-ion precursor (although some of the differences in stability listed are smaller than the experimental uncertainties). Part of the relative stability of the *t*-butyl cation presumably arises from the ability of α-alkyl substituents to stabilize sp^2-hybridized carbon relative to sp^3-hybridized carbon. However, this factor should also make *t*-butyl radicals relatively stable. Yet the difference between the stability of *t*-butyl and methyl radicals (as judged by enthalpies of homolysis of RH) is only 12 kcal/mole compared with the 79 kcal/mole difference in stability between the two carbonium ions.[3] Hence most of the stability of the *t*-butyl cation probably arises from the electron-donating ability of its three methyl

Table 7-1. Enthalpies of Heterolysis in the Gas Phase[3]

	$\Delta H_c°$ (kcal/mole) for heterolysis of			
R	RH	ROH	RCl	RI
Methyl	313	274	227	212
Ethyl	274	242	191	176
n-Propyl	268	235	185	171
i-Propyl	251	222	170	156
t-Butyl	234	208	157	140
Cyclohexyl	249	215	170	153
Vinyl	291		207	~194
Phenyl	300	275	219	202
Allyl	256	223	~173	159
Benzyl	247	215	~166	149
p-Methylbenzyl	240	206	~158	142
m-Methylbenzyl	244	210	162	146
Acetyl	227	223	157	138

groups. Much of this electron donation must arise from the polarizability of methyl groups (relative to that of hydrogen). (The greater stability of the *n*-propyl cation compared to the ethyl cation is probably almost entirely a polarizability effect.) However, it must also be relevant that alkyl substituents are electron donors, relative to hydrogen, when attached to sp^2-hybridized carbon, even though they need not be when attached to sp^3-hybridized carbon. This is revealed in the observations that replacement of a hydrogen atom in an alkane by an alkyl group gives a species with no dipole moment, but the same replacement in benzene or ethylene gives a dipolar species with the alkyl group at the positive end of the dipole.[4] The marked stability of allyl and benzyl cations relative to other primary carbonium ions is largely due to resonance. The large amount of energy required to produce vinyl and phenyl cations is associated with the relatively high electronegativity of the sp^2 orbitals with which these groups form bonds to whatever they are attached. Just as carbonium-ion formation from an sp^3-hybridized precursor gives an sp^2-hybridized cation, carbonium-ion formation from an sp^2-hybridized precursor preferably yields an sp-hybridized cation. This explains why the phenyl cation, which is secondary but can become linear at its positively charged carbon atom only by the introduction of a great deal of ring strain, is less stable than the vinyl cation, which is primary but which can become linear without strain. The positive charge in the acetyl cation is distributed onto the oxygen atom by resonance as well as by polarization.

$$CH_3\text{—}\overset{\oplus}{C}\text{=}O \leftrightarrow CH_3\text{—}C\equiv\overset{\oplus}{O}$$

There are several cases in which the relative stabilities of cations depend on what precursor they are formed from. The phenyl cation is formed 13 kcal/mole more easily than the methyl cation from the corresponding RH, but 1 kcal/mole less easily from the corresponding ROH. More strikingly, the acetyl cation is formed 20 kcal/mole more easily than the benzyl cation from the corresponding RH but 8 kcal/mole less easily from the corresponding ROH. These reversals in relative enthalpies of reaction, amounting to 14 and 28 kcal/mole, respectively, are attributed largely to resonance interaction of the unshared electron pairs on the oxygen atom of the hydroxy group with the pi systems of the phenyl and acetyl groups.

The heterolysis reactions covered by Table 7-1 would be much less endothermic in any solvent than they are in the gas phase. All the carbonium ions would be stabilized by solvation, but the largest amount of stabilization would be expected for those in which the positive charge is most concentrated and least shielded from close approach by the solvent.

7-1b. Carbonium-Ion Stability in Solution. More quantitative equilibrium measurements have been made on the stabilities of triarylmethyl cations than any other type of carbonium ion. The most common measure of this stability is the pK of the carbonium ion, where K is defined in Eq. 7-2. A plot of such pK values[5-11] versus the sums of the substituent constants (σ^+ for those substituents for which σ^+ constants exist, σ for the others)

$$K = \frac{[H^+][Ar_3COH]}{[Ar_3C^+]} \tag{7-2}$$

for the substituents on the three aromatic rings is shown in Fig. 7-1. The values of pK smaller than about zero had to be determined by measurements in water-sulfuric acid mixtures using the H_R function (see Section 5-2d). Although there is a clear tendency for pK to increase with increasing electron-donating power of the substituents, there would be some rather large deviations from any straight line that could be drawn through the entire set of points. If the Hammett equation is defined as a correlation of data on reactants of the Type ArX where Ar may be varied but X is

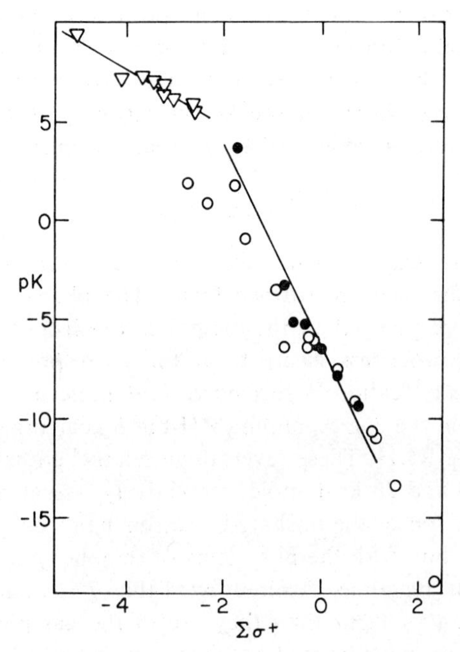

Figure 7-1. Plot of pK versus $\Sigma\ \sigma^+$ for ● monosubstituted triphenylmethyl cations, ▽ malachite green derivatives, and ○ other polysubstituted triphenylmethyl cations.

kept constant for the entire series of compounds studied, then in the present case we can apply the equation only to cases in which the substituents in two of the three rings are kept constant. Data for two such series are plotted in the figure. The solid circles are for triarylmethanols with substituents on only one ring; ρ for this set of points is -5.1. The triangular points are for malachite green derivatives, triarylmethanols in which two rings have *p*-dimethylamino substituents and the substituent in the third ring is varied. The ρ for this reaction series is -1.4. The principal reason for ρ being so much smaller for the malachite green derivatives is probably that in the malachite green cations a considerable fraction of the positive charge is on the dimethylamino groups, far away from any substituent in the third ring. It may also be that the resonance interactions involving the dimethylamino groups cause the two rings bearing these substituents to be more nearly coplanar than the third ring, whose ability to transmit electronic effects may be diminished somewhat by the greater deviation from coplanarity. (There is abundant evidence that the three aromatic rings in triarylmethyl cations are not coplanar but twisted so that the ion is like a three-bladed propeller.) The variation of substituents in all three rings can be treated as a simultaneous application of three linear free energy relationships to one set of data. Such a treatment[11] gives a much more nearly linear plot than that shown in Fig. 7-1.

The pK values for 10 benzhydrols (diphenylmethanols) with one meta or para substituent have been reported to give a satisfactory Hammett equation correlation with a ρ of -4.1 when σ^+ values are used but a better correlation when the Yukawa-Tsuno equation (which has the form of Eq. 7-3 in this case) is used.[12] It seems surprising at first glance that the

$$\log \frac{K}{K_0} = 3.43\,(\sigma + 1.54\;\Delta\sigma_R{}^+) \tag{7-3}$$

absolute magnitude of ρ is smaller for benzhydrols than for triphenylmethanols. However, Mindl and Večeřa[12] used a special H_R' function defined to fit the data on benzhydrols. Its value changed from -3 to about -12.0 over the same range of sulfuric acid concentrations that changed the H_R value of Deno and others from -3 to about -13.6. If the same acidity function had been used about the same ρ value would have been obtained for benzhydrols as for triphenylmethanols. Data on disubstituted benzhydrols were also treated.

Values of pK for a number of other carbonium ions are listed in Table 7-2. The most stable one listed is the sesquixanthydryl cation (**1**), for which the corresponding alcohol is almost as basic as ammonia in aqueous solution.[13] The stability of **1** must be due in part to the presence of three

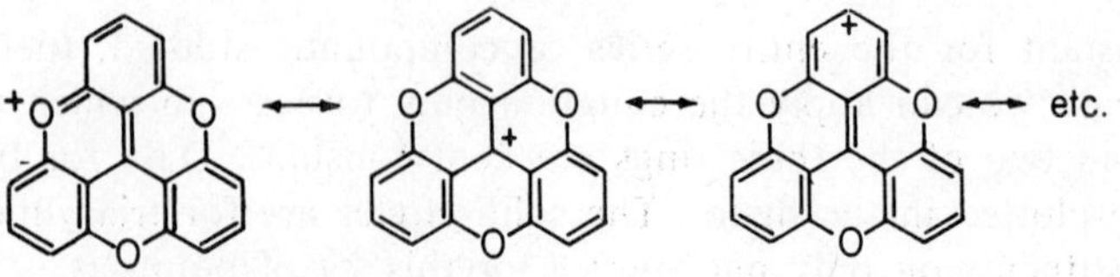

1

oxygen atoms onto which the positive charge may be distributed, but this must not be the only relevant factor because it is also present in the tris(*p*-methoxyphenyl)methyl cation, whose p*K* is much smaller. It must also be important that **1** can achieve the coplanar configuration required for maximum resonance stabilization without significant steric strain. There may also be significant quantum mechanical factors. Such factors are required to explain the relative instability of the fluorenyl cation (**2**).[7] Although this cation is planar and there are valence bond structures that distribute the positive charge onto each of its thirteen carbon atoms, it is no more stable than the diphenylmethyl cation, which could achieve coplanarity only at the cost of large steric interactions and whose positive charge can be dis-

Table 7-2. p*K* Values for Carbonium Ions

Cation	p*K*	Ref.
Sesquixanthydryl (**1**)	9.05	13
Tri-*n*-propylcyclopropenyl	7.2	18
Cycloheptatrienyl	4.7	16
Triphenylcyclopropenyl	3.1	18
Di-*n*-propylcyclopropenyl	2.7	18
2,3-(1′,2′-Naphtho)cycloheptatrienyl	2.2	19
2,3-Benzocycloheptatrienyl	1.7	19
Tris(*p*-methoxyphenyl)methyl	0.8	7
2,3-(2′,3′-Naphtho)cycloheptatrienyl	0.3	19
Xanthydryl	−0.2	8
Diphenylcyclopropenyl	−0.7	18
Tricyclopropylmethyl	−2.3	20
2,3,6,7-Dibenzocycloheptatrienyl	−3.7	19
1-Phenyl-2,3,6,7-Dibenzocycloheptatrienyl	−5.7	19
2,3,4,5-Dibenzocycloheptatrienyl	−5.8	19
Triphenylmethyl	−6.5	7, 8
Cyclopropenyl	−7.4	17
Diphenylmethyl	−13.3	7
Fluorenyl (**2**)	−14	7
2,3,4,5,6,7-Tribenzocycloheptatrienyl	−15	19

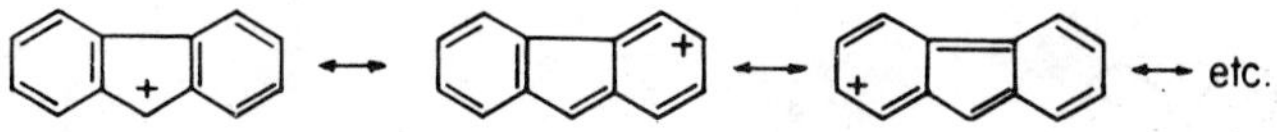

2

tributed onto only seven carbon atoms by significant valence bond structures. A simple Hückel-Molecular-Orbital (HMO) calculation of the difference in binding energy between carbonium ion and the corresponding alcohol gives the values 1.16 and 1.30β for the fluorenyl and diphenylmethyl cases, respectively.[14] [The bond integral β is a somewhat indeterminant parameter whose value is ordinarily within a factor of three of 1 eV (23 kcal/mole).] A more qualitative explanation for the relative instability of the fluorenyl cation is that it may be considered to be a derivative of the cyclopentadienyl cation, which does not obey the Hückel $4n + 2$ rule and, in fact, appears to be antiaromatic; that is, the conjugation of the two double bonds and the electron-deficient carbon in a ring appear to lead to less stabilization than they would in an acyclic compound.[15] The only monocyclic hydrocarbon cations with rings composed of sp^2 carbon atoms that are stable enough for their pK values to be determined are those having $4n + 2$ electrons. The cycloheptatrienyl[16] and cyclopropenyl[17] cations are two examples. Each is found by an HMO calculation to be stabilized by 2.0β relative to the corresponding alcohol.[14] The fact that the cycloheptatrienyl cation is more stable by a factor of about 10^{12} reflects the HMO method's neglect of the advantages of spreading a positive charge over seven atoms instead of three, the neglect of any differences in the strengths of the C—O bonds in the alcohols, and many other factors, no doubt. Phenyl substituents are seen to stabilize the cyclopropenyl cation, but not by as large a factor as *n*-propyl groups do, in spite of the fact that the triphenylcyclopropenyl cation can approach coplanarity considerably more closely than a triphenylmethyl cation can.[18] The fusion of a benzene ring to the cycloheptatrienyl cation to get the 2,3-benzocycloheptatrienyl cation gives four more atoms upon which the charge may be distributed by resonance. Nevertheless this structural change appears to disturb the simple aromatic system and decrease the stability for quantum mechanical reasons that are reflected in a decreased delocalization energy (1.77β) calculated by the HMO method and in a decreased pK[19] (Table 7-2). Fusion of several aromatic rings to the cycloheptatrienyl ring decreases the stability of the carbonium ion even more; increased difficulties in achieving coplanarity must contribute to these results.

The striking ability of the cyclopropyl substituent to stabilize carbonium ions is reflected in the pK of the tricyclopropylmethyl cation,[20] which is the most stable hydrocarbon cation known that lacks substituents with trigonally hybridized carbon atoms. Although the stabilizing effect of the

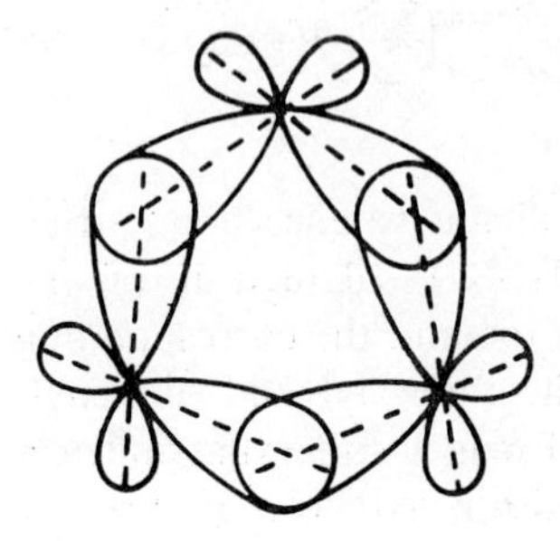

Figure 7-2. Overlap of the bonding orbitals (whose axes are shown by dashed lines) of the three carbon atoms of a cyclopropane ring.

cyclopropyl substituent has been rationalized in terms of valence bond resonance involving no-bonded structures, a clearer explanation may be given in terms of the special properties of the cyclopropane ring that result from the imperfect overlap of the orbitals in the bent bonds (Fig. 7-2). The reduced overlap makes the electrons in these orbitals more available for entry into nearby vacant orbitals. The observation that the two methyl groups in the 2-cyclopropyl-2-propyl cation give rise to two different absorption peaks in the proton magnetic resonance spectrum determined in FSO_3H-SO_2-SbF_5 solution tells us more about the way in which the cyclopropyl substituent stabilizes the carbonium ion.[21] It shows that there is a significant barrier to rotation around the single bond between the cyclopropyl group and the positively charged carbon atom and also rules out a structure like **3**. The most plausible structure for the carbonium ion is the

CH_3—$\overset{+}{C}$—CH_3 | C(H) | CH_2—CH_2

3

one in which the plane described by the three bonds to the positively charged carbon atom is a perpendicular bisector of the cyclopropane ring. In this conformation the vacant *p* orbital on the positively charged carbon atom can overlap fairly well with two of the three bent bonds of the cyclopropane ring.

The solution of saturated acyclic tertiary alcohols in concentrated sulfuric acid results in the formation of substituted cyclopentenyl cations, a variety of alkanes, and probably other species.[22] These products probably result from the reaction of the tertiary alkyl cation with the corresponding olefin. In order to make direct measurements on such cations, a medium so strongly acidic that the amount of olefin present is very small seems to be needed. Such superacidic solvent mixtures as FSO_3H-SbF_5 provide such media. Although no pK values seem to have been reported in such solvents, direct measurements (most commonly nmr) have been made on simple secondary and tertiary alkyl cations.[23,24] These measurements support the conclusion, based on solvolysis rate data and other evidence, that tertiary carbonium ions are more stable than secondary carbonium ions (relative to a common type of precursor, e.g., the corresponding chlorides or fluorides). The fact that the same conditions do not yield primary carbonium ions indicates that these are less stable.

7-1c. Carbon Basicity.[25] If R^+ is a carbonium ion then the equilibrium constant for Eq. 7-4 is a measure of the basicity of the base A^- toward R^+. Not many such equilibrium constants have been measured and it would be quite difficult to measure them in solution, at least in cases where R^+ is not

$$R^+ + A^- \rightleftharpoons RA \tag{7-4}$$

a particularly stable carbonium ion. This lack of data seems less tragic, however, when we remember that it is often not absolute but relative acidities and basicities that are needed. The equilibrium constant for Eq. 7-5, a measure of the basicity of B^- toward R^+, may be as unavailable as that for Eq. 7-4, but the quotient of these two equilibrium constants, that

$$R^+ + B^- \rightleftharpoons RB \tag{7-5}$$

is, the relative basicities of A^- and B^- toward R^+, is simply the equilibrium constant for Eq. 7-6. The determination of this equilibrium constant is

$$RB + A^- \rightleftharpoons RA + B^- \tag{7-6}$$

unaffected by the instability of R^+, of course. In comparing the strengths of various bases it is useful to select one as the standard with which all the others are compared. It seems convenient to choose the hydroxide ion for this purpose. Therefore let us define the equilibrium constant for Eq. 7-7 as the measure of the R^+ basicity of the base A^-. By analogy the proton

$$ROH + A^- \overset{K_A{}^R}{\rightleftharpoons} RA + OH^- \tag{7-7}$$

basicity of A^- will be measured by the equilibrium constant for Eq. 7-8.

$$HOH + A^- \overset{K_A}{\rightleftharpoons} HA + OH^- \tag{7-8}$$

Since there is a large body of data available on proton basicities and since there are often useful parallels between basicities toward protons and toward other acids, it is useful to have a numerical measure of how the R^+ basicities of various bases compare with their proton basicities. Such a measure is provided by the quotient K_A^R/K_A, which is the equilibrium constant for Eq. 7-9.

$$ROH + HA \overset{K_{HA}^{RA}}{\rightleftharpoons} RA + H_2O \qquad (7\text{-}9)$$

Values of K_A, K_A^R, and K_{HA}^{RA} for a number of R and A groups in water at 25° C are listed in Table 7-3. The first entry shows that although the proton basicity of methoxide ions is somewhat less than that of hydroxide ions the methyl basicity is somewhat greater. Combination of these two facts leads to a value of K_{HA}^{RA} that is larger than 1.0. This result, and therefore the conclusion that the methyl basicity of the methoxide ion exceeds its proton basicity, could have been predicted from some of the correlations of thermodynamic properties considered in Section 1-2. For the reaction shown in Eq. 7-9 in the case where R is methyl and A is methoxy a correlation in terms of additivity of bond contributions would predict no

Table 7-3. Relative Strengths of Various Bases toward Protons and Various Carbon Acids in Water at 25° C[25]

R	A	K_A[a]	K_A^R[a]	K_{HA}^{RA}[a]
Me	OMe	0.11	12	110
Me	OPh	9×10^{-7}	1.3×10^{-5}	14
Me	SH	1×10^{-9}	0.8	8×10^{8}
Me	SMe	2×10^{-6}	4×10^{4}	2×10^{10}
Me	SPh	3×10^{-10}	2	7×10^{9}
Me	CN	1.5×10^{-7}	1×10^{9}	7×10^{15}
MeCHOH	CN	1.5×10^{-7}	0.1	7×10^{5}
Ph	OMe	0.11	1.5	14
Ph	SMe	1.9×10^{-6}	1.1	6×10^{5}
Ac	SCH_2CH_2NHAc	3×10^{-7}	2×10^{-8}	0.06
Ac	OEt	0.9	6	7
Ac	$OC_6H_4NO_2$-*p*	1.3×10^{-9}	1.7×10^{-13}	1.3×10^{-5}
$HOCH_2$	$C(NO_2)_2Me$	6×10^{-12}	7×10^{-6}	1.1×10^{6}
$HOCH_2$	$C(NO_2)_2CN$	3×10^{-23}	2×10^{-17}	6×10^{5}
i-Pr	OPr-*i*	8	0.3	0.04
CF_3CH_2	OAc	3×10^{-12}	7×10^{-14}	0.03

[a] Corrected for symmetry effects.

change in thermodynamic properties, but a correlation in terms of bond contributions plus α-interactions predicts that the reaction will be exothermic by the stabilization energy of an interaction of two carbon atoms across an oxygen. Such a stabilizing interaction can be overwhelmed by steric effects, and that must be the reason why $K_{HA}{}^{RA}$ is less than 1.0 when R is isopropyl and A is isopropoxy.

In the case of bases in which the basic atom is not oxygen as it is in the reference base hydroxide ion, $K_{HA}{}^{RA}$ may deviate even further from 1.0. Thus, although the proton basicity of the bisulfide ion is only one-billionth that of the hydroxide ion its methyl basicity is almost the same, and although the thiomethoxide ion is a much weaker base than the hydroxide ion toward protons it is a much stronger base toward the methyl cation. That the carbon basicities of such sulfur bases would be much larger than their proton basicities (i.e., that $K_{HA}{}^{RA}$ would be much larger than 1.0) could have been predicted from a correlation of thermodynamic properties in terms of bond contributions or even from the Pauling bond energy equation, according to which a reaction involving the cleavage of two single bonds and the formation of two new single bonds (e.g., Eq. 7-10) will proceed in the direction that yields the bond joining the atoms with

$$\text{R—OH} + \text{H—SR}' \rightleftharpoons \text{R—SR}' + \text{H—OH} \qquad (7\text{-}10)$$

the largest difference in electronegativity (the bond between hydrogen and oxygen in the case of Eq. 7-10.) The carbon basicity of carbon bases ordinarily exceeds their proton basicity by an even larger factor than that observed with sulfur bases (in spite of the fact that the Pauling electronegativity of sulfur is no greater than that of carbon). Thus the methyl basicity of cyanide ion exceeds its proton basicity by around 10^{16}-fold. However, there are enormous differences in carbon basicities that depend on the nature of the specific carbon acid under consideration. Thus the basicity of the cyanide ion toward the methyl cation exceeds its basicity toward the α-hydroxyethyl cation by a factor of 10^{10}. Since the hydroxide ion was defined as the reference base the basicity of cyanide ion toward the α-hydroxyethyl cation, which is measured by the equilibrium constant for Eq. 7-11, must be greatly diminished by the strong stabilizing interaction

$$\text{MeCH(OH)}_2 + \text{CN}^- \rightleftharpoons \text{MeCH(OH)CN} + \text{OH}^- \qquad (7\text{-}11)$$

of two oxygen atoms across a carbon atom in $MeCH(OH)_2$ (cf. Table 1-3). This interaction is absent from the equilibrium that measures the methyl basicity of the cyanide ion. A somewhat different type of interaction may be blamed for the decrease in $K_{HA}{}^{RA}$ from 2×10^{10} for the methyl basicity of MeS^- to 6×10^5 for the phenyl basicity of MeS^- to 0.06 for the acetyl basicity for $AcNHCH_2CH_2S^-$. In this case there must be a weaker resonance interaction betweeh the aromatic ring and the unshared electron pairs of the atom attached to the ring in thioanisole than in phenol and a much weaker resonance interaction between the acetyl and RS groups in $AcSCH_2CH_2NHAc$ than between the acetyl and hydroxy groups in acetic acid. Jencks and co-workers observed than the acetyl basicities of 13 oxygen bases decrease more rapidly than their proton basicities when electron-withdrawing groups are introduced so that $K_{HA}{}^{RA}$ decreases steadily from 7 when the base is ethoxide ion to 1.3×10^{-5} when it is *p*-nitrophenoxide.[26,27] These results can be rationalized in terms of resonance interactions with the acetyl group but they may also be discussed in terms of Eqs. 3-6 and 3-13. The acetyl basicity relative to the proton basicity of RO^- is measured by the equilibrium constant for Eq. 7-12. According to Eq. 3-13 this equilibrium constant may be expressed as shown in Eq.

$$AcOH + HOR \rightleftharpoons AcOR + H_2O \qquad (7\text{-}12)$$

7-13, where τ, the proportionality constant for transmission of polar

$$\log K = \tau(\sigma_{Ac} - \sigma_H)(\sigma_H - \sigma_R) \qquad (7\text{-}13)$$

effects across oxygen, must be a positive number. Choice of hydrogen as the reference substituent so that σ_H is zero gives Eq. 7-14. Since the acetyl group is certainly an electron-withdrawing substituent relative to hydrogen,

$$\log K = -\tau \sigma_{Ac} \sigma_R \qquad (7\text{-}14)$$

the usual definition will make σ_{Ac} positive. Thus it follows that K will decrease as the electron-withdrawing power of R increases.

Hall studied a case in which variations in carbon basicity and in proton basicity ran rather closely parallel.[28] He measured equilibrium constants for reactions like Eq. 7-9 where R was hydroxymethyl (ROH was formaldehyde hydrate) and the five A groups studied were substituted dinitromethyl groups. Since the acidity constants of the five dinitromethanes were known, values of K_A, and therefore of $K_A{}^R$, could be calculated. On going from an HA of $HC(NO_2)_2Me$ to one of $HC(NO_2)_2CN$ the proton basicity of A^- decreases by more than 10^{10}, but so does the hydroxymethyl basicity. Thus the values of $K_{HA}{}^{RA}$ are all within the experimental uncertainty of 10^6. Dividing the $K_{HA}{}^{RA}$ for $HC(NO_2)_2X$ by that for $HC(NO_2)_2Y$ gives the equilibrium constant for Eq. 7-15, which is a form of Eq. 3-6 in which N is

$$HC(NO_2)_2X + HOCH_2C(NO_2)_2Y \rightleftharpoons HOCH_2C(NO_2)_2X + HC(NO_2)_2Y \qquad (7\text{-}15)$$

dinitromethylene, A is hydrogen, B is hydroxymethyl, C is X, and D is Y.

Therefore the equilibrium constant may be expressed as shown in Eq. 7-16.

$$\log K = \tau(\sigma_{H} - \sigma_{CH_2OH})(\sigma_{X} - \sigma_{Y}) \qquad (7\text{-}16)$$

Since the Taft substituent constants for hydrogen and hydroxymethyl (based on data arising from compounds in which the substituents are attached to carbon) are of about the same magnitude (0.49 and 0.56), it may not be surprising that log K is near zero and thus that all the K_{HA}^{RA} values are about the same.

7-2. BORON ACIDS AND STERIC HINDRANCE

One of the first extensive sets of measurements of equilibrium constants for adduct formation between Lewis acids and bases came from the studies of boron acids and amines by H. C. Brown and co-workers.[29] Table 7-4

Table 7-4. Equilibrium Constants for Addition to Trimethylboron in the Gas Phase at 100° C[29]

Amine	K^a
NH_3	0.22
$MeNH_2$	28
$EtNH_2$	14
n-$PrNH_2$	17
i-$PrNH_2$	2.7
t-$BuNH_2$	0.11
Me_2NH	47
Me_3N	2.1
Et_2NH	0.82
Et_3N	Small[b]
Quinuclidine	51
Pyridine	3.3
2-Picoline	Small[b]
3-Picoline	7.2
4-Picoline	9.5
Ethylenimine	35
Trimethylenimine	3100
Pyrrolidine	290
Piperidine	48

[a] Atm^{-1}.
[b] Too small to determine.

lists equilibrium constants for complexing of a number of amines with trimethylboron in the gas phase at 100° C. The increased basicity of most of the amines listed relative to ammonia probably has the same cause as the corresponding increased basicity toward protons, namely, the ability of the alkyl groups to donate electrons to and be polarized by the adjacent positively charged nitrogen atom. Although the basicity toward trimethylboron increases by more than 100-fold on going from ammonia to methylamine, the increase on going from methylamine to dimethylamine is less than two-fold, and going further to trimethylamine gives a 20-fold decrease in basicity. Apparently the base-strengthening effect of the methyl groups is being opposed by steric hindrance to an increasing extent. In the methylamine-trimethylboron adduct the mutually repelling methyl groups can move away from each other and toward the nitrogen-bound hydrogen atoms without greatly stretching the boron-nitrogen bond (but perhaps bending it significantly). As the number of methyl groups on nitrogen is increased this sort of freedom is lost and some sort of compromise between stretching the boron-nitrogen bond, bending the bonds to the methyl groups and their hydrogen atoms, and bringing methyl groups to within less than their van der Waals radii of each other is reached (at the expense of a significant amount of strain energy). Ethylamine may form an adduct with trimethylboron that is not significantly more strained than the methylamine adduct if it orients the methyl end of the ethyl group away from the trimethylboron as shown in **4**, which is a Newman projection down the carbon-

BMe_3
H H
H H
Me
4

nitrogen bond of the adduct. If this methyl group is equally stable in each of the three conformations of ethylamine, then its restriction to one conformation in the adduct should give an unfavorable contribution of $R \ln 3$, or 2.2 eu, to the entropy of reaction. Experimentally the entropy of adduct formation was found to be 2.4 eu more negative for ethylamine than for methylamine. Both carbon-bound methyl groups in the diethylamine adduct may be oriented away from the trimethylboron as shown in **5**, but only at the expense of mutual repulsions that are reflected in the significantly decreased equilibrium constant for adduct formation shown in Table 7-4. Studies of molecular models indicate that with triethylamine

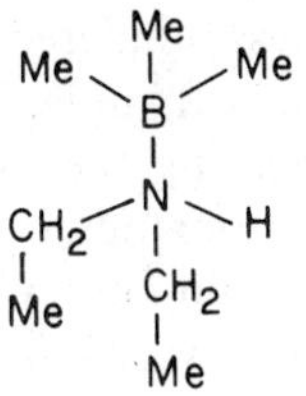

5 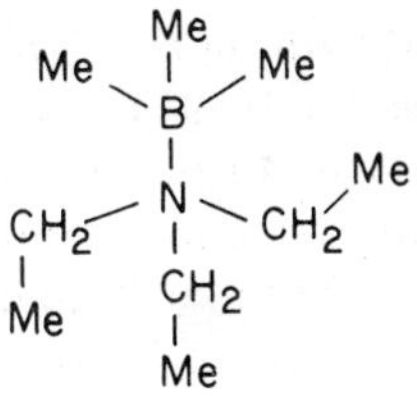 **6**

the strain involved in aiming all three methyl groups at each other is probably prohibitive and that one of the methyl groups must be gauche to the unshared electron pair on nitrogen, where it would crowd the methyl groups from boron upon adduct formation as shown in **6.** Thus the equilibrium constant for formation of this adduct was too small to measure.

Evidence that there is steric strain in the adduct formed by trimethylboron with secondary amines, including dimethylamine, is found in the equilibrium constants listed for piperidine, pyrrolidine, and trimethylenimine. The proton basicities of these three amines in aqueous solution are only about twice that of dimethylamine. Piperidine, whose bond angles are nearly tetrahedral, has about the same basicity as dimethylamine toward trimethylboron, but pyrrolidine and trimethylenimine, in which the two alkyl groups attached to nitrogen are tied back, away from the trimethylboron in the adduct, are much more basic toward trimethylboron. With ethylenimine this trend is interrupted, and the basicity toward trimethylboron is slightly less than that of dimethylamine. Even this is evidence for a large steric effect because the proton basicity of ethylenimine is less than that of dimethylamine by more than 500-fold, presumably because of the increased *p* character of the carbon-nitrogen bonds that accompanies the small bond angles, which results in increased *s* character in the nitrogen-hydrogen bonds.

Table 7-4 shows that 3-methyl and 4-methyl substituents increase the basicity of pyridine toward trimethylboron just as they increase the basicity toward protons. The 2-methyl substituent, which also increases the proton basicity, decreases the basicity toward trimethylboron almost undoubtedly because of steric hindrance. The magnitude of this decrease could not be measured by the method of determining equilibrium constants in the gas phase, but a similar result is seen in the enthalpies of addition to boron fluoride and trimethylboron in nitrobenzene solution listed in Table 7-5. It seems reasonable to assume that the enthalpy of addition of trimethylboron to 2-picoline would be about the same as to 4-picoline in the absence of steric effects (the proton basicities of the two are very similar). Hence the observed enthalpy of addition shows that the adduct of 2-picoline and trimethylboron is sterically strained by about 6 kcal/mole. The corresponding adduct with boron trifluoride, a less hindered (and stronger) Lewis acid, appears to be strained by about 2.2 kcal/mole. The strain in the 2-picoline adducts is no doubt relieved by bending the boron acid away from the offending 2-methyl group. When the obvious route of escape is blocked, as it is in 2,6-dimethylpyridine, the strain energy jumps to about 8 kcal/mole in the case of boron trifluoride and more than 15 kcal/mole in the base of trimethylboron. With a 2-*t*-butyl substituent the strain in the boron fluoride adduct is about 11 kcal/mole and with 2,6-di-*t*-butylpyridine the strain is too large to measure since no adduct is formed on treatment with boron fluoride.[30] (Cf. Section 6-4.)

Love, Cohen, and Taft determined equilibrium constants for complexing of trimethylboron in the gas phase with a series of primary aliphatic amines for which the polar character of the substituents was varied but the steric requirements kept essentially constant (near the reaction center, at least).[31] Table 7-6 lists their results and the ionization constants and enthalpies and entropies of ionization of the amines in aqueous solution. A log-log plot of the basicity constants toward trimethylboron versus the ionization constants gives a reasonably straight line except that the methoxy compounds are more than 10 times too basic toward trimethylboron to fit on the line.

Table 7-5. Enthalpies of Addition to Pyridine Derivatives

	$-\Delta H°$ (kcal/mole)	
Substituents(s)	BF_3[b]	BMe_3[c]
None	32.9	21.4
2-Me	31.2	16.1
3-Me	33.2	21.7
4-Me	33.4	22.0
2,6-Me_2	25.4	6.2
2-*t*-Bu	22.7	6.1

[a] Reactions of the gaseous acid with the dissolved base to give the dissolved complex.

[b] H. C. Brown, D. Gintis, and H. Podall, *J. Amer. Chem. Soc.*, **78,** 5375 (1976).

[c] H. C. Brown and D. Gintis, *J. Amer. Chem. Soc.*, **78,** 5378 (1956).

Table 7-6. Basicities of Primary Aliphatic Amines toward Trimethylboron in the Gas Phase and Protons in Aqueous Solution

		Ionization in Water at 25° C[b]		
Amine	Me_3B K[a] (atm^{-1})	10^8 K (M^{-1})	$\Delta H°$ (kcal/mole)	$\Delta S°$ (eu)
$EtNH_2$	430	43000	−0.1	−15.6
n-$PrNH_2$	520	34000	−0.2	−16.2
n-$BuNH_2$	720	40000	−0.4	−16.6
$MeO(CH_2)_3NH_2$	2800	8300		
$MeOCH_2CH_2NH_2$	980	2500		
$FCH_2CH_2NH_2$	72	620	1.6	−18.7
$F_2CHCH_2NH_2$	6.7	12	3.8	−19.0
$F_3CCH_2NH_2$	1.1	0.4	5.2	−20.9

[a] At 54° C. The entropy of reaction was −43.0 ± 0.2 for all the amines except the methoxy compounds, for which it was −45.3 ± 0.2.

[b] Calculated from the acidity constant data of Ref. 31 and K, $\Delta H°$, and $\Delta S°$ for the autoprotolysis of water.

These deviations must come from stabilization of the adduct by hydrogen bonding as shown in **7**. Internal hydrogen bonding in species of the type

```
   Me
     \
     O - - H  ⊕     ⊖
    /        \
 CH2          NH - BMe3
    \        /
     CH2—CH2
```

7

$MeO(CH_2)_nNH_3^+$ in aqueous solution should be relatively unimportant because of the great abundance of water, which should be a better hydrogen bonder than the oxygen atom of an ether with an electron-withdrawing substituent. When the equilibrium constants for complexing of trimethylboron with n-$BuNMe_2$, $MeOCH_2CH_2NMe_2$, and $MeO(CH_2)_3NMe_2$ were determined, the methoxy substituents were found to decrease the basicity, presumably because internal hydrogen bonding of the type shown in **7** was now impossible. If decreases of the same magnitude had appeared with the methoxy-substituted primary amines the points for these amines would have fit the line in the log-log plot. The slope of the line (0.54) or just examination of Table 7-6 shows that substituents change the basicity toward protons more than they change the basicity toward trimethylboron.

This is probably largely because protonation places a positive charge near the substituents, whereas trimethylboron addition merely places the positive end of a dipole near them. It is noteworthy that the log-log plot is reasonably linear in spite of the fact that the entropies of reaction are essentially constant in one reaction while they vary considerably in the other (cf. Section 6-7b). The increase in basicity toward trimethylboron on going from ethyl- to *n*-propyl- to *n*-butylamine suggests that polarizability effects are significant.

Brown and Holmes found the enthalpies of addition of pyridine to boron fluoride, boron chloride, and boron bromide in nitrobenzene solution to be −25.0, −30.8, and −32.0 kcal/mole, respectively.[32] This is certainly not the order that would be expected from the electronegativities of the halogens. The results were rationalized in terms of resonance stabilization of the boron halides, as shown in **8**, with the contribution of the double-bonded structures increasing in the order Br < Cl < F.

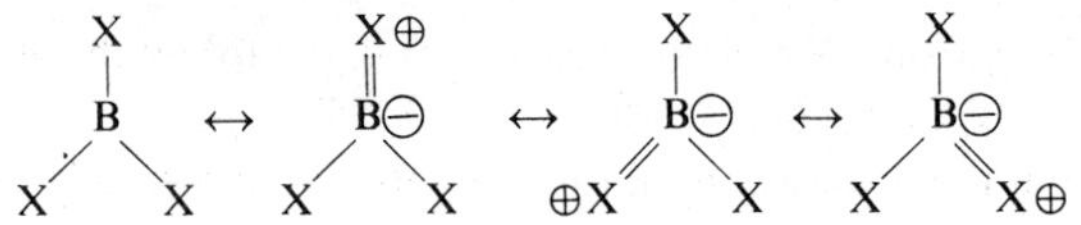

$$BX_3 \leftrightarrow (X^{\oplus}{=})B^{\ominus}X_2 \leftrightarrow (^{\oplus}X{=})B^{\ominus}X_2 \leftrightarrow X_2B^{\ominus}({=}X^{\oplus})$$

8

7-3. HARDNESS AND SOFTNESS OF LEWIS ACIDS AND BASES

7-3a. Class (*a*) and Class (*b*) Behavior of Lewis Acids. If acidity and basicity were a simple enough subject all acids could be arranged in a unique order of relative strengths that did not change with the nature of the base and all bases could be arranged in a unique order of relative strengths that was independent of the nature of the acid with which they were reacting. We have already noted now steric effects and solvent effects can produce reversals in relative acid strengths and relative base strengths. Thus steric effects explain why 2-methylpyridine is a stronger base than pyridine toward protons but a weaker base toward trimethylboron.[29] Solvent effects explain why the Brønsted basicity of pyridine is greater than that of acetate ions in water but less in methanol (Section 5-3). A number of reversals in relative acid strength and relative base strength that seemed clearly not to arise from steric or solvent effects were noted and rationalized after the Lewis definition of acidity and basicity became widely used. By the 1950s there were enough examples of this type to elicit broad generalizations from chemists in the field. Ahrland, Chatt, and Davies divided Lewis acids into class (*a*) and class (*b*).[33] Class (*a*) acids form a stronger bond to fluoride than to any other halide ion, a stronger bond to an oxygen base than to the analogous sulfur, selenium, or tellurium base, and a stronger bond to a nitrogen base than to an analogous phosphorus, arsenic, or antimony base. Class (*b*) acids form a stronger bond to the second element in a group, with the relative basicities of the higher elements in the group being variable. Thus toward class (*b*) acids the relative basicities of the halogens are usually F < Cl < Br < I; for a series of analogous Group V bases the usual order is N < P > As > Sb; and in Group VI, sulfur bases are stronger than the analogous oxygen bases, but the position of selenium and tellurium bases is variable.

The illustrative data most clearly uncomplicated by solvent effects refer to the gas phase. Boron fluoride acts as a class (*a*) acid in that its dimethyl ether adduct is less dissociated at 100° C[34] than its dimethyl sulfide adduct is at 60° C;[35] furthermore, the stabilities of its Group V adducts stand in the order Me_3N[36] > Me_3P[37] > Me_3As[38] > Me_3Sb.[38] Boron hydride, however, is a class (*b*) acid. Its dimethyl sulfide adduct[39] is more stable than its dimethyl ether adduct[40] or its dimethyl selenide adduct.[39]

Illustrative data in solutions are much more abundant.[41,42] The data in Table 7-7 classify Fe^{3+} and UO_2^{2+} as class (*a*) acids and Ag^+ and Hg^{2+} as class (*b*) acids, with In^{3+} and Cd^{2+} being nearer the borderline between the two classes. Based on these and many other observations, class (*b*) character was noted to occur when the acidic atom was in a roughly triangular region of the Periodic Table stretching from magnanese to copper in

Table 7-7. Lewis Basicity of Halide Ions in Aqueous Solution [a]

	log K_1			
Acid	F^-	Cl^-	Br^-	I^-
Fe^{3+}	6.04	1.41	0.49	
UO_2^{2+}	4.77	−0.10	−0.20	
In^{3+}	3.78	2.36	2.06	1.64
Cd^{2+}	0.57	1.59	1.76	2.08
Ag^+	0.36	3.04	4.38	8.13
Hg^{2+}	1.03	6.74	8.94	12.87

[a] Measurements were made near room temperature; some of the data refer to fairly high ionic strengths, but an attempt was made to find cases in which the conditions were the same for all the K_1 values determined for a given acid.[41,42]

Period IV, from molybdenum to cadmium in Period V, and from tungsten to polonium in Period VI, and, in some cases, when the acidic atom was boron or carbon. The borders of this region are not sharp. Whether some of the elements near the border exhibit class (b) behavior may depend on their state of oxidation, what other ligands are already attached to them, etc.

Pearson has extensively discussed reversals in relative acidity and basicity such as those illustrated in Table 7-7 in terms of the adjectives "hard" and "soft" suggested by Busch.[43–46] Since fluorine, oxygen, and nitrogen are the least polarizable elements in their groups their derivatives are referred to as hard bases, with the more polarizable bases derived from phosphorus, sulfur, etc., being referred to as soft. Class (*a*) acids tend to be less polarizable than class (*b*) acids and are called hard acids, with the class (*b*) acids being soft. It then follows that deviations in the relative strength of a series of acids observed upon going from one base to a second base may result from hard acids forming stronger complexes with a hard base or soft acids forming stronger complexes with a soft base than would otherwise be expected. This leads one to hope that all equilibrium constants for Lewis acid-base reactions may be correlated in terms of two factors, strength and hardness (or softness), for each acid and base, in those cases where steric effects may be neglected. Whether this hope will be realized is not certain. However, if it is true that hardness (and softness) is related to (1) the degree of ionic character of the sigma bonding, (2) the extent of pi bonding, (3) electron-correlation energy changes, (4) solvation effects, and other factors, as has been suggested, it seems a lot to hope that hardness will be treatable as a simple, rather than a complex, quantity. Like many other concepts, for example, "ion-solvating power of a solvent," hardness and softness may prove to be a rather "fuzzy" concept, but nevertheless a useful one in discussing chemical phenomena.

There is evidence that the bond formed between a soft acid and a soft base in many cases is a multiple bond; in addition to the sigma bond formed from the unshared electron pair of the base there is also pi bonding. Some of the softest acids are metal ions in which the outer *d* orbital is largely or completely filled with electrons. When such a metal ion coordinates with a base that has vacant *d* orbitals or a pi electron system that is capable of accepting more electrons, a *d* orbital on the metal may overlap with an orbital on the base, delocalizing the *d* electrons and donating electron density to the base.[47] There is a synergistic interaction between such "back bonding," which, by increasing the electron density on the base, makes it more easily share the unshared electron pair with which it is forming a sigma bond to the acid, and the sigma bonding, which decreases the electron density on the base, making it accept electrons from the *d* orbitals of the acid to a greater extent. The net result is that coordination with the soft acid does not change the net charge on the base as much as would coordination with an acid that could not donate electrons to the base by such pi bonding.

Taube and co-workers have described a case in which electron donation from the *d* orbitals of a metal to the pi system of a ligand is so pronounced that the complex is stabilized by substituents that make the pi system more electron deficient in spite of the fact that such substituents must decrease the availability of the unshared electron pair that makes the ligand a base.[48,49] The acid in question was $Ru(NH_3)_5H_2O^{2+}$, whose remarkable ability to coordinate with elemental nitrogen in aqueous solution—a fact also attributed to efficient back-bonding—did not simplify the study. When the water in this acid is replaced by a pyrazine molecule the basicity of the remaining nitrogen atom in the pyrazine molecule is greater than in pyrazine. The increase in basicity, as shown in Table 7-8, amounts to 2.2 log units when allowance for statistical effects is made. Some mechanism of electron donation is required to explain the observation that attachment of an atom with a double positive charge to one nitrogen atom of pyrazine increases the basicity of the other nitrogen atom. Algebraic manipulation

Table 7-8. Acidity of Some Azine Derivatives[48–50]

Acid	pK_a
$N(C_4H_4)NH^+$ (pyrazinium)	0.6
$(NH_3)_5\overset{++}{Ru}N(C_4H_4)NH^+$	2.5
$NC{-}C_5H_4NH^+$ (4-cyanopyridinium)	1.90
$(NH_3)_5\overset{++}{Ru}NC{-}C_5H_4NH^+$ (4-)	2.72
$NC{-}C_5H_4NH^+$ (3-cyanopyridinium)	1.36
$(NH_3)_5\overset{++}{Ru}NC{-}C_5H_4NH^+$ (3-)	1.75

of the acidity constants shows that, after statistical corrections, the equilibrium constant for complex formation from $Ru(NH_3)_5H_2O^{2+}$ and pyrazinium ions is $10^{2.2}$ times as large as the equilibrium constant for complex formation from $Ru(NH_3)_5H_2O^{2+}$ and unprotonated pyrazine. Other entries in Table 7-8 show that $Ru(NH_3)_5{}^{2+}$, when coordinated to the nitrogen atom of a nitrile substituent, acts as an electron donor to the ring nitrogen atom of pyridine and that it does so more efficiently in 4-cyanopyridine, where it is conjugated with the ring nitrogen atom. These observations further show that the cyano groups are more basic toward $Ru(NH_3)_5(H_2O)^{2+}$ when they are attached to protonated pyridine rings than when attached to unprotonated pyridine rings. Further evidence for pi bonding between ruthenium and an azine ring is found in the values of the equilibrium constants for displacement of water from $Ru(NH_3)_5H_2O^{2+}$ by pyridine and ammonia.[51] These values, 2.4×10^7 and 3.5×10^4 M^{-1} (statistically corrected), respectively, show that although pyridine is only about 10^{-4} times as basic as ammonia toward protons (and less basic toward most metal ions), it is 700 times as basic toward $Ru(NH_3)_5H_2O^{2+}$. The preceding and additional observations have been discussed in detail by Taube.[49]

7-3b. Correlation of Enthalpies of Lewis Acid-Base Reactions in Inert Solvents. Drago and co-workers correlated a large number of enthalpies of Lewis acid-base reactions in alkane or carbon tetrachloride solution or in the gas phase in terms of the four-parameter relationship shown in Eq. 7-17.[52–54] Each acid is given two parameters, E_A and C_A, and each

$$-\Delta H = E_A E_B + C_A C_B \qquad (7\text{-}17)$$

base is given two parameters, E_B and C_B. The terms E and C were chosen to stand for "electrostatic" and "covalent," respectively. The four parameter values that could be assigned arbitrarily without affecting the ability of the equation to correlate the data were chosen by setting E_A and C_A for the acid iodine equal to 1.00, E_B for *N,N*-dimethylacetamide equal to 1.32, and C_B for diethyl sulfide equal to 7.40. Most of the 280 values of ΔH used in obtaining parameter values referred to processes like hydrogen bonding and coordination with iodine, which do not involve very complete coordination of the acid with any electron pair of the base. It is interesting that the formation of iodine complexes, which are often referred to as charge-transfer complexes, can be treated by a correlation that includes many reactions that are more clearly simple Lewis acid-base additions. Some of the parameter values obtained are listed in Table 7-9.

Equation 7-17 can be thought of as a method of expressing the concept of hard and soft acids and bases quantitatively. It is not clear from most qualitative discussions whether the softness of an acid or base is best

Table 7-9. Parameter Values for Eq. 7-17[54]

Acid	C_A	E_A	Base	C_B	E_B
I_2	1.00	1.00	Pyridine	6.40	1.17
ICl[a]	0.830	5.10	NH_3	3.46	1.36
PhSH[a]	0.198	0.987	$MeNH_2$	5.88	1.30
p-t-BuC_6H_4OH	0.387	4.06	Me_2NH	8.73	1.09
PhOH	0.442	4.33	Me_3N[b]	11.54	0.808
p-FC_6H_4OH	0.446	4.17	MeCN	1.34	0.886
p-ClC_6H_4OH	0.478	4.34	$ClCH_2CN$	0.530	0.940
m-FC_6H_4OH	0.506	4.42	Me_2NCN	1.81	1.10
m-$CF_3C_6H_4OH$	0.530	4.48	$AcNMe_2$	2.58	1.32
CF_3CH_2OH	0.434	4.00	AcOEt	1.74	0.975
$(CF_3)_2CHOH$	0.509	5.56	Me_2CO	2.33	0.987
Pyrrole[a]	0.295	2.54	Et_2O[b]	3.25	0.963
HNCS	0.227	5.30	1,4-Dioxane	2.38	1.09
BF_3[a,b]	1.62	9.88	Tetrahydrofuran	4.27	0.978
Me_3B[b]	1.70	6.14	Me_2SO	2.85	1.34
Me_3Al	1.43	16.9	Et_2S	7.40	0.339
Me_3Ga	0.881	13.3	Me_2Se	8.33	0.217
Me_3SnCl[b]	0.0296	5.76	Pyridine *N*-oxide	4.52	1.34
SO_2[a]	0.808	0.920	Me_3P[a]	6.55	0.838
$SbCl_5$[b]	5.13	7.38	Benzene	0.707	0.486
$CHCl_3$	0.150	3.31	$(Me_3N)_3PO$[a]	1.33	1.73

[a] Tentative value calculated from a limited range of data.
[b] Sensitive to steric effects.

equated to its C value (and its hardness to $1/C$) or to its C/E ratio but the latter seems closer to the qualitative concept. The E_A and E_B parameters could perhaps be called the general acid strength and general base strength, respectively.

Since entropy effects in acid-base reactions tend to arise largely from solvation effects the relative values of ΔH in the gas phase and in inert solvents that were used to calculate the parameters in Table 7-9 tend to be moderately near the relative values of ΔG, so that Eq. 7-17 offers some hope for correlating equilibrium constants. However, it is not clear how well the equation will survive exposure to new experimental data. In subsequent work, for example, the enthalpy of addition of tetrahydrofuran to antimony pentachloride in carbon tetrachloride solution was found to be 19.3 kcal/mole.[55] The value calculated from the parameters in Table 7-9 is 29.12 kcal/mole. Perhaps this is a steric effect. In any case, there is considerable uncertainty as to the nature and extent of solvent effects and the deviation from Eq. 7-17 that may result from them.[55,56]

7-4. ACIDITY OF METALS IN COMPOUNDS AND IONS

7-4a. Acidity of Covalent Halides. In their reviews on quantitative measurements of the Lewis acidity of covalent metal halides the Satchells described a number of reaction series in which log K for the coordination of derivatives of aniline or benzamide with a given metal halide (Eq. 7-18) is linearly related to the pK_a of the protonated bases in aqueous solution

$$B + MX_n \overset{K}{\rightleftharpoons} B{-}MX_n \quad (7\text{-}18)$$

(Eq. 7-19).[57] The results obtained in some of these correlations are sum-

$$\log K = a\, pK_a + b \quad (7\text{-}19)$$

marized in Table 7-10. In several cases ortho-substituted compounds and *N*-substituted compounds deviated significantly from the correlation, and their equilibrium constants were not used in calculating a and b.

As an extreme approximation we might say that since a is 0.81 for the coordination of anilines with zinc chloride in ether, the nitrogen-zinc bonds must have about 81% as much covalent character as the nitrogen-hydrogen bond in the corresponding anilinium ion. However, this approximation neglects a number of important factors. For example, the basicities toward zinc chloride in *ether* are compared with the basicity toward protons in *water*. Many observations, such as the solvent dependence of Hammett ρ constants (cf. Table 6-1), show that Brønsted basicity is less sensitive to substituent effects in water than in poorer ion-solvating media. Hence if the Brønsted pK_a values in Eq. 7-19 had referred to basicities toward protons in ether smaller values of a should have been obtained. On the other hand, Eq. 7-19 is comparing Lewis basicity to give a neutral (but dipolar) adduct with Brønsted basicity to give a cationic adduct. The reaction of anilines with a Brønsted acid to give a neutral adduct, such as an ion pair or a hydrogen-bonded complex, should be less sensitive to substituent effects than reaction to give free anilinium ions. Hence if the Brønsted pK_a values in Eq. 7-19 had referred to the formation of a neutral adduct, larger values of a should have been obtained. Still another factor that affects the interpretation of a is the implicit assumption that the zinc atom is bonded to nitrogen in the adduct only by a sigma bond, with no pi interactions. Although this assumption is fairly plausible in the present case it will not be in certain other Lewis acid-base reactions that we shall discuss.

Table 7-10. Correlations of Brønsted Basicity and Lewis Basicity toward Covalent Halides[a]

Acid	Bases	Solvent	Temperature (°C)	a^b	b^b
$ZnCl_2$	$ArNH_2$	Et_2O	20	0.81	−0.97
$ZnBr_2$	$ArNH_2$	Et_2O	20	0.70	−0.62
$ZnCl_2$	$ArNH_2$[c]	Me_2CO[d]	25	0.63	0.08
$ZnBr_2$	$ArNH_2$[c]	Me_2CO[d]	25	0.81	0.04
ZnI_2	$ArNH_2$[c]	Me_2CO[d]	25	0.64	0.15
$ZnCl_2$	$ArCONH_2$	Et_2O	25	0.79	4.69
BF_3	$ArNH_2$	Et_2O	25	0.96	0.29
BF_3	$ArCONH_2$	Et_2O	25	1.11	5.54
BF_3	$ArCONH_2$	THF[e]	26	1.25	4.47
$AlCl_3$	$ArNH_2$[c]	Et_2O	25	0.60	0.75
$AlBr_3$	$ArNH_2$[c]	Et_2O	25	0.75	1.15
$GaCl_3$	$ArNH_2$[c]	Et_2O	25	0.99	−0.01
$GaBr_3$	$ArNH_2$[c]	Et_2O	25	1.01	0.05
$SnCl_4$	$ArNH_2$[c]	Et_2O	25	0.65	0.80
$SnBr_4$	$ArNH_2$[c]	Et_2O	25	1.07	−0.98
$PhSnCl_3$	$ArNH_2$[c]	Et_2O	25	0.80	−1.40

[a] Data from Ref. 57; K in M^{-1}.

[b] These are the values of a and b obtained by fitting the data to Eq. 7-19.

[c] All the anilines studied had at least one nitro substituent.

[d] These results may be complicated by the formation of imines from the anilines and the solvent, a possibility that seems not to have been considered.

[e] Tetrahydrofuran.

The larger values of b obtained for benzamides than for anilines show that the basicity of a benzamide toward zinc chloride or boron trifluoride in ether is significantly greater than that of an aniline with the same Brønsted basicity in aqueous solution. Much of these differences in b values is probably a solvent effect. Since protonation of an amide, which takes place almost entirely on the oxygen atom, gives a cation with a considerably more diffused charge than that on a protonated amine, the Brønsted basicity of an amide should increase, relative to that of an amine, on going from water to ether as a solvent. Hence if the Brønsted pK_a in Eq. 7-19 had referred to ether solution, it is not clear whether the b values would be larger for amides or amines. Any remaining differences in b values could refer to any preferences that the Lewis acids in question have for coordinating with an oxygen rather than a nitrogen atom, or with an unsaturated atom rather than a saturated atom. (There is good evidence that these Lewis acids, like protons, coordinate almost exclusively with the oxygen atom of an amide.)

None of the solvents listed in Table 7-10 are very acidic but all of them contain a basic oxygen atom. In these solvents the reactions may be written more accurately as nucleophilic displacement equilibria, as exemplified in Eq. 7-20, rather than as simple adduct formation as may be implied by

$$Et_2O{-}BF_3 + ArNH_2 \rightleftharpoons ArNH_2{-}BF_3 + Et_2O \qquad (7\text{-}20)$$

Eq. 7-18. The quotient of the equilibrium constants for Eq. 7-20 for the two amines $ArNH_2$ and $Ar'NH_2$ is simply the equilibrium constant for Eq. 7-21, which measures the relative basicities of $ArNH_2$ and $Ar'NH_2$

$$ArNH_2 + Ar'NH_2{-}BF_3 \rightleftharpoons Ar'NH_2 + ArNH_2{-}BF_3 \qquad (7\text{-}21)$$

toward boron fluoride. The quotient of the equilibrium constant for Eq. 7-20 and the equilibrium constant for the formation of the analogous aluminum chloride adduct is the equilibrium constant for Eq. 7-22, which,

$$Et_2O{-}BF_3 + ArNH_2{-}AlCl_3 \rightleftharpoons Et_2O{-}AlCl_3 + ArNH_2{-}BF_3 \qquad (7\text{-}22)$$

however, does not tell us anything so simple about the relative acidities of aluminum chloride and boron fluoride. If the basicity of $ArNH_2$ exceeds that of ether toward aluminum chloride by exactly the same amount as it does toward boron fluoride the equilibrium constant for Eq. 7-22 will be 1.0; that is, $ArNH_2$ will have the same equilibrium constant for reaction with boron fluoride as for reaction with aluminum chloride in ether solution. After admitting that there must be some interaction between the solvent and the unshared electron pair of an amine, we could object that Eq. 7-20 is also an oversimplification and that the $ArNH_2$ should have been written as a complex with the solvent. If this were done, Eq. 7-21 would become Eq. 7-23, whose equilibrium constant would not be simply

$$ArNH_2{-}Et_2O + Ar'NH_2{-}BF_3 \rightleftharpoons Ar'NH_2{-}Et_2O + ArNH_2{-}BF_3 \qquad (7\text{-}23)$$

a measure of the relative basicities of $ArNH_2$ and $Ar'NH_2$ toward boron fluoride, but would also depend on their relative basicities toward the solvent. Such complications may be avoided completely only by relying on measurements in the gas phase. Nevertheless, it is reasonable to try to draw conclusions from solution phase equilibrium constants like those in Table 7-10 by arguing that the acid-base interaction between ether and the unshared electron pair in an aniline derivative is so small that the difference between such interactions for two aniline derivatives may be neglected safely in comparison to the difference between their energies of acid-base interaction with boron fluoride. An extrapolation of this argument is that, other factors being equal (which they never are), for two bases significantly more basic than the solvent, the one that forms the stronger bond to a given acid will have the larger equilibrium constant for formation of an adduct with that acid. This approximation, which treats the equilibrium constant as measuring the strengths of the acid-base interactions rather than the differences between these interactions and the acid-solvent interactions, will become less reliable with increasing basicity of the solvent, relative to that of the bases being studied, and with increasing chemical differences between the solvent and the bases being studied. According to this extrapolation, the strongest Lewis acids in Table 7-10 will be those for which a has the largest value for a given series of bases in a given solvent. The a values listed, most of which have an uncertainty of 0.1 or less, suggest that boron fluoride is a stronger acid toward amides than zinc chloride is and that stannic bromide, gallium bromide, gallium chloride, and boron fluoride are stronger acids than zinc chloride, phenyltin trichloride, aluminum bromide, zinc bromide, stannic chloride, and aluminum chloride toward anilines. The extrapolation would also lead to the conclusion that the strongest Lewis acids will be those for which b has the largest value for a given series of bases in a given solvent, provided that the solvent is sufficiently similar to the bases. It is doubtful that any of the solvents listed are similar enough to the bases studied to warrant application of this generalization to the relatively small differences in b observed for various pairs of acids reacting with a given family of bases in a given solvent. Part of this doubt comes from the differences between the conclusions that would be drawn and those that arose tentatively from the relative sizes of the a values.

7-4b. Metal-Ion Basicity versus Brønsted Basicity of Monodentate Ligands. Equilibrium constants for the coordination of metal ions with bases in solution provide a major pool of data for seeking and testing generalizations concerning structural effects on Lewis acidity and basicity. Most of the data reported before 1969 has been compiled by Sillén and Martell,[41,42] Some of the trends that have been observed were discussed by Irving and the Rossottis[58,59] and by Chipperfield.[60] One of these trends, which has already been described for other types of Lewis acids, is the tendency for equilibrium constants for members of a given family of bases to increase with increasing Brønsted basicity. Bruehlman and Verhoek, for example, determined equilibrium constants for coordination of a number of amines with silver ions in 0.5 M potassium nitrate solution.[61] A plot of log K versus the pK_a values of the protonated amines under the same conditions showed that the points for diethylamine, morpholine, and piperidine described one line (of slope 0.31) and the points for pyridine derivatives and four primary amines of the type RCH_2NH_2 fell near a second line of similar slope. However, since all the primary amines were considerably more basic than any of the pyridines it is hard to be sure that the equilibrium constant for a pyridine would necessarily be so near that for a primary amine of the same basicity that the two types of bases are really most reasonably treated as belonging to the same family. The linearity of the plots shows that the data may be correlated by Eq. 7-19 in which a is the slope and b, the intercept, is the value of log K for a base whose conjugate acid has a pK_a of 0. However, since the bases that have been studied in the present cases are all much stronger than this, the value of b results from a long extrapolation. For this reason it seems more useful to transform Eq. 7-19 to Eq. 7-24, in which a has the same meaning as before but c is

$$\log K = a(pK_a - 7) + c \tag{7-24}$$

the value of log K for a base whose conjugate acid has a pK_a of 7.0. Values of a and c calculated from the data of a number of workers[61–67] are listed in Table 7-11.

Table 7-11. Correlations of Brønsted Basicity and Lewis Basicity toward Metal Ions in Aqueous Solution[a]

Acid	Bases	a	c	Reference
Ag^+	$(RCH_2)_2NH$[b]	0.31	1.73	61
Ag^+	RCH_2NH_2	0.29	2.28	61
Ag^+	Pyridines	0.28	2.47	61
Ni^{2+}	Pyridines	0.27	2.32	62
Cd^{2+}	Pyridines	0.32[c]	1.80[c]	62,63
Zn^{2+}	Pyridines	0.35[c]	1.67[c]	62,64
Cu^{2+}	Pyridines	0.45	3.26	62
CoL_5[d]	Pyridines	0.21[e]	3.58[e]	65
CoL_5[d]	RCH_2NH_2	0.38[e]	2.22[e]	65
CoL_5[d]	RCH_2S^-	0.18[e]	4.92[e]	66
Fe^{3+}	ArO^-	0.94[c]	5.99[c]	67

[a] At 25° C in the presence of 0.5 M potassium nitrate unless otherwise noted. K's in M^{-1}, correlated by Eq. 7-24.

[b] Including cyclic amines.

[c] In the presence of 0.1 M sodium perchlorate.

[d] Methylaquocobaloxime.

[e] In the presence of 1.0 M potassium chloride.

In all cases the value of *a* is between 0 and 1.0, indicating that structural effects change the basicity toward the metal ion in the same direction as they change the Brønsted basicity but that they change the metal-ion basicity less. The reaction of ferric ions with meta- and para-substituted phenoxide ions shows the largest *a* (0.94). This must be partly because of the large charge on the metal ion. Even if the iron-oxygen bond is not fully enough covalent to make the charge on oxygen change by one unit when it coordinates to iron, the presence of the positively charged ferric ion only slightly further from the substituents makes electron-donor substituents help and electron-withdrawer substituents hinder the coordination reaction more than they would otherwise. However, the charge on the ion is obviously not the only factor of importance. Pyridines show essentially the same *a* value for coordination with single charged silver ions as for coordination with doubly charged nickel ions. Primary amines show a smaller *a* value for coordination with silver ions than for coordination with methylaquocobaloxime (abbreviated CoL_5), an electrically neutral complex that may be thought of as arising from the complexing of a cobaltic ion with five basic ligand atoms in addition to the water molecule that is replaced in the equilibria that were measured.

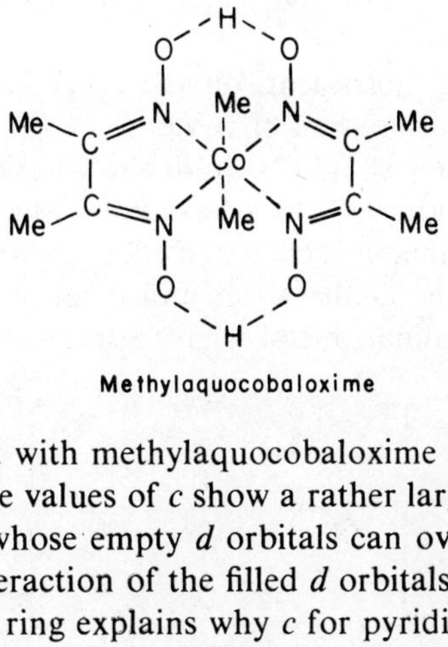

Methylaquocobaloxime

The results obtained with methylaquocobaloxime classify this species as a soft acid. The relative values of *c* show a rather larger tendency for coordination with sulfur, whose empty *d* orbitals can overlap with the filled *d* orbitals on cobalt. Interaction of the filled *d* orbitals on cobalt with the pi system of the pyridine ring explains why *c* for pyridines is larger than that for primary amines. The donation of electrons to the ligand by pi bonding at the same time that they are withdrawn by sigma bonding results in less change in the charge on the ligand atom than would have accompanied sigma bonding alone. This explains why the *a* values for coordination with methylaquocobaloxime increase in the order $RCH_2S^- <$ pyridines $< RCH_2NH_2$. In view of the fact that the behavior of Ag^+ is usually distinctly that of a soft acid, it is not clear why the values of *a* and *c* listed for its reactions with pyridines and primary amines are so near each other.

The nature of the bases that deviated from the correlations in Table 7-11 (and were therefore not used in calculating *a* and *c*) can be as informative as the correlations themselves. For example, a 2-methyl or 2-amino substituent in the pyridine ring causes the equilibrium constant for complexing with Cu^{2+} ions to be 15- to 30-fold too small to fall on the correlation line described by 3- and 4-substituted pyridines.[62] The 2-methyl substituent causes the equilibrium constant for complexing with Ni^{2+} and Cd^{2+} ions to be too small to measure.[62,63] Such results, which presumably arise from steric hindrance, are not observed with silver ions, whose equilibrium constants for complexing with 2-methylpyridine and 2,4-dimethylpyridine fall on the same correlation line as those for pyridines without 2-substituents.[61,62] The difference in behavior must arise from the fact that Ag^+ is the only metal ion covered by Table 7-11 that is dicoordinate. The equilibrium constants for the silver ion referred to involve the reaction of the $Ag(OH_2)_2^+$ ion, in which the three nonhydrogen atoms are collinear, with the base **B** to give $BAgOH_2^+$ ions, in which the two valence bonds to silver are again collinear. In such a complex a 2-methyl substituent encounters little crowding. The other metals listed are all at least tetracoordinate. Thus the methyl substituent in a $2\text{-}MeC_5H_4N\text{-}Cu(OH_2)_5^{2+}$ ion, for example should be crowded by at least one of the five water molecules in the first coordination sphere of the metal ion.

A 2-hydroxymethyl substituent, on the other hand, makes the equilibrium constants for coordination of pyridine to Cu^{2+} and Ni^{2+} ions about 10 times as large as they should be to fit the correlations obtained with 3- and 4-substituted pyridines.[62] This deviation, which also does not occur with Ag^+ ions, is presumably a chelation effect, in which the hydroxymethyl substituent replaces one of the water molecules in the inner coordination sphere of the hexacoordinate metal to give a species such as **9**. An analogous

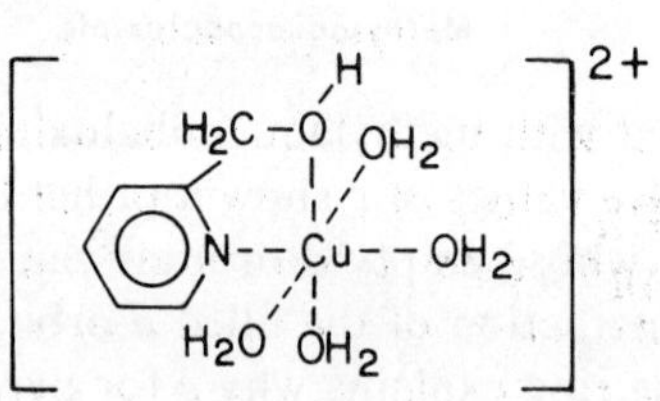

9

complex of the silver ion would be incompatible with the collinearity of the two bonds to silver. A more detailed discussion of chelation effects on equilibrium constants for complexing will be given in Section 9-3.

Although the equilibrium constants correlated in Table 7-11 refer to the coordination of only one molecule of base with a given metal ion, most of the metals listed are capable of coordinating with several molecules of base.

Comparisons of equilibrium constants for coordination of the second, third, etc., molecule (or ion) of base are complicated. In comparing the equilibrium constant for the coordination of the first molecule of a number of different bases with a given metal ion, the Lewis acid (the aquo metal ion) is the same for each. The equilibrium constant for coordination of a second base molecule refers to a different Lewis acid (e.g., AgB^+, AgB'^+, etc.) for each base, Hence the overall equilibrium constant for the coordination of two molecules of base with a metal ion depends upon how the first molecule that coordinates affects the Lewis acidity of the metal ion as well as upon the Lewis basicity of the base. If this substituent effect of coordinated bases is a linear function of the number that have coordinated, log K_1^{chem} for coordination of the first base molecule should differ from log K_2^{chem} (for the second base molecule) by the same amount that log K_2^{chem} differs from log K_3^{chem}, etc. Statistical corrections have not been made in all the comparisons of successive equilibrium constants for coordination that have been published, and in many cases the application of Eqs. 1-7 and 1-8 cannot be made correctly; the dicoordinated complex may be a mixture of cis and trans isomers of unknown composition, and the two isomers would have different symmetry numbers. However, since K_σ (see Section 1-1) will tend to increase monotonically with the number of base molecules that have coordinated, symmetry effects will not mask a systematic trend in K_n^{chem} values unless it is a rather small one. There are a number of cases, such as the coordination of Al^{3+} with one to six fluoride ions or the coordination of Ni^{2+}, Co^{2+}, or Mg^{2+} with one to six molecules of ammonia, in which log K_n has been seen to decrease almost linearly with increasing n.[60] Silver ions, however, have a larger equilibrium constant for combining with the second ligand than the first in a number of cases.[41,42,62] These facts do not appear to be understood, but it would be interesting to correlate the data in terms of interaction energies (Section 1-2d).

7-4c. Metal-Ion Basicity Versus Bronsted Basicity of Bidentate Ligands. The bases covered by Table 7-11 were all monodentate ligands. With polydentate ligands there are additional reasons for deviations from a correlation between basicities toward a metal ion and basicities toward a proton. This may be illustrated by the plot of log K_1 for coordination of substituted salicylaldehyde anions with Cu^{2+} versus the pK_a values of the salicylaldehydes[68,69] shown in Fig. 7-3. Neglecting the compounds with 3- and 6-substituents (which would be ortho to the hydroxy group and the aldehyde group, respectively), there is a good linear correlation for all the other compounds except the two for which strong resonance interaction with the reaction center would be expected. The 4-methoxy substituent should

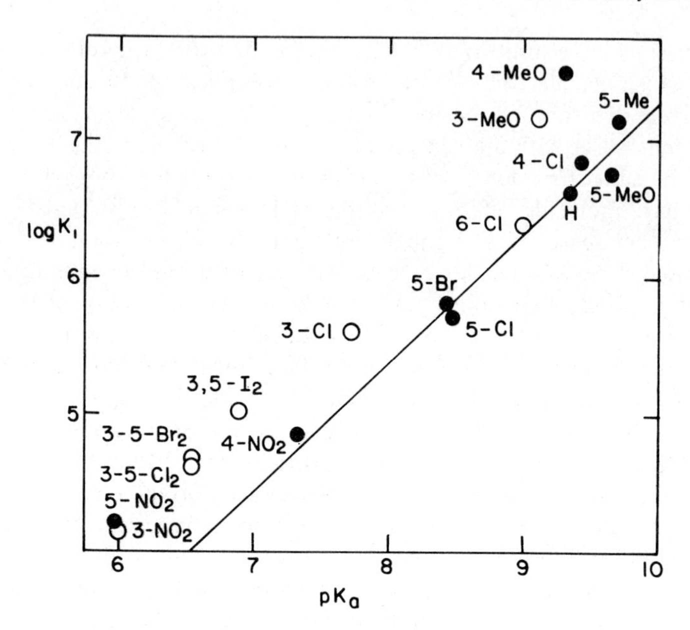

Figure 7-3. Plot of the logarithm of the equilibrium constant for the coordination of one substituted salicylaldehyde anion with a Cu^{2+} ion versus the pK_a of the salicylaldehyde, both in 50 vol % aqueous dioxane with 0.3 *M* sodium perchlorate.

interact strongly with the aldehyde group, especially when it has coordinated with the metal ion. This electron-donating ability of the 4-methoxy substituent, which would make **10** a significant contributor to the total structure of the complex, should increase the basicity of the aldehydic

10

oxygen atom. Since the Cu^{2+} ion probably forms almost as strong a bond to the aldehydic oxygen atom as to the phenolic oxygen atom this increase in basicity increases log K_1 markedly. However, since protonation of the 4-methoxysalicylaldehyde anion involves the formation of a strong covalent bond to the phenolic oxygen atom and only a relatively weak hydrogen bond to the aldehydic oxygen atom, there is a disproportionately small effect on pK_a. All the other substituents affect the basicities of the two oxygen atoms rather similarly except for the 5-nitro substituent, whose resonance electron-withdrawing ability greatly decreases the basicity of the phenolic oxygen atom without decreasing the basicity of the aldehydic oxygen atom proportionately. Hence the basicity of the 5-nitrosalicylaldehyde anion as measured by pK_a significantly underestimates the ability of the anion to use both its oxygen atoms simultaneously to form strong bonds to an acid.

If the deviations by the points in Fig. 7-3 for the 4-methoxy and 5-nitro compounds arise from pK_a being very largely a measure of the basicity of the phenolic oxygen atom of a salicylaldehyde anion but log K_1 being a measure of the basicities of both oxygen atoms, a better correlation might have been obtained if log K_1 had been plotted against some measure of the proton basicity of the aldehydic oxygen atom as well as that of the phenolic oxygen atom. Unfortunately no reliable measure of the Brønsted basicity of the aldehydic oxygen atoms of substituted salicylaldehydes is available. However, in the case of derivatives of 8-hydroxyquinoline (whose trivial name is oxine), which forms metal-ion complexes like **11**, there are measures

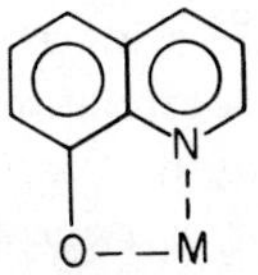

N
O——M

11

of the basicities of both atoms that form bonds to the metal. The available equilibrium constants for the formation of such complexes are scattered over so many different solvents, ionic strengths, metal ions, temperatures, and oxine derivatives, and the question of which structural changes would cause deviations from a correlation even in a monodentate ligand is so complicated that a detailed interpretation of the data will not be attempted here. Nevertheless, there is good evidence that log K_1 for coordination with metals is better correlated with the sum of pK_{NH}, the acidity constant of the protonated oxine, and pK_{OH}, the acidity constant of the electrically neutral oxine, than it is with either pK alone.[59,68,70–72]

For several different metal ions log K_1 for the coordination with the conjugate bases of β-diketones of the type $RCOCH_2COR'$ (where R and R′ need not be different) has been found to be linearly correlated with the pK_a of the ketones in those cases where R and R′ are phenyl, 2-furyl, or 2-thienyl.[73] When either R or R′ is methyl, K_1 is only about half as large as it should be to fit this correlation, and when it is trifluoromethyl there are also deviations. These deviations may arise in part from the nature of the pK_a values, which were determined on materials that were mixtures of both the keto and enol forms of the β diketone and, in the cases of the trifluoromethyl compounds, must also have contained large fractions of 1,1-diols [$RCOCH_2C(OH)_2CF_3$] in the solutions in which the measurements were made. Since the enolate anions use only their oxygen atoms to coordinate with the metal ions, log K_1 would most plausibly be expected to be correlated with the Brønsted basicities of these oxygen atoms, that is, with the pK_a of the enol form. It is possible that if the enol pK_a values were available the points for the methyl and trifluoromethyl compounds would agree better with the correlation observed for the compounds with aromatic R and R′ groups but the agreement would still not be very good. The set of pK_a values required to give a good correlation with the log K_1 values for one metal ion is not the same as that required for a good correlation with another metal ion. It *could* be found that when pK_a values for the enols are used even the compounds with aromatic R and R′ groups give a poor correlation. It seems more likely, however, that the enol contents of these compounds are all about the same, so that all the pK_a(enol) values differ from pK_a(observed) by the same amount, or that the enol contents vary with the nature of R and R′ in such a way that the differences between pK_a(enol) and pK_a(observed) are a linear function of pK_a(enol). If either of these things is true a linear correlation of log K_1 with pK_a(observed) will mean that there is also a linear correlation with pK_a(enol). Although the absolute value of a, the slope of a correlation with pK_a(observed), may not be very meaningful, the relative values of a obtained for two different metal ions will be. For this reason a number of such a values are listed in Table 7-12. For some of the metal ions listed only two β-diketones, dibenzoylmethane (pK_a 13.75) and 2-furoyl-2-thenoylmethane (pK_a 12.30), were studied[74] and the correlation *assumed* to exist. The log K_1 values for coordination with dibenzoylmethide ions are also listed. Since the log K_1 values were not corrected for coordination by the anion of the metal salt, it should be noted that, as shown by the results obtained with Ni^{2+}, the metals used as their chlorides would have given larger log $K_1(Bz_2CH_2)$ values and smaller a values and those used as the perchlorates smaller log $K_1(Bz_2CH_2)$ values and larger a values than if the nitrates had been used. Considering these facts, there is a clear but imperfect tendency for the selectivity of the metal ions among various β-diketone anions, which is measured by a, to decrease with decreasing selectivity between a given diketone anion and water, which is measured by log $K_1(Bz_2CH_2)$. Perhaps one reason why this trend is clearer in the present case than was the analogous trend for the data summarized in Table 7-11 is that the basic atoms in the ligands presently under consideration are all oxygen just as the basic atom in water is.

Table 7-12. Correlations of log K_1 for Coordination of β-Diketone Anions with Metal Ions[a]

Metal Ion	a[b]	log K_1[c] (Bz_2CH_2)	Metal Ion	a[b]	log K_1[c] (Bz_2CH_2)
Be^{2+}	0.61	13.6	Zn^{2+}	0.43	10.2
Cu^{2+}	0.74[d]	13.0[d]	Pb^{2+}	0.45	9.8
Fe^{2+}	0.52	11.2	Mn^{2+}	0.35	9.3
Ni^{2+}	0.38[e]	11.1[e]	Cd^{2+}	0.30	8.7
Ni^{2+}	0.43	10.8	Mg^{2+}	0.30	8.5
Ni^{2+}	0.47[d]	10.5[d]	Ca^{2+}	0.31	7.2
Co^{2+}	0.39	10.4	Sr^{2+}	0.31	6.4

[a] Refers to compounds of the type $RCOCH_2COR'$, where R and R′ are phenyl, 2-furyl, or 2-thienyl, in 75% dioxane-25% water (by volume) using 0.01 *M* metal nitrate solutions unless otherwise noted.[73,74] Correlated with pK_a(observed) values.

[b] The value of a in Eq. 7-19 or 7-24.

[c] The value of log K_1 for dibenzoylmethane.

[d] Metal chloride used.

[e] Metal perchlorate used.

7-5. ACIDITY OF DIAZONIUM CATIONS

Most aromatic diazonium salts, when titrated with sodium hydroxide in aqueous solution, are rapidly equilibrated with the corresponding *syn*-diazohydroxides and *syn*-diazotates.

$$ArN_2^+ + OH^- \rightleftharpoons \underset{\substack{|\\ Ar}}{N}{=}N{-}OH \overset{OH^-}{\rightleftharpoons} \underset{\substack{|\\ Ar}}{N}{=}N{-}O^- \qquad (7\text{-}25)$$

Ordinarily the *syn*-diazohydroxide is more acidic than the diazonium ion from which it is formed and therefore it is never present in directly detectable concentrations. Hence the titration gives only one end point, from which an equilibrium constant K_{12} may be calculated (Eq. 7-26) but not

$$K_{12} = \frac{[ArN_2O^-]}{[ArN_2^+][OH^-]^2} \qquad (7\text{-}26)$$

the equilibrium constants for the individual steps of the reaction. Equilibrium with the *trans*-diazohydroxides and diazotates, which are considerably more stable, has been obtained, and the equilibrium constants for each step of this reaction (Eq. 7-27) have been determined in several cases.

$$ArN_2^+ + OH^- \rightleftharpoons \begin{matrix} Ar{-}N \\ \quad \diagdown \\ \qquad N{-}OH \end{matrix} \overset{OH^-}{\rightleftharpoons} \begin{matrix} Ar{-}N \\ \quad \diagdown \\ \qquad N{-}O^- \end{matrix} \tag{7-27}$$

Determination of 16 values of K_{12} for the formation of *syn*-diazotates gave a fairly good Hammett equation correlation, using ordinary σ values, and a ρ value of 6.3.[75] The reaction is particularly interesting in that one could imagine that both σ^+ and σ^- values would give a better correlation than ordinary σ values. The fact that σ values for *p*-nitro, cyano, and acetyl substituents give a much better correlation than σ^- values do shows that the *p*-nitro diazotate, for example, is not markedly stabilized by the contribution of structure **12.** Similarly, the fact that σ^+ values give a much poorer correlation than σ values for the *p*-methyl and *p*-halo substituents shows that the *p*-fluorodiazonium ion is probably not greatly stabilized by

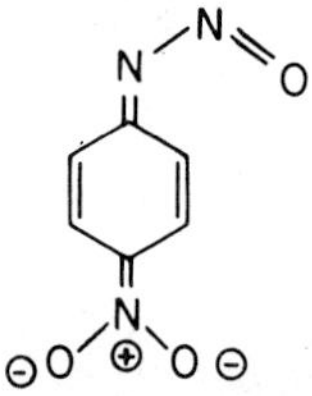

12 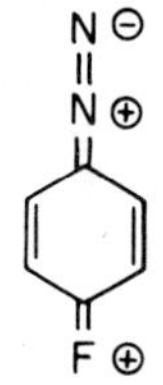 **13**

contributions of structure **13.** Since it could be argued that steric destabilization of the coplanar form of the *syn*-diazotate anion would minimize the contribution of a structure like **12** to the total structure of the anion, it is relevant that the ordinary σ value for the *p*-nitro substituent also gives a much better fit than σ^- does to the Hammett equation correlation of the acidities of *anti*-diazohydroxides, for which ρ has been found to be 1.2.[76]

If ρ for the acidity of *syn*-diazohydroxides is the same as for the *anti* isomers, ρ for the combination of aryl diazonium cations with hydroxide ions to give *syn*-diazohydroxides must be 5.1 (since the observed ρ of 6.3 is the sum of the ρ's for the two reactions that go together to give Eq. 7-25. This is not far from the values 5.5 and 4.7 that have been found for the combination of diazonium ions with sulfite ions[77] and cyanide ions,[78] respectively, to give *syn* adducts. Any interpretation of these relative values is complicated by the probability that both steric and electronic effects are important. The most strongly electron-donating substituent for which an equilibrium constant was obtained is *p*-methoxy, which was studied only in the reaction with sulfite ions. Its deviation from the Hammett equation line suggests significant contributions of a structure analogous to **13** to the total structure of the *p*-methoxybenzenediazonium ion, but not enough such contributions to make σ^+ give a better correlation than σ.

PROBLEMS

1. In the gas phase at 25° C, di-*t*-butyl peroxide, *t*-butyl hydroperoxide, *t*-butyl alcohol, and water have enthalpies of formation of −84.7, −58.6, −74.9, and −57.8 kcal/mole, respectively, and entropies of 116.1, 87.3, 76.8, and 45.1 cal/degree-mole, respectively. What is the *t*-butyl basicity of the *t*-butylperoxide anion relative to its proton basicity, both relative to hydroxide ion as the standard base, in the gas phase at 25° C? Correct the value of $K_{HA}{}^{RA}$ you calculate to remove any symmetry effects. If the corrected value is not 1.0, explain why in terms of the structures of the reactants and products.

2. The straight line in Fig. 7-3 is a plot of Eq. 7-24 using an *a* value of 0.94 and a *c* value of 4.44. Discuss the magnitudes of these two values relative to the values of *a* and *c* listed in Table 7-11.

3. For each of the reactions

$$Al^{3+} + F^- \rightarrow AlF^{2+}$$
$$Hg^{2+} + Cl^- \rightarrow HgCl^+$$

$\Delta G°$ is in the range -8.8 ± 0.4 kcal/mole. In one case, however, $\Delta H°$ is favorable, being −5.5 kcal/mole, and in the other case it is unfavorable (1.1 kcal/mole).[79] Tell which case is which and why you think so.

4. Suggest an explanation for why BH_3 is a softer acid than BF_3.

5. Use the data in Table 7-9 to estimate values of E_A and C_A for *m*-nitrophenol.

REFERENCES

1. A. J. Parker, *Proc. Chem. Soc.*, 371 (1961).
2. D. Bethell and V. Gold, *Carbonium Ions*, Academic Press, New York, 1967.
3. K. W. Egger and A. T. Cocks, *Helv. Chim. Acta*, **56**, 1516, 1537 (1973).
4. L. E. Sutton, in *Determination of Organic Structures by Physical Methods*, E. A. Braude and F. C. Nachod, Eds., Academic Press, New York, 1955, Chap. 9.
5. R. J. Goldacre and J. N. Phillips, *J. Chem. Soc.*, 1724 (1949).
6. M. Gillois and P. Rumpf, *C. R. Acad. Sci., Paris*, **238**, 591 (1954).
7. N. C. Deno, J. J. Jaruzelski, and A. Schriesheim, *J. Amer. Chem. Soc.*, **77**, 3044 (1955); N. C. Deno and W. L. Evans, *J. Amer. Chem. Soc.*, **79**, 5804 (1957).

8. E. M. Arnett and R. D. Bushick, *J. Amer. Chem. Soc.*, **86,** 1564 (1964).
9. R. Breslow, L. Kaplan, and D. LaFollette, *J. Amer. Chem. Soc.*, **90,** 4056 (1968).
10. I. I. Schuster, A. K. Colter, and R. J. Kurland, *J. Amer. Chem. Soc.*, **90,** 4679 (1968).
11. C. D. Ritchie, W. F. Sager, and E. S. Lewis, *J. Amer. Chem. Soc.*, **84,** 2349 (1962).
12. J. Mindl and M. Večeřa, *Collect. Czech. Chem. Commun.*, **36,** 3621 (1971); **37,** 1143 (1972).
13. J. C. Martin and R. G. Smith, *J. Amer. Chem. Soc.*, **86,** 2252 (1964).
14. A. Streitwieser, Jr., *Molecular Orbital Theory*, Wiley, New York, 1961, Sec. 12.2.
15. Cf. R. Breslow, *Chem. Eng. News*, **43** (26), 90 (1965); *Accounts Chem. Res.*, **6,** 393 (1973); R. Breslow and S. Mazur, *J. Amer. Chem. Soc.*, **95,** 584 (1973).
16. W. v. E. Doering and L. H. Knox, *J. Amer. Chem. Soc.*, **76,** 3203 (1954).
17. R. Breslow and J. T. Groves, *J. Amer. Chem. Soc.*, **92,** 984 (1970).
18. R. Breslow, H. Höver, and H. W. Chang, *J. Amer. Chem. Soc.*, **84,** 3168 (1962).
19. G. Naville, H. Strauss, and E. Heilbronner, *Helv. Chim. Acta.*, **43,** 1221 (1960).
20. N. C. Deno, H. G. Richey, Jr., J. S. Liu, D. N. Lincoln, and J. O. Turner, *J. Amer. Chem. Soc.*, **87,** 4533 (1965).
21. C. U. Pittman, Jr., and G. A. Olah, *J. Amer. Chem. Soc.*, **87,** 2998 (1965).
22. N. C. Deno, *Progr. Phys. Org. Chem.*, **2,** 129 (1964); *Chem. Eng. News*, **42** (40), 88 (1964).
23. G. A. Olah and C. U. Pittman, Jr., *Advan. Phys. Org. Chem.*, **4,** 305 (1966).
24. G. A. Olah, *Chem. Eng. News*, **45** (13), 77 (1967); *Science*, **168,** 1298 (1970).
25. J. Hine and R. D. Weimar, Jr., *J. Amer. Chem. Soc.*, **87,** 3387 (1965).
26. W. P. Jencks and M. Gilchrist, *J. Amer. Chem. Soc.*, **86,** 4651 (1964).
27. J. Gerstein and W. P. Jencks, *J. Amer. Chem.* Soc., **86,** 4655 (1964).
28. T. N. Hall, *J. Org. Chem.*, **29,** 3587 (1964).
29. H. C. Brown, D. H. McDaniel, and O. Häfliger, in *Determination of Organic Structures by Physical Methods*, E. A. Braude and F. C. Nachod, Eds., Academic Press, New York, 1955, pp. 634-643.
30. H. C. Brown and B. Kanner, *J. Amer. Chem. Soc.*, **75,** 3865 (1953); **88,** 986 (1966).
31. P. Love, R. B. Cohen, and R. W. Taft, *J. Amer. Chem. Soc.*, **90,** 2455 (1968).
32. H. C. Brown and R. R. Holmes, *J. Amer. Chem. Soc.*, **78,** 2173 (1956).
33. S. Ahrland, J. Chatt, and N. R. Davies, *Quart. Rev.* (London), **12,** 265 (1958).
34. H. C. Brown and R. M. Adams, *J. Amer. Chem. Soc.*, **64,** 2557 (1942).
35. H. L. Morris, N. I. Kulevsky, M. Tamres, and S. Searles, Jr., *Inorg. Chem.*, **5,** 124 (1966).
36. A. B. Burg and A. A. Green, *J. Amer. Chem. Soc.*, **65,** 1838 (1943).

37. H. C. Brown, *J. Chem. Soc.*, 1248 (1956).
38. R. H. Harris, Ph. D. Thesis, Purdue University, 1952, as quoted in the first citation of Ref. 57.
39. W. A. G. Graham and F. G. A. Stone, *J. Inorg. Nucl. Chem.*, **3,** 164 (1956).
40. H. I. Schlesinger and A. B. Burg, *J. Amer. Chem. Soc.*, **60,** 290 (1938).
41. L. G. Sillén and A. E. Martell, *Stability Constants of Metal-Ion Complexes*, Special Publication No. 17, The Chemical Society, London, 1964.
42. L. G. Sillén and A. E. Martell, *Stability Constants of Metal-Ion Complexes, Supplement No. 1*, Special Publication No. 25, The Chemical Society, London, 1971.
43. R. G. Pearson, *J. Amer. Chem. Soc.*, **85,** 3533 (1963).
44. R. G. Pearson, *Science*, **151,** 172 (1966).
45. R. G. Pearson, *Chem. Brit.*, **3,** 103 (1967).
46. R. G. Pearson, in *Survey of Progress in Chemistry*, Vol. 5, A. F. Scott, Ed., Academic Press, New York, 1969, Chap. 1.
47. L. Pauling, *The Nature of the Chemical Bond*, 2nd ed., Cornell University Press, Ithaca, N.Y., 1945, Sec. 31.
48. P. Ford, D. F. P. Rudd, R. Gaunder, and H. Taube, *J. Amer. Chem. Soc.*, **90,** 1187 (1968).
49. H. Taube, in *Survey of Progress in Chemistry*, Vol. 6, A. F. Scott, Ed., Academic Press, New York, 1973, Chap. 1.
50. R. E. Clarke and P. C. Ford, *Inorg. Chem.*, **9,** 495 (1970).
51. R. E. Shepherd and H. Taube, *Inorg. Chem.*, **12,** 1392 (1973).
52. R. S. Drago and B. B. Wayland, *J. Amer. Chem. Soc.*, **87,** 3571 (1965).
53. R. S. Drago, *Chem. Brit.*, **3,** 516 (1967).
54. R. S. Drago, G. C. Vogel, and T. E. Needham, *J. Amer. Chem. Soc.*, **93,** 6014 (1971).
55. G. Olofsson and I. Olofsson, *J. Amer. Chem. Soc.*, **95,** 7231 (1973).
56. M. S. Nozari and R. S. Drago, *J. Amer. Chem. Soc.*, **94,** 6877 (1972).
57. D. P. N. Satchell and R. S. Satchell, *Chem. Rev.*, **69,** 251 (1969); *Quart. Rev.* (London), **25,** 171 (1971).
58. H. Irving and H. S. Rossotti, *Acta Chem. Scand.*, **10,** 72 (1956).
59. F. J. C. Rossotti in *Modern Coordination Chemistry*, J. Lewis and R. G. Wilkins, Eds., Interscience, New York, 1960, Chap. 1.
60. J. R. Chipperfield, in *Advances in Linear Free Energy Relationships*, N. B. Chapman and J. Shorter, Eds., Plenum Press, New York, 1972, Chap. 7.
61. R. J. Bruehlman and F. H. Verhoek, *J. Amer. Chem. Soc.*, **70,** 1401 (1948).
62. M. S. Sun and D. G. Brewer, *Can. J. Chem.*, **45,** 2729 (1967).
63. A. G. Desai and M. B. Kabadi, *J. Indian Chem. Soc.*, **38,** 805 (1961).
64. A. G. Desai and M. B. Kabadi, *J. Inorg. Nucl. Chem.*, **28,** 1279 (1966).

65. K. L. Brown, D. Chernoff, D. J. Keljo, and R. G. Kallen, *J. Amer. Chem. Soc.*, **94,** 6697 (1972).
66. K. L. Brown and R. G. Kallen, *J. Amer. Chem. Soc.*, **94,** 1894 (1972).
67. A. G. Desai and R. M. Milburn, *J. Amer. Chem. Soc.*, **91,** 1958 (1969).
68. J. G. Jones, J. B. Poole, J. C. Tomkinson, and R. J. P. Williams, *J. Chem. Soc.*, 2001 (1958).
69. K. Clarke, R. A. Cowen, G. W. Gray, and E. H. Osborne, *J. Chem. Soc.*, 245 (1963).
70. J. C. Tomkinson and R. J. P. Williams, *J. Chem. Soc.*, 1153 (1958).
71. H. Irving and H. S. Rossotti, *J. Chem. Soc.*, 2910 (1954).
72. G. Gutnikov and H. Freiser, *Anal. Chem.*, **40,** 39 (1968).
73. L. G. Van Uitert, W. C. Fernelius, and B. E. Douglas, *J. Amer. Chem. Soc.*, **75,** 457 (1953).
74. L. G. Van Uitert, W. C. Fernelius, and B. E. Douglas, *J. Amer. Chem. Soc.*, **75,** 2736 (1953).
75. E. S. Lewis and H. Suhr, *Chem. Ber.*, **91,** 2350 (1958).
76. J. S. Littler, *Trans. Faraday Soc.*, **59,** 2296 (1963).
77. E. S. Lewis and H. Suhr, *Chem. Ber.*, **92,** 3031 (1959).
78. E. S. Lewis and H. Suhr, *Chem. Ber.*, **92,** 3043 (1959).
79. G. Schwarzenbach, *Pure Appl. Chem.*, **24,** 307 (1970).

8

Equilibrium in Additions to and Migrations of Multiple Bonds

8-1. EQUILIBRIUM IN CARBONYL ADDITION REACTIONS

The addition of almost anything to a carbon-oxygen double bond gives a new group that is less electron withdrawing and more space filling than the original carbonyl group. For this reason we would expect equilibrium constants for carbonyl addition reactions to be increased by electron withdrawing, and decreased by electron donating and by bulky substituents on the carbonyl compound. In addition, we expect a decrease in the equilibrium constant to result from the loss of resonance stabilization that would accompany addition to a carbonyl group that is conjugated with an unshared pair of electrons or a system of π electrons.

Taft and Kreevoy correlated the *relative* equilibrium constants for addition of hydrogen to carbonyl groups by treating the equilibrium constants determined by Adkins and co-workers for the Meerwein-Ponndorf-Oppenauer equilibrium.[1,2] The equilibrium constants for the reduction of 17 ketones (none of which contained substituents conjugated with the carbonyl group) by isopropyl acohol (Eq. 8-1) gave a rather scattered plot

$$R-\overset{O}{\overset{\|}{C}}-R' + Me_2CHOH \rightleftharpoons R-\overset{OH}{\overset{|}{C}}H-R' + Me_2CO \qquad (8\text{-}1)$$

when the simple Taft equation was used. It was noted that a markedly improved correlation could be obtained by assuming that the starting ketones were stabilized by hyperconjugation and that the magnitude of this stabilization was simply proportional to the number of hydrogen atoms available for hyperconjugation. This assumption may be expressed in terms of Eq. 8-2, in which n is the number of α-hydrogen atoms in the ketone

$$\log K = \rho^*(\sigma_R^* + \sigma_R'^*) + (6 - n)h \tag{8-2}$$

(the "6" is the number in the reference compound, acetone), and the best values of ρ^* and h were determined by the method of least squares to be -4.2 and 0.35, respectively. Before accepting this correlation as strong evidence for the necessity of invoking hyperconjugation to explain the data, an alternative explanation should be considered. Ritchie showed that the Taft equation with the σ^* values listed in Table 3-6 and used by Taft and Kreevoy displayed a type of internal inconsistency that could be removed if the σ^* values for saturated alkyl groups and hydrogen were set equal to zero (Section 3-6b). If such σ^* values are used in correlating the data of Adkins and co-workers the points for the three aldehydes fall near each other but far from the line described by the points for the 14 ketones. The deviation of the aldehydes is an example of the rather common tendency for the hydrogen substituent to produce deviations from data for other substituents. The correlation of the ketone data obtained using σ^* values of zero for alkyl groups is about as good as that obtained using Eq. 8-2 with ordinary σ^* values and the hypothesis of hyperconjugation. Hence it may be that Eq. 8-2 works largely because the hyperconjugation term provides a way of compensating for the differences among the σ^* constants for the various alkyl groups. Unfortunately, interpretation of the data is further complicated by the possibility that steric effects may be significant and the fact that the values of σ^* for the substituents used do not cover as wide a range as would be desired. (The only nonhydrocarbon substituent was the methoxymethyl group.)

By comparing the equilibrium constants for the reduction of such carbonyl compounds as crotonaldehyde, benzaldehyde, and acetophenone with the values that may be calculated from Eq. 8-2, Kreevoy and Taft concluded that conjugation of the carbonyl group with a phenyl group or a carbon-carbon double bond leads to a resonance stabilization of about 6 kcal/mole.[3]

The fraction of a carbonyl compound that is hydrated at equilibrium in aqueous solution (Eq. 8-3) is a particularly significant number in view of the importance of water as a solvent. The simplest aldehyde, formaldehyde,

$$\text{R—}\overset{\text{O}}{\overset{\|}{\text{C}}}\text{—R}' + \text{H}_2\text{O} \rightleftharpoons \text{R—}\underset{\text{OH}}{\underset{|}{\overset{\text{OH}}{\overset{|}{\text{C}}}}}\text{—R}' \tag{8-3}$$

is about 99.95% hydrated in aqueous solution at 25° C,[4] and when electron-withdrawing substituents are present, as in the case of chloral, for example, the extent of hydration is probably even larger.[5] In contrast, the simplest ketone, acetone, is only about 0.14% hydrated,[6] and aromatic and α,β-unsaturated ketones are probably even less hydrated. Bell[5] correlated the data for the hydration of 11 aldehydes and ketones in water at 25° C in terms of a relationship that, when written for the reaction in the direction shown in Eq. 8-3 and expressed in terms of dimensionless equilibrium constants, becomes Eq. 8-4, in which the E_s values are Taft's steric sub-

$$\log K = -2.70 + 2.6\,(\sigma_R{}^* + \sigma_R{}'^*) + 1.3\,(E_s{}^R + E_s{}^{R'}) \qquad (8\text{-}4)$$

stituent constants. A more precise correlation could probably have been obtained by adding another term to allow for hyperconjugation, but it is not clear how meaningful this would have been. The smaller absolute value of ρ^* (2.6) obtained in the hydration reaction than that (4.2) obtained in the hydrogenation reaction is largely a reflection of the fact that the —$C(OH)_2$— group is a considerably stronger electron withdrawer than —CHOH— is.

In the addition of hydrogen cyanide to meta- and para-substituted benzaldehydes, where complications arising from steric effects should be negligible and those arising from hyperconjugation very small, the equilibrium constants of Baker and co-workers in 95% ethanol at 20° C[7] give a ρ of 1.49.[8]

The equilibrium constants for addition of hydrogen cyanide to carbonyl compounds listed in Table 8-1 include examples in which steric effects are clearly important. Wheeler's study of alkyl cyclohexanones shows that 2-alkyl substituents decrease the equilibrium constant to an extent that increases as the size of the alkyl group increases.[9] Alkyl substituents also decrease the stability of the adduct from the 3-position, where they can be involved in an axial-axial interaction with either the hydroxy or the cyano group. In the 4-position, however, there is no significant effect. The fact that 3,3-dimethylation destabilizes the cyanohydrin of cyclobutanone more than 2,2,4,4-tetramethylation does was rationalized in terms of the nonplanar structure characteristic of most cyclobutane derivatives.[10] In such a structure, as shown in **1** there will be an axial-axial interaction between either the hydroxy or cyano group and one of the 3-substituents.

Bond angle strain is probably the main factor that causes the equilibrium constant for addition to cyclobutanone to be larger than that for simple aliphatic ketones. As Brown and co-workers pointed out, the average internal bond angle in a cyclobutane ring must be 90° if it is planar and less than 90° if it is nonplanar; therefore, an sp^2-hybridized carbon atom in the ring will tend to have its ring bond angle strained by about 30°, whereas an sp^3-hybridized carbon atom will have only about 20° of strain. Hence

Table 8-1. Equilibrium Constants for Addition of Hydrogen Cyanide to Carbonyl Compounds[a]

Carbonyl Compound	K (M^{-1})
Acetaldehyde	7100[b,c]
Acetone	28[d,e]
Methyl ethyl ketone	33[d,e]
Methyl isopropyl ketone	52[d,e]
Methyl *t*-butyl ketone	29[d,e]
Isopropyl *t*-butyl ketone	0.12[e,f]
Dioctyl ketone	14
Cyclohexyl methyl ketone	56[d,e]
Acetophenone	0.67[d,e]
Benzaldehyde	200[d,g]
Cyclobutanone	108[h]
3,3-Dimethylcyclobutanone	4.7[h]
2,2,4,4-Tetramethylcyclobutanone	37[h]
Cyclopentanone	34[h]
2,2,4,4-Tetramethylcyclopentanone	5.3[h]
Cyclohexanone	1700[i]
2-Methylcyclohexanone	930[i]
2-Ethylcyclohexanone	280[i]
2-Isopropylcyclohexanone	59[i]
2-*t*-Butylcyclohexanone	7.6[i]
3-*t*-Butylcyclohexanone	156[i]
4-*t*-Butylcyclohexanone	1700[i]
Cycloheptanone	7.7
Cyclooctanone	1.2
Cyclononanone	0.59
Cyclodecanone	Too small to measure
Cycloundecanone	0.89
Cyclododecanone	3.2
Cyclotridecanone	3.8
Cyclotetradecanone	17
Cyclopentadecanone	9.0
Cyclohexadecanone	11
Cycloheptadecanone	8.3
Cyclooctadecanone	10
Cyclononadecanone	10

[a] Data from Ref. 13 in 95% ethanol at 22–23° C unless otherwise stated.

[b] In water at 25° C; data on other compounds suggest that K would be larger, by as much as 50%, in 95% ethanol.

[c] W. F. Yates and R. L. Heider, *J. Amer. Chem. Soc.*, **74**, 4153 (1952).

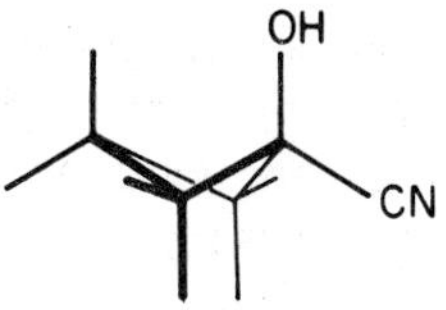

OH

CN

1

reactions in which an sp^2 carbon atom in a cyclobutane ring is transformed to an sp^3 carbon atom will tend to be favored relative to the corresponding reaction of an analogous acyclic compound.[11,12] Acting in the opposite direction is the tendency for the transformation of an sp^2 to an sp^3 carbon atom in a cyclobutane ring to increase the bond eclipsing. Although this factor must be present it does not usually seem to be dominant in reactions that interchange sp^2 and sp^3 carbon atoms in a cyclobutane ring. In the case of cyclohexanone, however, bond eclipsing is often a major factor. Addition to cyclohexanone gives a species in which there is nearly perfect staggering of bonds and nearly optimum bond angles. In the starting material these two requirements for maximum stability cannot be met simultaneously. Hence the equilibrium constant for cyanohydrin formation will be larger than in the case of an acyclic compound in which optimum staggering and optimum bond angles can be closely approached in both reactants and products. The *internal* ring strains from various sources that have been discussed are sometimes referred to as *I strain*.[11,12] In the case of a cyclopentane ring, rehybridization of a ring atom from sp^3 to sp^2 should increase the angle strain since the internal angles in a five-member ring can average no more than 108°. The same rehybridization should tend to decrease the eclipsing strain. In the planar form of either cyclopentane or cyclopentanone every bond in the ring is in its highest energy conformation. However, since the rotational barrier in acetone is smaller than that in ethane or propane the amount of torsional strain in planar cyclopentanone should be less than that in planar cyclopentane. If cyclopentane is not more capable of relieving this strain by bond angle distortions than cyclopentanone is, this difference in strain energies will also be present for the actual nonplanar molecules. The fact that cyclopentanone adds hydrogen

---

[d] Interpolated between data at 20 and 35° C.

[e] D. P. Evans and J. R. Young, *J. Chem. Soc.*, 1310 (1954).

[f] At 35° C; the value at 22–23° C would probably be about twice as large.

[g] Data from Ref. 7.

[h] Data from Ref. 10.

[i] Data from Ref. 9.

cyanide to about the same extent at equilibrium as acetone and other simple aliphatic ketones may be rationalized by assuming that the activation of cyclopentanone arising from bond angle strain is of about the same magnitude as the deactivation arising from torsional energy effects. In certain other reactions of cyclopentane derivatives the torsional energy effects appear to be dominant.

The studies of Prelog and Kobelt[13,14] on cycloalkanones containing 6- to 20-membered rings show that the equilibrium constants for cyanohydrin formation are much smaller than for a simple aliphatic ketone in the C_7 through C_{13} cases, the minimum being reached in the case of cyclodecanone, where the constant was too small to measure. These decreases in K are believed to result from simple steric hindrance and, in many cases, unfavorable torsional energy effects. The larger cycloalkanones contain enough atoms in the ring to give almost as much conformational freedom as in an aliphatic ketone. Hence it is not surprising that their equilibrium constants for cyanohydrin formation are not much different from that for dioctyl ketone.

Equilibrium constants for additions to carbonyl compounds depend not only on the structure of the carbonyl compound but also on that of the species that is adding to it. Sander and Jencks measured and tabulated a large number of equilibrium constants for addition of various species to formaldehyde, *p*-chlorobenzaldehyde, and pyridine-4-carboxaldehyde.[15] From their data, much of which is listed in Table 8-2, it may be seen that the values of K for addition to *p*-chlorobenzaldehyde vary with the structure of the addend in essentially the same way as do the values of K for addition to pyridine-4-carboxaldehyde. Since the aldehyde group in each of these compounds is attached directly to an aromatic ring and steric factors are essentially constant, this similarity in behavior, which may be illustrated by the straight line of slope 1.00 obtained in a log-log plot of the equilibrium constants, is not surprising. Comparison of the K values for either of the compounds with those for formaldehyde reveals some differences in trends, at least some of which are probably due to steric hindrance. Thus, for example, methanol adds to formaldehyde three times as well as isopropyl alcohol does, but there is more than a sevenfold difference in the ability of these two alcohols to add to the more sterically encumbered pyridine-4-carboxaldehyde. Another difference may be seen if the values of log K for formaldehyde are plotted against those for either of the two aromatic aldehydes. The points tend to approximate two lines, one for the oxygen and sulfur addends and one, displaced by about 1.3 log units toward larger log K values for formaldehyde, for nitrogen addends. If the free energies of formation of all the species involved under the reaction

Table 8-2. Equilibrium Constants for Additions to Aldehydes[a]

	K (M^{-1})		
Addend	CH_2O	p-ClC_6H_4CHO	NC_5H_4-4-CHO[b]
$(H_2N)_2CO$	5.8×10^4 [c]		1.3[c]
Morpholine	1.8×10^6	0.40	32
Piperidine	3.6×10^6	0.45	
$EtNH_2$	3.9×10^6		43
$MeNH_2$	3.4×10^6		87
$H_2NCONHNH_2$		3.1	250
H_2NNH_2		7.0[c]	
H_2NOH		24	1500
H_2O	41		0.023
CF_3CH_2OH			<0.05
i-PrOH	4.1×10^2		<0.07
EtOH	8.6×10^2		0.27
MeOH	1.3×10^3		0.50
$(CH_3)_2C{=}NOH$			0.80
MeOOH		0.35	12.1
H_2O_2	3.5×10^4	0.44[c]	19.8[c]
$HOCH_2CH_2SH$	1.4×10^6	2.3	193
HCN		300	
HSO_3^-	1×10^{10} [d]	11,400	

[a] From Ref. 15. In water at 20–25° C. Corrected for any competing hydration of the aldehyde.
[b] Pyridine-4-carboxaldehyde.
[c] Obtained by dividing the observed value by a statistical factor of 2.
[d] Calculated from data given by A. Skrabal and R. Skrabal, *Monatsh. Chem.*, **69**, 11 (1936) and P. E. Sørensen and V. S. Andersen, *Acta Chem. Scand.*, **24**, 1301 (1970).

conditions could be correlated in terms of bond contributions plus interactions among adjacent bonds, as in the correlation of enthalpies of formation for which contributions are listed in Table 1-3, the straight line plot of log K values for pyridine-4-carboxaldehyde versus those for p-chlorobenzaldehyde would follow automatically, but it would also follow that a plot of log K values for either of these aromatic aldehydes versus those for formaldehyde would give three lines (some of which might accidentally coincide) for the oxygen, nitrogen, and sulfur addends. The location of the observed lines would require that the interaction of an aromatic carbon and a nitrogen atom attached to the same carbon be less stabilizing than that of the aromatic carbon and an oxygen or sulfur atom attached to the same carbon. There appears to be too little thermochemical data available to evaluate all these contributions, or even the analogous contributions to thermodynamic properties in the gas phase.

In comparing the relative abilities of various species to add to a given aldehyde it may be noted that all the additions studied by Sander and Jencks were reactions in which a hydrogen atom is added to the oxygen atom of the carbonyl group and the rest of the species added to the carbon atom. Hence K for addition of HX to RCHO divided by K for addition of HY to RCHO is simply the equilibrium constant for reaction 8-5. Such equilibrium constants may be discussed in terms of carbon basicity (Sec-

$$\mathrm{HX} + \underset{\substack{|\\ \mathrm{OH}}}{\mathrm{RCHY}} \rightleftharpoons \mathrm{HY} + \underset{\substack{|\\ \mathrm{OH}}}{\mathrm{RCHX}} \tag{8-5}$$

tion 7-1c). Thus the equilibrium constant for addition of HX to an aldehyde will increase with increasing carbon basicity of X^-. For example, other factors being equal, the equilibrium constant should decrease with increasing electronegativity of the atom in X that is bonded to hydrogen in HX and to carbon in RCH(OH)X. In agreement with this generalization, K values for simple alcohols in Table 8-2 are seen to be smaller than those for primary amines or for a mercaptan or hydrogen cyanide. In discussing carbon basicity in the present connection it is important to remember that in the formation of RCH(OH)X it is the basicity of X^- toward a carbon atom with an α-hydroxy substituent that is being considered. Hence in terms of correlations of thermodynamic properties using bond and interaction contributions (Section 1-2d), the interaction between this oxygen atom and the atom by which X is attached to carbon should be considered, as was done in discussing the equilibrium constant for Eq. 7-11. In the absence of parameters for correlation of free energies in aqueous solution, let us use contributions to enthalpies of formation in the gas phase. From the bond contributions of Table 1-3, a nitrogen adduct gains an advantage of 4.28 kcal/mole (corresponding to a 1400-fold effect on K at 25° C) over an oxygen adduct. However, interaction terms of -13.09 and about -10 kcal/mole for Γ_{OCO} and Γ_{OCN}, respectively (Table 1-3), decrease this advantage to about 1.1 kcal/mole. With any addends other than water and ammonia there are other interactions to consider. Thus when methanol adds to an aldehyde a new interaction of two carbon atoms across an oxygen is created. The value of -5.90 kcal/mole for Γ_{COC} corresponds to stabilization and thus qualitatively agrees with the larger K values observed for methanol than for water, but it greatly overestimates the magnitude of the difference. The relatively large values of K shown in Table 8-2 for hydrazine and hydrogen peroxide are qualitatively consistent with the rather large negative values of Γ_{CNN} (−5.8 kcal/mole) and Γ_{COO} (−8.63 kcal/mole), but there are probably significant differences between parameters referring to the gas phase and parameters referring to aqueous solution, especially in the case of oxygen and nitrogen compounds, where hydrogen bonding and other polar interactions may be so important.

8-2. ENTHALPIES OF ADDITION TO CARBON-CARBON MULTIPLE BONDS

In discussing the relative stabilities of various olefins, as in other relative-stability discussions, it is important to keep the reference point(s) in mind. Statements about relative stabilities are most useful when they are most general, but an increase in the scope of a statement is often accompanied by a decrease in its validity.

Consider, for example, the enthalpies of addition of bromine[16] listed in Table 8-3. These additions tend to become increasingly exothermic as the number of alkyl groups attached to the carbon-carbon double bond is increased. Such data might be used as the basis for a rather broad generaliza-

Table 8-3. Enthalpies of Addition to Olefins[a]

	$\Delta H°$ (kcal/mole)	
Olefin	Br_2[b]	H_2[c]
Ethylene	29.1	32.8
Propylene	29.4	30.1
1-Butene	29.6	30.3
cis-2-Butene	30.2	28.6
trans-2-Butene	29.1	27.6
Isobutene		28.4
2-Methyl-2-butene	30.4	26.9
2,3-Dimethyl-2-butene		26.6
1-Heptene		30.1

[a] In the gas phase at 82° C.
[b] Ref. 16.
[c] Refs. 17–19.

tion that the stability of olefins with respect to saturated compounds decreases with an increase in the number of alkyl groups on the carbon-carbon double bond. However, the enthalpies of hydrogenation[17–19] shown could lead to an exactly reversed generalization. The addition of hydrogen becomes less exothermic as the number of alkyl groups attached to the carbon-carbon double bond is increased. Hence it is desirable to narrow the excessively broad generalizations that might have been made. We might say that the stability of olefins relative to their dibromides decreases with an increase in the number of alkyl groups on the double bond. Then on the basis of the enthalpies of hydrogenation we might generalize: the stability of olefins relative to the corresponding alkanes increases with an increase in the number of alkyl groups on the double bond. Although there is no ironclad guarantee that such a generalization will be applicable beyond the group of compounds for which we have data, it provides a useful basis for making predictions in new cases. Consider, for example, the question of whether 1-octene or *trans*-2-octene is more stable. From the fact that the enthalpies of hydrogenation of propylene, 1-butene, and 1-heptene are all equal, within the experimental uncertainty, we may infer that any compound of the type $RCH{=}CH_2$, where R is a saturated straight-chain alkyl group, will have an enthalpy of hydrogenation of about 30.2 kcal/mole and hence that 1-octene will have an enthalpy of hydrogenation of this magnitude. Inasmuch as the enthalpy of hydrogenation of *trans*-2-butene differs from that of propylene by about the same amount that the enthalpy of hydrogenation of propylene differs from that of ethylene, we may infer that the effect of a second alkyl group on the enthalpy of hydrogenation will be about the same as that of the first, if the second alkyl group is placed on the other carbon atom trans to the first alkyl group so as to minimize steric interactions. From these two inferences it follows that the enthalpy of hydrogenation of *trans*-2-octene should be about the same as that of *trans*-2-butene. We now have measures (or estimates of measures) of the stabilities of 1-octene and *trans*-2-octene, relative to the corresponding alkane, which is the same compound for each of the two olefins. It therefore follows that if the enthalpies of hydrogenation are as estimated, 30.2 and 27.6 kcal/mole, respectively, then 1-octene is 2.6 kcal/mole less stable than *trans*-2-octene.

If an organic chemist states that *trans*-2-butene is more stable than 1-hexene he might mean "*trans*-2-butene is more stable relative to butane than 1-hexene is relative to hexane" (it is), or he might mean (if he was discussing the dehydration of alcohols) "trans-2-butene is more stable relative to 2-butanol than 1-hexene is relative to 2-hexanol" (it almost certainly is), or possibly "*trans*-2-butene is more stable relative to 2-

butanol than 1-hexene is relative to 1-hexanol" (it probably is not). If the organic chemist meant what he said literally he would be referring to the reaction

$$3\ trans\text{-}CH_3CH{=}CHCH_3 \rightleftharpoons 2CH_2{=}CHC_4H_9\text{-}n \qquad (8\text{-}6)$$

which has not been studied experimentally but for which the thermodynamics of reaction may be calculated from the thermodynamics of formation of the reactant and product. In this case, as in any reaction in which there is a change in the number of molecules, the free energy change is a rather arbitrary measure of the relative stabilities of reactants and products because the free energy change for the reaction may be made either positive or negative depending on the choice of the standard state. $\Delta H°$ for reaction 8-6 is -11.9 kcal/mole in the gas phase at 25° C, and thus in terms of an enthalpy definition of stability 1-hexene is more stable than *trans*-2-butene.

The fact that the addition of bromine to olefins becomes more exothermic as the number of alkyl groups on the carbon-carbon double bond increases whereas the reverse is true for the addition of hydrogen may be explained in terms of the stabilization that accompanies the attachment of bromine and carbon to the same saturated carbon (cf. Section 1-2d); that is, tertiary alkyl bromides are more stable than the isomeric secondary bromides, which are in turn more stable than primary bromides.

It would probably be possible to select olefins of such structures as to violate the generalization that increasing alkylation of the carbon-carbon double bond increases the exothermicity of the addition of bromine. There are certainly exceptions to the generalization that the most stable of several isomeric olefins with the same carbon skeleton will be the one with the largest number of alkyl groups on the double bond. For example, the enthalpies of hydrogenation of 2,4,4-trimethyl-1-pentene and 2,4,4-trimethyl-2-pentene show that the latter is less stable by about 1.2 kcal/mole,[20] a result that is supported by direct equilibration experiments.[21] Brown and Berneis attributed the decreased stability of the latter olefin to steric repulsions between the *t*-butyl and methyl groups that are cis to each other.[22]

```
        H           CH3
          \        /
   CH3     C ==== C
      \   /        \
        C           CH3
      /   \
   CH3     CH3
```

Since sp^2-hybridized carbon is electron withdrawing relative to sp^3-hybridized carbon, electron-withdrawing substituents would be expected to destabilize alkenes relative to the corresponding alkanes. Evidence that this is the case has been provided by Taft and Kreevoy, who correlated the enthalpies of hydrogenation of olefins of the type *trans*-RCH=CHR′ relative to that of *trans*-2-butene; that is, they correlated the enthalpies of reaction 8-7.[1] For substituents for which there was no conjugation of

$$\underset{R}{\overset{H}{}}\!\!\!>C{=}C<\!\!\!\underset{H}{\overset{R'}{}} + \underset{Me}{}CH_2CH_2\overset{Me}{} \rightarrow \underset{R}{}CH_2CH_2\overset{R'}{} + \underset{Me}{\overset{H}{}}\!\!\!>C{=}C<\!\!\!\underset{H}{\overset{Me}{}} \qquad (8\text{-}7)$$

aromatic rings, multiple bonds, or unshared electrons with the reaction site, the data were correlated by Eq. 8-8, in which n is the number of

$$\Delta H_{(8\text{-}7)} = -2.41\,(\sigma_R{}^* + \sigma_R{}'^*) + 0.44\,(6 - n) \qquad (8\text{-}8)$$

hydrogen atoms available for hyperconjugation with the double bond and the value 6 refers to the fact that there are 6 such hydrogens in the reference compound, *trans*-2-butene. As in the case of the similar correlation of the equilibrium constants for reduction of carbonyl compounds, the data may be treated alternatively in terms of a set of substituent constants in which all saturated alkyl groups and hydrogen have σ^* values of zero. When this is done and hyperconjugation is ignored, the data for hydrogen substituents deviate widely, but those for other substituents give about as good a correlation as that obtained with Eq. 8-8. When polar effects were allowed for by use of Eq. 8-8, aromatic rings, carbon-carbon double bonds, and carbonyl groups that were conjugated with the double bond being reduced were found to give from 4 to 7 kcal/mole of resonance stabilization.[3]

Other factors that influence the stabilities of carbon-carbon multiple bonds may be deduced from the enthalpies of hydrogenation[17–20,23,24] listed in Table 8-4. The first three entries show that alkyl substituents stabilize acetylenes (relative to the corresponding olefins) to about the same extent that they stabilize olefins (relative to the corresponding alkanes). The next four entries show that when two double bonds are on the same carbon atom they are relatively unstable, when they are conjugated they are relatively stable, and when they are separated farther they are about as stable as double bonds in monoolefins. The heat of hydrogenation of ethyl vinyl ether shows that the ethoxy substituent stabilizes the double bond greatly, presumably because of conjugation of the unshared electrons on oxygen with the double bond. When it is realized that the highly electronegative oxygen substituent should destabilize the double bond markedly by a polar effect, it is seen that this conjugation effect must be quite important. This stabilization is seen to be smaller (per double bond) in divinyl

Table 8-4. Enthalpies of Hydrogenation[a]

	$-\Delta H^{\circ}$[b] (kcal/mole)		$-\Delta H^{\circ}$[c] (kcal/mole)
Acetylene	42.2	*cis*-Cyclooctene	23.0
Propyne	39.6	*trans*-Cyclooctene	32.2
2-Butyne	37.6[d]	*cis*-Cyclononene	23.6
Allene	41.2	*trans*-Cyclononene	26.5
1,3-Butadiene	26.7	*cis*-Cyclodecene	20.7
1,4-Pentadiene	30.7	*trans*-Cyclodecene	24.0
1,5-Hexadiene	30.4	1,2-Dimethylcyclopropene	43.3
Ethyl vinyl ether	26.7	1-Methylcyclobutene	28.5
Divinyl ether	30.5	1,2-Dimethylcyclobutene	26.5
Vinyl acetate	31.1	Methylmethylenecyclopropane	38.3
2-Ethoxypropene	25.1	Methylenecyclobutane	29.4
Styrene	27.6[e]	Methylenecyclopentane	26.9
Indene	24.1[e]	Methylenecyclohexane	27.8
Cyclopentene	26.9	Methylenecycloheptane	26.3
1,3-Cyclopentadiene	24.0	Bicyclo[2.2.1]heptene	33.1
Cyclohexene	28.6	Bicyclo[2.2.1]heptadiene	35.0
1,3-Cyclohexadiene	26.7	Bicyclo[2.2.2]octene	28.2
Cycloheptene	26.4	Bicyclo[2.2.2]octadiene	28.0

[a] For addition of one mole of hydrogen.
[b] Data from Refs. 17–19 in the gas phase at 82° C.
[c] Data from Refs. 20, 23, and 24 in acetic acid at 25° C. The reactions would be about 1.0 ± 0.5 kcal/mole more exothermic in the gas phase at 82° C.
[d] For reduction to *trans*-2-butene.
[e] For reduction of the nonaromatic double bond.

ether, in which interaction of the unshared electrons with either double bond renders them less available for interaction with the other double bond, and still smaller in vinyl acetate, in which the unshared electrons must be greatly involved in conjugation with the strongly electron-withdrawing carbonyl group.

The reduction in angle strain associated with changing the hybridization of an atom in a small ring from sp^2 to sp^3 [11,12] is probably the dominant factor in making 1,2-dimethylcyclopropene, in which there are two sp^2-hybridized ring atoms (or, more precisely, two atoms of a type that would be sp^2-hybridized if the ring structure were not further modifying their hybridization), so unstable. The instability of methylmethylenecyclopropane, in which there is only one sp^2-hybridized ring atom, is less but still pronounced. This angle strain effect is still present in the cyclobutane series but, as expected, it is much smaller than in the cyclopropane series.

It also appears that certain steric effects in the cyclobutane series differ significantly from the analogous effects in acyclic compounds. Comparison of such enthalpies of hydrogenation as those of ethylene, propylene, and *trans*-2-butene (Table 8-3) suggests that replacement of a hydrogen atom attached to a double bond by a primary alkyl group increases the stability of the alkene (relative to that of the corresponding alkane) by about 2.6 kcal/mole if there is no other alkyl group on the same carbon atom or on the other carbon atom of the double bond and cis to the new alkyl group. The additional methylation of the double bond that occurs on going from *trans*-2-butene to 2-methyl-2-butene decreases the enthalpy of hydrogenation by only 0.7 kcal/mole, presumably because of steric repulsions between the new methyl group and the two old methyl groups that are near it. If we now go from 2-methyl-2-butene to 2,3-dimethyl-2-butene the decrease in enthalpy of hydrogenation is seen to be only 0.3 kcal/mole. In this case the new methyl group again crowds two old methyl groups, but now the old methyl groups have less freedom to move away from the new methyl group. Such crowding, which is a function of angles α and β in structure **2**, should be considerably smaller in 1,2-dimethylcyclobutene (**3**), in which the approximately 90° ring bond angle makes angles α and β

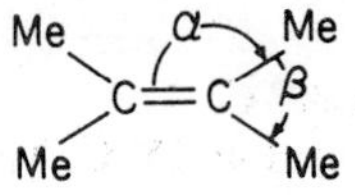

Me, Me, α, Me, β, Me — C=C

2

Me, α, Me, C=C, β, H_2C—CH_2

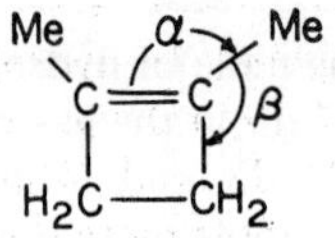

3

much larger. It is therefore not surprising that the introduction of another methyl group into 1-methylcyclobutene to give 1,2-dimethylcyclobutene is accompanied by a 2.0 kcal/mole decrease in the enthalpy of hydrogenation.

With certain cycloalkenes, such as the cyclodecenes, the cycloalkanes appear to be more strained (because of torsional strains and steric repulsions) than the cycloalkenes.

8-3. EQUILIBRIUM IN THE MIGRATION OF CARBON-CARBON DOUBLE BONDS

The enthalpies of hydrogenation discussed in the preceding section provide useful information concerning the effects of substituents on the stabilities of unsaturated compounds but they suffer from three disadvantages.

Data are not available for a very wide variety of substituents. The data may be used to calculate equilibrium constants only by neglecting entropy effects or by estimating them in some way. Many of the data are uncertain by several tenths of a kilocalorie per mole, an uncertainty that could lead to a fairly large error in an equilibrium constant. For these reasons equilibrium constants for reactions involving migrations of double bonds along carbon chains provide valuable new information concerning the effects of substituents on the stabilities of carbon-carbon double bonds.

Let us first consider double-bond migration reactions in which steric effects have been minimized. In a reaction like Eq. 8-9 there should be no significant steric interactions across the double bond with ordinary X and Y

$$trans\text{-}XCH_2CH{=}CHY \rightarrow trans\text{-}XCH{=}CHCH_2Y \qquad (8\text{-}9)$$

groups. If we assume that the molar free energies of the reactants and products may be correlated in terms of bond contributions plus interactions among bonds leading to a common atom, such as the enthalpy contributions listed in Table 1-3, the statistically corrected free energy change for Eq. 8-9 may be expressed as shown in Eq. 8-10. If we define a double-

$$\Delta G^{chem} = B(C{-}Y) + B(C_d{-}X) + \Gamma_{C_dCY} - B(C{-}X) - B(C_d{-}Y) - \Gamma_{C_dCX} \qquad (8\text{-}10)$$

bond stabilization parameter D such that D_X is the free energy change for the reaction of vinyl X with propylene to give allyl X and ethylene (Eq. 8-11), this defines D_X as shown in Eq. 8-12 and permits ΔG^{chem} to be

$$CH_2{=}CHX + CH_2{=}CHCH_3 \rightarrow CH_2{=}CH_2 + CH_2{=}CHCH_2X \qquad (8\text{-}11)$$

$$D_X = B(C_d{-}H) + B(C{-}X) - B(C_d{-}X) - B(C{-}H) + \Gamma_{C_dCX} \qquad (8\text{-}12)$$

expressed as shown in Eq. 8-13. This equation is an expression of the idea

$$\Delta G^{chem} = D_Y - D_X \qquad (8\text{-}13)$$

that the equilibrium constant for Eq. 8-9 is a measure of the relative double-bond stabilizing abilities of the substituents X and Y. It should be tested by application to a large number of equilibrium constants for reactions like Eq. 8-9 carried out in the same solvent (or all in the gas phase) at the same temperature. Unfortunately, no such body of data exists but, since there is reason to think that entropy effects and solvent effects will not be very large, data over a range of temperatures (25–125° C) and in several different solvents as well in the gas phase have been treated.[25] When all primary alkyl groups were assigned the same D value (since several were found to have essentially the same double-bond stabilizing ability) and thus treated as a single substituent, there were seven substituents (not counting hydrogen, the reference substituent, which may be seen from Eq. 8-12 to have been assigned a D value of zero) that figured in two or more reactions for which equilibrium constants were available. Least squares treatment of the 20 equilibrium constants for these reactions gave D values from which the 20 ΔG^{chem} values could be calculated with a standard deviation of 0.37 kcal/mole. This provides a useful correlation, but there seemed to be some systematic trends in the deviations that were not easily explained as entropy or solvent effects or experimental error. For the reactions shown in Eqs. 8-14 and 8-15, the ΔG^{chem} values are

$$trans\text{-}RCH_2CH{=}CHOMe \rightarrow trans\text{-}RCH{=}CHCH_2OMe \qquad (8\text{-}14)$$

$$trans\text{-}RCH_2CH{=}CHCO_2Me \rightarrow trans\text{-}RCH{=}CHCH_2CO_2Me \qquad (8\text{-}15)$$

about 1.21 and 0.75 kcal/mole, respectively. These, then, would be the values of $D_{OMe} - D_R$ and $D_{CO_2Me} - D_R$, respectively. From them a value of -0.46 kcal/mole may be calculated for $D_{CO_2Me} - D_{OMe}$, the ΔG^{chem} value for Eq. 8-16, but this value differs markedly from the experimental

$$trans\text{-}MeOCH_2CH{=}CHCO_2Me \rightarrow trans\text{-}MeOCH{=}CHCH_2CO_2Me \qquad (8\text{-}16)$$

value of -2.39 kcal/mole. This discrepancy may be largely accounted for if we allow for interactions across the double bond as described in the following paragraph.

By analogy to Eq. 3-11 let us assume that the free energy of interaction of the XCH_2 and Y groups in the reactant in Eq. 8-9 is equal to $\tau_V\sigma_{XCH_2}\sigma_Y$, where the σ's are substituent constants for the groups in question and τ_V is a proportionality constant for interactions across a trans vinylene group. (Since we are to be correlating ΔG values rather than log K values it is convenient to absorb the factor $2.3RT$, which was kept separate in Eq. 3-11, into the τ value in the present case.) This and an analogous assumption for the product transforms Eq. 8-13 to Eq. 8-17. Since there is evidence

$$\Delta G^{chem} = D_Y - D_X + \tau_V(\sigma_X\sigma_{CH_2Y} - \sigma_Y\sigma_{CH_2X}) \qquad (8\text{-}17)$$

that Hammett substituent constants for para substituents may be used in describing interactions across trans vinylene groups (Section 3-5), these substituent constants were used to test Eq. 8-17.[25] Since there is evidence that τ values are not *highly* sensitive to the nature of the solvent, τ_V was taken as having the same value in several organic solvents and in the gas phase, and this value was treated as a disposable parameter. When the 20 values of ΔG^{chem} that had been correlated in terms of Eq. 8-13 were now correlated in terms of Eq. 8-17, values of D and τ_V were obtained from which the ΔG^{chem} values could be calculated with a standard deviation (0.20 kcal/mole) only about half as large as that obtained previously. The values of D are listed in Table 8-5, not only for the seven substituents for which two or more equilibria were available, but also for several substituents that appeared in only one equilibrium. The optimum value of τ_V was 13.4 with a standard deviation of 2.4 kcal/mole. This value is plausible enough to support the reality of the interaction term in Eq. 8-17. From a correlation of acidities of *trans*-3-substituted acrylic acids, τ_V may be shown to be around 8.9 kcal/mole in aqueous solution at 25° C. It should be somewhat larger in the gas phase or in the organic solvents used.

Table 8-5. Double-Bond Stabilization Parameters for Various Substituents[a]

Substituent	D (kcal/mole)
OMe	5.2
Ph	4.9
F	3.3[b]
n-C_nH_{2n+1}	3.2
SMe	3.2
CO_2Me	3.2
NO_2	2.9[b,c]
CH_2OMe	2.6[b]
i-Pr	2.5[b]
CN	2.3
CH_2CO_2Me	2.1[b]
Cl	1.8[b]
SOMe	0.7
Br	0.3[b]
H	0.0[d]
SO_2Me	−0.4[b]

[a] From Ref. 25; as defined by Eq. 8-17 with a τ_V value of 13.4 kcal/mole.

[b] Based on only one equilibrium constant.

[c] Based on an equilibrium in which less than 0.3% of the minor component was present; therefore relatively uncertain.

[d] By definition.

For a qualitative explanation of how interactions across the double bond explain why ΔG for Eq. 8-16 is so much more negative than ΔG for Eq. 8-15 minus ΔG for Eq. 8-14, we should first note that since the σ values for CH_2OMe and CH_2CO_2Me are 0.04 or less, the interactions across the double bonds in the products in Eqs. 8-14 and 8-15 and in both the reactants and products of Eq. 8-16 are relatively small. However, the reactant in Eq. 8-14 is destabilized considerably by the interaction of the electron-donating methoxy and alkyl groups. Allowance for this destabilization would give a value for $D_{OMe} - D_R$ considerably larger than 1.21 kcal/mole. The reactant in Eq. 8-15, on the other hand, is markedly stabilized by the interaction of the electron-donating alkyl group and the electron-withdrawing carbomethoxy group. Allowance for this stabilization would make $D_{CO_2Me} - D_R$ considerably smaller than 0.75 kcal/mole. It is therefore not surprising that ΔG for Eq. 8-16 is much more negative than -0.46 kcal/mole.

The tendency for electron-withdrawing substituents to destabilize double bonds noted in the preceding section may be seen in the relative values of D listed in Table 8-5 for substituents that are attached via saturated carbon atoms: $n\text{-}C_nH_{2n+1} > CH_2OMe > CH_2CO_2Me$. Other factors are also important, however. The methoxy and fluoro substituents, in spite of being attached via highly electronegative atoms, are among the best double-bond stabilizers. This is probably a result of the electron delocalization arising from the overlap of the filled $2p$ orbitals of the substituents with the $2p\pi$ systems of the carbon-carbon double bonds. Such overlap should be less efficient when the filled orbital of the substituent is $3p$, as in the case of methylthio and chloro substituents, and indeed, the D values for these substituents are smaller than those for methoxy and fluoro in spite of the fact that the methylthio and chloro substituents are attached via less electronegative atoms. Overlap of the $2p\pi$ system of the substituent with that of the carbon-carbon double bond must add to the double-bond stabilizing abilities of the phenyl, carbomethoxy, nitro, and cyano substituents. The fact that the methylsulfonyl and methylsulfinyl groups are much poorer double-bond stabilizers than the nitro and cyano groups, which are equally strong electron withdrawers, shows that interaction of these substituents with a carbon-carbon double bond leads to much less electron delocalization.

The D values in Table 8-5 are also in agreement with a number of qualitative and semiquantitative observations. For example, the isomerizations of allyl methyl ether and allyl methyl sulfide to the corresponding propenyl methyl compounds may be calculated to be more than 99% complete at equilibrium at room temperature, and both have been found to be so.[26] However, there are also experimental observations that may appear to be at variance with the *D* values. Most such observations deal with compounds that are not of the type *trans*-$XCH_2CH{=}CHY$. In the case of the halobullvalenes, for example, the chloro and bromo compounds exist entirely (within the experimental uncertainty) as the vinyl compounds (**5** and **6**), whereas the fluoro compound exists largely as the bridgehead isomer (**4**).[27] The *D* values in Table 8-5 reflect a tendency for all the halogens to prefer to be vinyl substituents, in the order F > Cl > Br. Although this apparent conflict between halogen substituent effects in the bullvalene series and in allyl-propenyl equilibria could mean that one of the observations is in error, it does not necessarily do so. Treating the two types of equilibria in terms of bond contributions and pairwise interactions of bonds leading to a common atom and neglecting interactions across double bonds, ΔG^{chem} for isomerization of allyl X to propenyl X may be calculated to exceed ΔG^{chem} for isomerization of **4** to **5** or **6** by the amount $2\,\Gamma_{C_dCX} - \Gamma_{CC_dX}$.

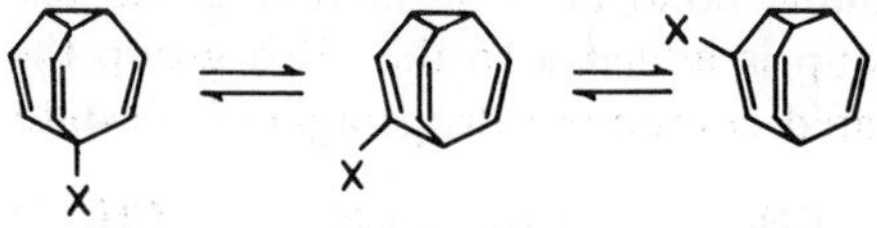

4 ⇌ **5** ⇌ **6**

Data that permit the independent calculation of these interactions do not appear to exist, but the interactions in Table 1-3 are large enough and varied enough to permit the possibility that variations in the way that halogen substituents affect bullvalene and allyl-propenyl equilibria may be explained in terms of variation in $2\,\Gamma_{C_dCX} - \Gamma_{CC_dX}$. The fact that this is true reminds us that the largely unknown factors that influence the relative magnitudes of such interactions must influence the position of equilibrium in many double-bond migration reactions as well as in many other processes. Until these factors are better understood any rationalization of *D* values in terms of polar and resonance effects, such as that we have given, will be an incomplete explanation of the facts at best.

There are many cases in which steric effects on the position of equilibrium in double bond migration seem to be important. For example, only about 0.2% of the unconjugated isomer is present in the equilibrium between the 1-nitropropenes and 3-nitropropene in an alkane solvent at 30° C.[28] Introduction of a new methyl substituent β to the nitro group

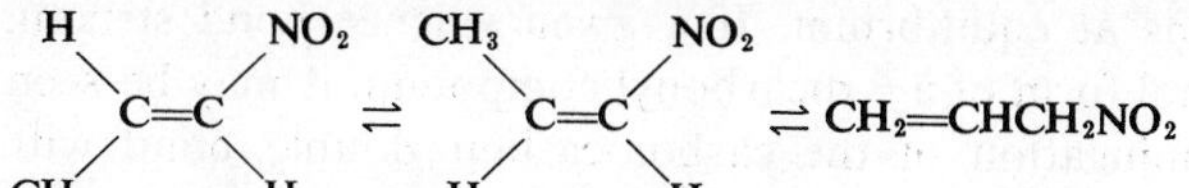

$$\underset{96.6\%}{(H)(CH_3)C{=}C(NO_2)(H)} \rightleftharpoons \underset{3.3\%}{(CH_3)(H)C{=}C(NO_2)(H)} \rightleftharpoons \underset{0.2\%}{CH_2{=}CHCH_2NO_2}$$

should stabilize the double bond whether it is α,β or β,α to the nitro group.

The stabilization might be expected to be greater when the double bond is α,β because of the stabilizing interaction of the methyl and nitro groups across the double bond that would result. Experimentally, however, the relative stability of the conjugated α,β unsaturated isomer is less in the equilibrium between 1-nitro-2-methylpropene and 3-nitro-2-methylpropene[29,30] than in the equilibrium between *cis*-1-nitropropene and 3-

$$(CH_3)_2C{=}CHNO_2 \rightleftharpoons CH_2{=}C(CH_3)CH_2NO_2$$

82% 18%

nitropropene. The new methyl substituent must have a buttressing effect that aggravates the destabilizing interaction between the nitro group attached to the double bond and the methyl group cis to it. When still another methyl group is added α to the nitro group the relative stability of the α,β unsaturated compound drops again,[30] despite the fact that the

$$(CH_3)_2C{=}C(CH_3)NO_2 \rightleftharpoons CH_2{=}C(CH_3){-}CH(CH_3)NO_2$$

66% 34%

α-methyl group can act directly as a stabilizing substituent on the double bond only when the double bond is α,β. The interactions between the nitro group and the methyl groups α and cis-β to it not only produce strains in bond angles and bond distances and destabilizing van der Waals interactions, but also rotate the nitro group out of the plane of the carbon-carbon double bond and thus decrease the stabilizing resonance interaction.

8-4. EQUILIBRIUM IN ENOLIZATION

Many of the quantitative studies of equilibrium in enolization have dealt with β-dicarbonyl compounds, which are commonly enolized to such an extent that both the keto and enol forms are present in measurable concentrations at equilibrium. If a given valence bond structure is written for the enol form of a β-dicarbonyl compound, it may be seen to be stabilized by conjugation of the carbon-carbon double bond with the remaining carbonyl group and by conjugation of the electron-donating hydroxy group with the electron-withdrawing carbonyl group across the double bond. In addition most such enols exist in a cyclic form, in which there is a stabilizing hydrogen bond between the hydroxy group and the carbonyl group. When this is so there are two valence bond structures that may be written. X-ray and electron diffraction studies on the enol forms of 1,1,2,2-tetraacetylethane,[31] hexafluoroacetylacetone,[32] and acetylacetone[33] give evidence that the hydrogen bond is symmetrical and that the two valence bond structures thus contribute equally to the total structure of the enol, which exists in only one form. On the other hand, X-ray diffraction studies

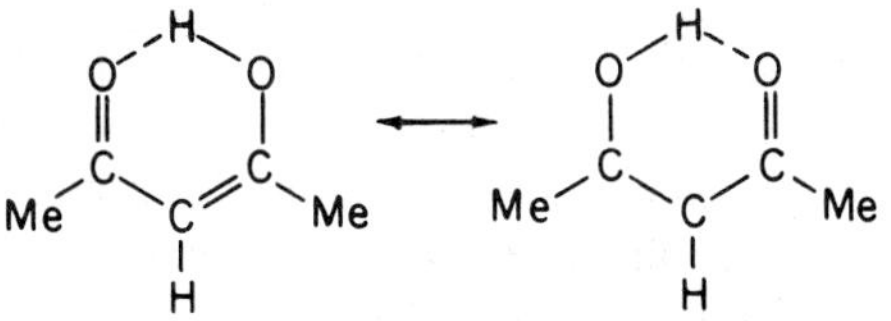

on dibenzoylmethane[34] and benzoylacetone[35] support structures with unsymmetrical hydrogen bonds, as do nmr studies on hydroxymethylene ketones, in which the fractions of the two types of tautomers present at equilibrium appear to depend markedly on the nature of R and R′.[36] It is possible that whether these cyclic hydrogen-bonded enols have one or two

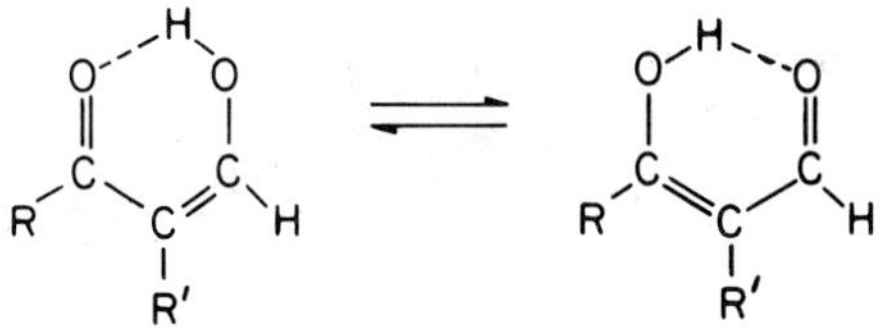

potential energy minima for the oxygen-bond hydrogen depends on the exact structure of the compound and perhaps on the medium in which it is located. It is also possible, however, that one of the two sets of experimental data has been misinterpreted. Additional studies, using neutron diffraction or perhaps microwave spectral techniques, might locate the enolic hydrogen atom more precisely.

Regardless of whether these internal hydrogen bonds are symmetrical or not they are strong and have an important influence on solvent effects on enolization, which are often fairly large. Acetylacetone, for example, is about 92% enolized in the vapor phase or in hexane solution but only about 15% enolized in aqueous solution.[37] There must be strong stabilizing hydrogen-bonding interactions between the carbonyl groups of the diketo form of acetylacetone and the solvent in aqueous solution, which would be lost on going to the vapor phase or to hexane as a solvent. The enol form, which has a cyclic structure, is hydrogen bonded internally and thus undergoes considerably less hydrogen bonding with the solvent in aqueous solution. Since the enol thus loses less stabilization due to hydrogen bonding on going to the vapor phase, its stability relative to that of the diketo form increases.

When internal hydrogen bonding is made sterically impossible, as in the case of 1,3-cyclohexanedione and its derivatives, the solvent effects are reversed. In aqueous solution at 20° C 5,5-dimethyl-1,3-cyclohexanedione is 95% enolized[38] but in dilute solution in cyclohexane, in which it cannot hydrogen bond to the solvent, it is less than 2% enolized.[39]

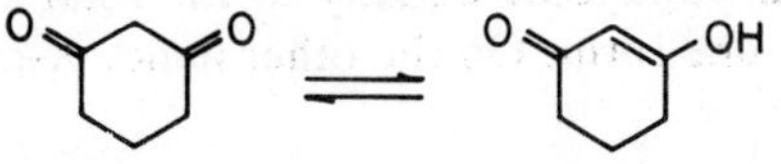

Polar effects on the enolization of β-diketones have been investigated by use of the Hammett equation. For meta- and para-substituted 2-benzoylcyclohexanones,[40] 2-benzoylcyclopentanones,[41] and benzoylacetones,[42] ρ values of 0.70, 0.93, and 0.63, respectively, were obtained. Spectral evidence was taken to indicate that the enols produced are largely those with their carbon-carbon double bonds conjugated with the aromatic rings, as in **7**. It is probably not suprising that electron-withdrawing substituents increase the extent of enolization. The ring to which Ar is attached in **7** would be expected, from substituent constants for related groups, to be less electron withdrawing than the CH_3COCH_2CO substituent to which it is attached in the diketone.

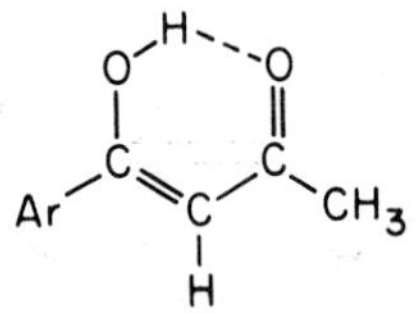

7

Table 8-6 lists data on the enolization of α-substituted derivatives of ethyl acetoacetate.[37,43] Some of the changes in substituents change polar, steric, and resonance effects all at once. There are also solvent effects that result from the fact that the measurements were made on the neat esters. However, since the solvents are all similar the solvent effects are probably not large. Although the methyl substituent has a *D* value (Table 8-5) that corresponds to its being a much better double-bond stabilizer than hydrogen, it is seen in Table 8-6 to lower the extent of enolization slightly. This is probably partly a steric effect, resulting from the necessity of having the X group in the cyclic enol form of $CH_3COCHXCO_2C_2H_5$ coplanar with, and on a carbon atom adjacent to, the ethoxy- and methyl-bearing carbon atoms in the ring. This interpretation is supported by the

Table 8-6. Enol Contents of Esters of the Type $CH_3COCHXCO_2C_2H_5$

X	% Enol[a]
H	8
Me	5
Et	~1
Ph	30[b]
F	15
Cl	15
Br	5
CF_3	89
CN	93

[a] For the neat esters at 33° C; data are from Ref. 43, unless otherwise stated.
[b] At 25° C.[37]

fact that when X is an alkyl group larger than methyl there is even less enol present.[43] The fact that the trifluoromethyl substituent, which is slightly larger than a methyl substituent, leads to much more enolization suggests that electron-withdrawing substituents encourage enolization. It is not clear whether there is any resonance interaction in the case of the trifluoromethyl substituent, but resonance interaction would certainly be expected in the case of the cyano substituent. Since steric factors should prevent the phenyl substituent from being coplanar with the hydrogen-bond-containing ring of the enol, resonance interactions should be markedly reduced. Perhaps the weakly electron-withdrawing power of the phenyl group combines with the small remaining resonance effect to give the increase in enolization noted with the phenyl substituent. The amount of enol observed with the fluorine substituent is probably smaller than would be expected in view of the rather large value of D_F but is more in line with the observations on fluorobullvalene.[27]

Most simple monocarbonyl compounds are so little ionized at equilibrium that it is difficult to determine the enol content reliably. The Kurt Meyer titration, in which an excess of halogen is added to react with the enol that is present and then the amount of halogen remaining determined, suffers from the disadvantage that any impurity present that reacts with halogen will be included in the enol determination. As Allinger and coworkers pointed out, some of the enol contents determined by this method are too high by more than a millionfold.[44] The Kurt Meyer titration has been greatly improved by developing a method for the reliable determination of very small concentrations of halogen. A highly purified sample of ketone is first treated with excess halogen, the excess determined, the ketone allowed to stand long enough to reestablish the keto-enol equilibrium, excess halogen again added, etc.[45] If the amount of halogen used is the same each time (or each time after the first) then it seems unlikely that the halogen was reacting with impurity (except in the first determination, if that value was high). This method yielded contents of 0.004 and 0.0002% for cyclopentanone and cyclohexanone, respectively, in aqueous solution at 25° C. In view of this evidence that the enol of cyclohexanone is less stable than that of cyclopentanone, it would be expected that the hydrolysis of 1-methoxycyclohexene (a model for 1-cyclohexenol) to cyclohexanone and methanol would be more exothermic than the hydrolysis of 1-methoxycyclopentene to cyclopentanone and methanol. Actually the reverse is true.[46] It is not clear whether the methyl ethers of the enols are poor models for the enols, whether there are large entropy and/or solvent effects, experimental error, or what. The enol constant of acetone was found to be too small to measure ($<10^{-4}\%$) by Bell and Smith, but Dubois and Toullec[47] have described kinetic results suggesting that about $1.5 \times 10^{-6}\%$ enol is present at equilibrium.

Whereas with simple monocarbonyl compounds the enols are usually so unstable as to be difficult to detect, the reverse is true with most phenols, which are so highly enolized that the keto form cannot be detected. This, of course, is a result of the loss of aromaticity that accompanies ketonization of a phenol. The polyhydroxy phenols have a particularly great tendency to ketonize because the stability resulting from aromatic resonance is pitted against the favorable energy change resulting from the ketonization of *several* enol groups. However, even in the case of phloroglucinol, which reacts readily with hydroxylamine to give the trioxime of 1,3,5-cyclohexanetrione,[48] the concentration of the keto form is so low that it has not yet been detected directly.[48,49] The dianion of phloroglucinol, however, appears to exist largely in the nonaromatic form, which permits the negative charge to be delocalized onto all the oxygen atoms.[50,51]

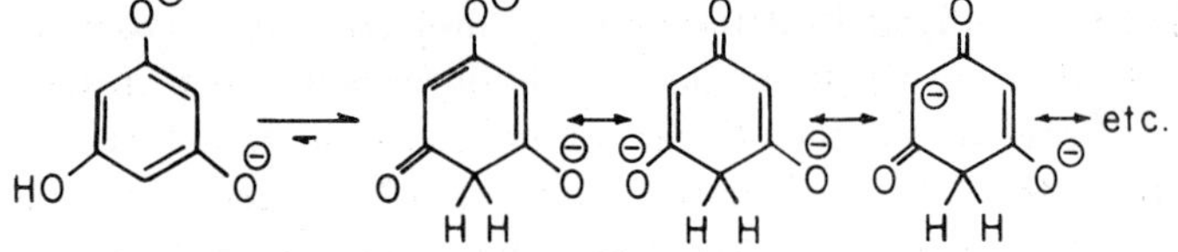

The tautomerism of phenols has been the subject of several reviews.[52,53]

PROBLEMS

1. Give a derivation, including statements as to what assumptions are used, for the following relationship for the equilibrium constant for Eq. 8-5.

$$\log K = \frac{8(X_Y - X_X)}{RT}$$

In this relationship X_X and X_Y are the electronegativities of X and Y or the atoms by which X and Y are attached.

2. From the data in Tables 8–3 and 8-4 estimate the equilibrium constant for the transformation of 1-butyne to 2-butyne in the gas phase at 82° C. Discuss the assumptions made in the estimate.

3. Which of the following two reactions would have the larger equilibrium constant? Why?

$$CH_2{=}CHCH_2OH \rightleftharpoons CH_3CH{=}CHOH$$
$$CH_2{=}CHCH_2O^- \rightleftharpoons CH_3CH{=}CHO^-$$

4. Use the data in Table 8-6 to name at least one compound of the type $CH_3COCHXCO_2C_2H_5$ for which the keto form is more acidic than the enol form (in the neat esters as solvent). Name at least one for which the reverse is true and give your reasoning in both cases.

REFERENCES

1. R. W. Taft, Jr., and M. M. Kreevoy, *J. Amer. Chem. Soc.*, **79**, 4011 (1957).
2. H. Adkins, R. M. Elofson, A. G. Rossow, and C. C. Robinson, *J. Amer. Chem. Soc.*, **71**, 3622 (1949).
3. M. M. Kreevoy and R. W. Taft, Jr., *J. Amer. Chem. Soc.*, **79**, 4016 (1957).
4. P. Valenta, *Collect. Czech. Chem. Commun.*, **25**, 853 (1960).
5. R. P. Bell, *Advan. Phys. Org. Chem.*, **4**, 1 (1966).
6. J. Hine and R. W. Redding, *J. Org. Chem.*, **35**, 2769 (1970).
7. J. W. Baker et al., *J. Chem. Soc.*, 191 (1942); 1089 (1949); 2831 (1952).
8. H. H. Jaffé, *Chem. Rev.*, **53**, 191 (1953).
9. O. H. Wheeler, *J. Org. Chem.*, **29**, 3634 (1964).
10. O. H. Wheeler and E. G. de Rodriguez, *J. Org. Chem.*, **29**, 718 (1964).
11. H. C. Brown, R. S. Fletcher, and R. B. Johannesen, *J. Amer. Chem. Soc.*, **73**, 212 (1951).
12. H. C. Brown, *J. Chem. Soc.*, 1248 (1956).
13. V. Prelog and M. Kobelt, *Helv. Chim. Acta*, **32**, 1187 (1949).
14. V. Prelog, *J. Chem. Soc.*, 420 (1950).

15. E. G. Sander and W. P. Jencks, *J. Amer. Chem. Soc.*, **90,** 6154 (1968).
16. J. B. Conn, G. B. Kistiakowsky, and E. A. Smith, *J. Amer. Chem. Soc.*, **60,** 2764 (1938); **61,** 216 (1939).
17. M. A. Dolliver, T. L. Gresham, G. B. Kistiakowsky, and W. E. Vaughan, *J. Amer. Chem. Soc.*, **59,** 831 (1937).
18. M. A. Dolliver, T. L. Gresham, G. B. Kistiakowsky, E. A. Smith, and W. E. Vaughan, *J. Amer. Chem. Soc.*, **60,** 440 (1938).
19. J. B. Conn, G. B. Kistiakowsky, and E. A. Smith, *J. Amer. Chem. Soc.*, **61,** 1868 (1939).
20. Cf. R. B. Turner, D. E. Nettleton, Jr., and M. Perelman, *J. Amer. Chem. Soc.*, **80,** 1430 (1958).
21. E. D. Hughes, C. K. Ingold, and V. J. Shiner, *J. Chem. Soc.*, 3827 (1953).
22. H. C. Brown and H. L. Berneis, *J. Amer. Chem. Soc.*, **75,** 10 (1953).
23. R. B. Turner and coworkers, *J. Amer. Chem. Soc.*, **79,** 4116, 4122, 4133 (1957); **80,** 1424 (1958).
24. R. B. Turner, W. v. E. Doering, and co-workers, *J. Amer. Chem. Soc.*, **90,** 4315 (1968).
25. J. Hine and N. W. Flachskam, *J. Amer. Chem. Soc.*, **95,** 1179 (1973).
26. C. C. Price and W. H. Snyder, *J. Amer. Chem. Soc.*, **83,** 1773 (1961); *J. Org. Chem.*, **27,** 4639 (1962).
27. J. F. M. Oth, R. Merényi, H. Röttele, and G. Schröder, *Tetrahedron Lett.*, 3941 (1968).
28. G. Hesse and V. Jäger, *Justus Liebigs Ann. Chem.*, **740,** 79 (1970).
29. H. Schechter and J. W. Shepherd, *J. Amer. Chem. Soc.*, **76,** 3617 (1954).
30. G. Hesse, R. Hatz, and H. König, *Justus Liebigs Ann. Chem.*, **709,** 79 (1967).
31. J. P. Schaefer and P. J. Wheatley, *J. Chem. Soc., A*, 528 (1966).
32. A. L. Andreassen, D. Zebelman, and S. H. Bauer, *J. Amer. Chem. Soc.*, **93,** 1148 (1971).
33. A. H. Lowrey, C. George, P. D'Antonio, and J. Karle, *J. Amer. Chem. Soc.*, **93,** 6399 (1971).
34. D. E. Williams, *Acta Cryst.*, **21,** 340 (1966).
35. D. Semmingsen, *Acta Chem. Scand.*, **26,** 143 (1972).
36. E. W. Garbisch, Jr., *J. Amer. Chem. Soc.*, **85,** 1696 (1963); **87,** 505 (1965).
37. J. B. Conant and A. F. Thompson, Jr., *J. Amer. Chem. Soc.*, **54,** 4039 (1932).
38. G. Schwarzenbach and E. Felder, *Helv. Chim. Acta*, **27,** 1044 (1944).
39. A. Yogev and Y. Mazur, *J. Org. Chem.*, **32,** 2162 (1967).
40. R. D. Campbell and H. M. Gilow, *J. Amer. Chem. Soc.*, **82,** 2389 (1960).
41. R. D. Campbell and W. L. Harmer, *J. Org. Chem.*, **28,** 379 (1963).
42. D. J. Sardella, D. H. Heinert, and B. L. Shapiro, *J. Org. Chem.*, **34,** 2817 (1969).
43. J. L. Burdett and M. T. Rogers, *J. Amer. Chem. Soc.*, **86,** 2105 (1964).

44. N. L. Allinger, L. W. Chow, and R. A. Ford, *J. Org. Chem.*, **32,** 1994 (1967).
45. R. P. Bell and P. W. Smith, *J. Chem. Soc.*, *B*, 241 (1966).
46. K. Arata, M. S. Thesis, The Ohio State University, 1970.
47. J. E. Dubois and J. Toullec, *Tetrahedron*, **29,** 2859 (1973).
48. V. C. Farmer and R. H. Thomson, *Chem. Ind.* (London), 86 (1956).
49. K. W. F. Kohlrausch, *Ber.*, **69,** 527 (1936).
50. H. Köhler and G. Scheibe, *Z. Anorg. Allgem. Chem.*, **285,** 221 (1956).
51. R. J. Highet and T. J. Batterham, *J. Org. Chem.*, **29,** 475 (1964).
52. R. H. Thomson, *Quart. Rev.* (London) **10,** 27 (1956).
53. V. V. Ershov and G. A. Nikiforov, *Russian Chem. Rev.*, **35**, 817 (1966).

9

Equilibrium in Ring Formation

9-1. FORMATION OF VARIOUS TYPES OF RINGS

There are a number of equilibrium constants for cyclization reactions in the literature but they are widely scattered, and in only a few cases have measurements been made on a series of compounds with systematic variations in structure. One such case is Eberson and Welinder's study of the formation of anhydrides from vicinal dicarboxylic acids in aqueous solution.[1] Their equilibrium constants, which are listed in Table 9-1 in dimensionless form, that is, with the constant water concentration absorbed into the equilibrium constant, cover a range of more than a millionfold. Alkyl substituents on the two carbons between the carboxy groups are seen to increase the stability of the anhydride relative to that of the acid in almost every case. This is believed to be largely the result of a steric effect. In the case of maleic acid, for example, the C═C—H angle in the acid is about 118° and that in the anhydride is about 128.5°. If the corre-

sponding bond angles in the dialkylmaleic acids are similar, the formation of the anhydride will relieve some of the strain between the two alkyl groups. This steric explanation is in agreement with the tendency for the anhydride-stabilizing ability of the alkyl groups to increase with their bulk. Polar factors may also be significant, however. The greater electron-withdrawing power of anhydride than carboxy substituents would be expected to make the carbon-carbon double bond more electron deficient in the anhydride than in the acid. Hence an electron-donating substituent, such as an alkyl group, should stabilize the anhydride relative to the acid.

The greater tendency for anhydride formation from the maleic acids than from the corresponding succinic acids must arise in part from the destabilizing interaction between the two carboxy groups that are cis to each other in the maleic acids. The fact that fumaric acid is more stable than maleic acid shows that this interaction is destabilizing. The analogous

Table 9-1. Equilibrium Constants for Transformation of Vicinal Dicarboxylic Acids to Anhydrides[a]

Acid	K[b]
Succinic	7×10^{-6}
Methylsuccinic	4×10^{-5}
2,2-Dimethylsuccinic	1×10^{-4}
dl-2,3-Dimethylsuccinic	2.5×10^{-4}
Tetramethylsuccinic	0.17
meso-2,3-Diethyl-2,3-dimethylsuccinic	1.0
dl-2,3-Diethyl-2,3-dimethylsuccinic	3.4
dl-2,3-Di-*t*-butylsuccinic	6
Tetraethylsuccinic	10
Maleic	~0.01[c]
Dimethylmaleic	5.3[d]
Methylethylmaleic	4.3[d]
Diethylmaleic	3.2[d]
Diisopropylmaleic	25
1,2-Diisopropyl-*cis*-1,2-cyclopropanedicarboxylic	0.3
Norbornane-*endo*-*cis*-2,3-dicarboxylic	0.6
2-Methylnorbornane-*endo*-*cis*-2,3-dicarboxylic	0.6
Bicyclo[2.2.2]octane-*cis*-2,3-dicarboxylic	0.3
Phthalic	~0.01[c]
3,6-Dimethylphthalic	0.2
3,6-Diethylphthalic	1.5

[a] In water at 60° C unless otherwise stated.
[b] Dimensionless equilibrium constants from Ref. 1.
[c] Estimated from rate data.
[d] At 20° C.

destabilizing interaction in succinic acid should be smaller even in a gauche conformation because the two C—CO_2H bonds would not be coplanar, and it may be reduced further by having the molecule assume the trans conformation as it is known to do in the crystalline state.

Hurd and Saunders studied the effect of ring size on the stabilities of cyclic hemiacetals relative to the corresponding acyclic hydroxy aldehydes.[2] They compared the extinction coefficients of aldehydes of the type $HO(CH_2)_nCHO$ with those of the corresponding methoxy aldehydes in dioxane-water (75:25%). The extinction coefficient of the free hydroxy aldehyde was assumed to be the same as that of the corresponding free methoxy aldehyde. The smaller extinction coefficient obtained experimentally for each hydroxy aldehyde was then assumed to result from the existence of part of the aldehyde in the form of its cyclic hemiacetal.

$$HO(CH_2)_nCHO \rightleftharpoons \underbrace{(CH_2)_nCHOH}_{O} \qquad (9\text{-}1)$$

On this basis the percentages of hemiacetal present for the various aldehydes shown in Table 9-2 may be calculated. The remaining material would be either aldehyde or aldehyde hydrate (1,1-diol). If the amount of aldehyde hydrate present is significant, which seems likely, and if the ratios of hydrate to free aldehyde for the hydroxy aldehydes are considerably different from those for the corresponding methoxy aldehydes, which seems unlikely, the percentages of hemiacetal listed in Table 9-2 will be significantly in error. The five- and six-membered rings are seen to be more stable than the larger rings. Consideration of the sources and plausible magnitudes of experimental errors suggests that the difference in stabilities between the seven- and nine-(and perhaps even ten-)membered rings is not

Table 9-2. Cyclic Hemiacetal Formation from ω-Hydroxy Aldehydes in Dioxane-Water (75:25%) at 25° C[a]

Aldehyde	Ring Size	% Hemiacetal
$HO(CH_2)_3CHO$	5	89
$HO(CH_2)_4CHO$	6	94
$HO(CH_2)_5CHO$	7	15
$HO(CH_2)_7CHO$	9	20
$HO(CH_2)_8CHO$	10	9

[a] Data from Ref. 2.

significant. The difference between the five- and six-membered rings is probably significant, however. Nuclear magnetic resonance, a more suitable method, which was not available at the time of Hurd and Saunders' study, has been used to study similar equilibria in the case of sugars, where the pyranose forms are ordinarily more stable than the furanose forms, and where cyclic hemiacetals of other ring sizes are rarely stable enough to be observed directly. The many studies that have been made on sugars demonstrate the importance of other factors, some steric and some polar, beyond ring size that influence the relative stabilities of furanose and pyranose forms.[3]

In addition to the preceding equilibria, in which an acyclic compound is transformed to a cyclic compound by the formation of one new bond, there are a number of reactions in which two or more of the ring bonds are formed in the process whose equilibrium constant is measured. In the formation of cyclic ketals from acetone and various diols, Anteunis and Rommelaere obtained the results shown in Table 9-3.[4] Alkyl substituents increase the equilibrium constants but not to as great an extent as they did in the case of anhydride formation, perhaps because of destablizing steric interactions with the methyl groups from the acetone.

Table 9-3. Equilibrium Constants for the Formation of Cyclic Ketals from Acetone and Diols at 25° C[a]

Diol	K^b
1,2-Ethanediol	0.14
3-Chloro-1,2-propanediol	0.21
1,2-Propanediol	0.48
1,2,3-Propanetriol	0.73
1,3-Propanediol	0.026
2-Methyl-1,2-propanediol	0.58
meso-2,3-Butanediol	0.81
2,2-Dimethyl-1,3-propanediol	0.27
cis-1,2-Cyclohexanediol	0.14

[a] Determined by nmr measurements on reaction mixtures containing only the four species involved in the equilibrium plus an acid catalyst.[4]

[b] Obtained in most cases by interpolation from data at other temperatures.

Evidence for another type of steric effect in the formation of cyclic acetals has been found by Newman and Dickson in a study of the reaction of 1,1-bis(hydroxymethyl)cycloalkanes with 1- and 2-naphthaldehyde and 1- and 2-naphthyl methyl ketone.[5] The equilibrium constants obtained

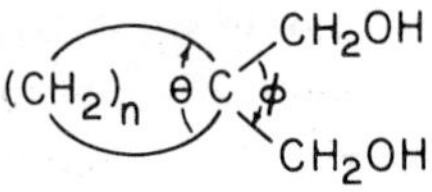

$(CH_2)_n$ θ C φ CH_2OH, CH_2OH $\qquad$ $(CH_2)_n$ C $(CH_2O)_2$CHAr

1

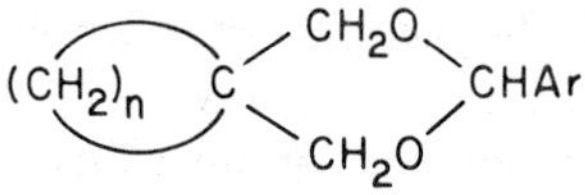

with 1-naphthaldehyde and the three-, four-, five-, and six-membered ring diols are listed in Table 9-4. About the same relative values were obtained with the other aldehyde and the ketones; namely, the equilibrium constants increase monotonically with the size of the ring in the diol. It is suggested that decreases in the size of angle θ in a compound like **1** are accompanied by increases in angle ϕ. Schleyer pointed out that the strengths of the hydrogen bonds in these diols, as measured by infrared spectra, give evidence for such an effect, amounting to a difference of about 5° between the values of ϕ for the cyclopropane and cyclohexane derivatives.[6] There is other evidence for such a relationship between θ and ϕ. Some such evidence was a basis for the "Thorpe-Ingold Valency Deflection Hypothesis."[7,8] The enlarged ϕ destabilizes the new ring being formed in the acetalization reaction.

The equilibrium constants for many cyclization reactions are too large or too small to be measured directly. However, even in cases like this the relative magnitudes of the equilibrium constants for possible competing cyclization reactions may be determined by studying the equilibrium constant(s) for interconversion of the possible cyclic products. And, of

Table 9-4. Equilibrium Constants for Acetal Formation from 1-Naphthaldehyde and 1,1-*Bis*(hydroxymethyl) cycloalkanes[a]

Cycloalkane	*K*
Cyclopropane	0.94
Cyclobutane	11
Cyclopentane	26
Cyclohexane	61

[a] In dioxane at 34.3° C.[5]

course, if the species involved are sufficiently closely related to compounds whose thermochemical properties have been determined, the equilibrium constants may be calculated with a reasonable degree of reliability. Consider, for example, the cyclization of 1-hexene to give cyclohexane or methylcyclopentane. The group contributions to the enthalpies of formation (Table 1-4) show contributions of 6.3 and 0.0 kcal/mole for the cyclopentane and cyclohexane rings, respectively. This 6.3 kcal/mole advantage of the less strained six-membered ring may seem to favor the formation of cyclohexane so much as to make it almost the exclusive product. However, there are several compensating factors. First, methylcyclopentane contains a branched carbon chain, which ordinarily leads to increased stability. Second, the entropy contribution for a cyclopentane ring is 8.5 eu more favorable than that for a cyclohexane ring, in which the freedom of movement of a larger number of atoms has been restricted. Third, cyclohexane is highly symmetrical, and highly symmetrical species are disfavored at equilibrium. From the appropriate group contributions from Table 1-4 and symmetry contributions from Eqs. 1-11 and 1-12, enthalpies of formation of -25.48 and -29.70 kcal/mole and entropy contents of 81.14 and 71.76 eu may be calculated for methylcyclopentane and cyclohexane, respectively. These lead to a free energy difference of -0.77 kcal/mole and a content of 21% methylcyclopentane and 79% cyclohexane in the gas phase equilibrium mixture at 25° C. According to vapor pressure data and equilibrium measurements in the liquid phase[9] the gas phase equilibrium mixture is actually about 16% methylcyclopentane at 25° C. The deviation of this experimental observation from the calculated value is probably significant, but it is clear that the calculations are reliable enough to be useful. The group contributions lead to an equilibrium constant of about 10^9 for the formation of either of the cycloalkanes from 1-hexene. It would be extremely difficult to determine such an equilibrium constant by direct measurements.

The various ring contributions to the enthalpies of formation listed in Table 1-4 are measures of the ring strain. However, it should be clear that the magnitude of a ring-strain contribution will depend on how ring strain is defined, that is, on how the contribution is calculated. For example, suppose we are using a group contribution scheme for calculating the enthalpies of formation of hydrocarbons and are not including separate terms for the destabilization resulting from gauche interactions. Then, to the extent to which we are trying to fit data on hydrocarbons in which there are gauche interactions, the group contributions will include an allowance for the gauche interactions; that is, they will correspond to higher enthalpy contents than an analogous set of group contributions designed for use with separate contributions for gauche interactions. Most sets of parameters for empirical correlations represent some sort of compromise between simplicity and ease of calculation on one hand and a high degree of precision in fitting the data on the other hand. The group contributions in Table 1-4 were calculated by using a separate term for *obligatory* gauche interactions such as that in 2-methylbutane, where there is a gauche interaction even in the most stable conformer. No allowance was made for the fact that in the case of straight-chain alkanes a large fraction of the molecules may be in a gauche conformation at 25° C even though there is a gauche-free conformation of lower enthalpy content available. This results from the fact that there is only one trans conformer but two gauche conformers possible with respect to rotation around any $RCH_2—CH_2R$ bond and the fact that although the probability of the conformation around any one $RCH_2—CH_2R$ bond being gauche may be only about one-third, if there are a large number of such bonds it is highly likely that one or more will be gauche. Hence an alternative, more complicated way of treating the data would be in terms of "single conformation" parameters.[10] In this approach *n*-butane is treated as a mixture of two compounds, the gauche and trans forms. Since that part of the enthalpy content of *n*-butane that arises from the presence of some gauche isomer is now being allowed for explicitly, the single conformation group contributions should not, as a whole, give as large an enthalpy content as they otherwise would. It happens that the difference between the single conformation contributions and the contributions in Table 1-4 resides almost entirely in the $CH_2(C)_2$ contribution, which is reduced from −4.95 to −5.13 kcal/mole.

Cyclohexane is 1.35 kcal/mole less stable than it is calculated to be from the single-conformation group contributions. This strain energy for cyclohexane differs from the 0.0 kcal/mole cyclohexane contribution listed in Table 1-4 because the Table 1-4 contribution is a measure of the strain relative to acyclic hydrocarbons in which there are significant fractions of gauche conformers. If the optimum bond angles for all acyclic saturated carbon compounds were 109.5°, it might at first seem surprising that cyclohexane, which can achieve such bond angles with perfect staggering around every carbon-carbon bond, is significantly strained. However, there are at least two good reasons why such ring strain would be expected. First, the equilibrium $C—CH_2—C$ angles found in propane and higher alkanes are around 112.4°, whereas the C—C—C bond angles in cyclohexane are only 111.5°, and even in achieving this smaller deviation from 109.5° the bond staggering has been made slightly imperfect.[11] Second, there is a type of gauche interaction between any carbon atom in cyclohexane and the carbon atom across the ring from it. This interaction is not the same as the interaction between the two methyl groups in gauche butane because the two hydrogen atoms whose close approach contributes most to the gauche methyl-methyl interaction are not present in cyclohexane. Nevertheless, the gauche interactions in cyclohexane have been estimated to be large enough that when added to the angle strain and the torsional strain they explain the strain energy of cyclohexane.[10] Adamantane, in which the C—C—C bonds are held even nearer 109.5°, has 6.48 kcal/mole of strain on the basis of single-conformation group contributions.[10]

Benson and co-workers[12] list contributions for a number of rings, both carbocyclic and heterocyclic, in addition to those covered by Table 1-4.

9-2. COMPARISON OF CYCLIZATION AND ANALOGOUS NONCYCLIZATION REACTIONS

In addition to comparing various cyclization reactions with each other, it is also interesting to compare cyclization reactions with analogous processes that do not involve cyclization. Page has discussed both rate and equilibrium comparisons in a review of the energetics of neighboring group participation.[13] One important difference between the two types of processes is that cyclization reactions involve an increase in the number of molecules relative to analogous noncyclization reactions. The formation of a cyclic anhydride and water from a dicarboxylic acid, for example, is the formation of two molecules from one, whereas the formation of an acyclic anhydride from the corresponding acid is the formation of two molecules from two. This means that the two equilibrium constants being compared will not have the same dimensions and that the ratio of the two equilibrium constants will have the dimensions of concentration (or its reciprocal). Thus if we are expressing concentrations in molarity, that is, using a standard state of 1 *M*, and if the ratio of the equilibrium constant for a cyclization reaction to the equilibrium constant for an analogous noncyclization reaction is greater than 1 *M*, then there will always be another concentration unit (and standard state) in which the ratio is smaller than 1. Similarly, if the ratio is smaller than 1 *M* there will always be another concentration unit in which the ratio is larger than 1. For this reason it may not seem very meaningful to say that the equilibrium constant for cyclization is larger or smaller than the equilibrium constant for the analogous noncyclization reaction. However, in a case in which the ratio is larger than 1 for any standard state that would be attainable under the pressure at which the reactions were studied,* perhaps there would be

*Note that a standard state of 50 kg/ml would not be attainable under conditions where equilibrium involving an organic compound has ever been studied.

some meaning in saying that the cyclization equilibrium constant was larger.

The translational and rotational entropy of the additional molecule formed in a cyclization reaction gives the cyclization reaction an advantage, which often amounts to 40–50 eu (based on a standard state of 1 *M*),[14] over an analogous reaction that does not involve cyclization. However, since each molecule, if nonlinear, has $3n - 6$ vibrational modes, where n is the number of atoms it contains, the cyclization reaction will lose six vibrational modes relative to the noncyclization reaction. This will have an effect that depends on the frequency of the vibrations but has been estimated to be around 15 eu (in the direction unfavorable for the cyclization reaction) in a typical case.[14] In addition there will be an unfavorable entropy effect in the cyclization reaction resulting from the increased extent of hindrance of rotations around carbon-carbon and other single bonds, which usually amounts to 3–4 eu per rotation.[14]

In order to illustrate the net result of these entropy effects it would be desirable to have appropriate data on a number of specific reactions. A difficulty in comparing a cyclization reaction and an analogous noncyclization reaction is that in any such comparison there are always special factors that relate to the two specific reactions being compared rather than to the general case. Examples of such special factors will be discussed in connection with data on specific reaction. The largest body of appropriate data relates to the formation of chelate complexes and "analogous" nonchelate complexes from metal ions. This will be discussed in the next section. Unfortunately, there are very few other cases in which equilibrium constants have been determined for both a cyclization and an analogous noncyclization reaction and even fewer in which the enthalpy and entropy of reaction have been determined for each reaction. The equilibrium constant for the formation of acetic anhydride from acetic acid in aqueous solution at 25° has been found to be about $3 \times 10^{-12}\ M^{-1}$ (with the water concentration absorbed into the equilibrium constant).[15] This is smaller by a factor of about $3 \times 10^{5}\ M$ than the equilibrium constant for the formation of succinic anhydride from succinic acid in water at 25° C.[16] However, this difference in equilibrium constants appears to be largely an enthalpy effect. If the enthalpy of hydrolysis of acetic anhydride in aqueous solution is about the same as it is for the pure liquids[17] it is about 5 kcal/mole more exothermic than the hydrolysis of succinic anhydride.[16] This difference in enthalpies of reaction suggests that there are electronic factors that make the coplanar conformation of the anhydride group that is present in succinic anhydride particularly stable. (The carbon and oxygen atoms of succinic anhydride are coplanar within 0.02 Å[18];

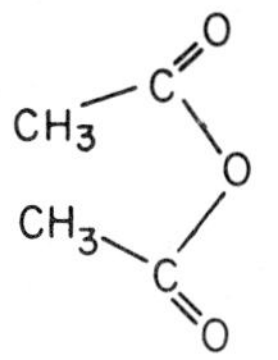

this is much more nearly coplanar than most five-membered ring compounds.) Such a conformation for acetic anhydride would be greatly destabilized by methyl-methyl interactions, however. From the enthalpies and free energies of reactions, the entropy of formation of succinic anhydride from the acid may be calculated to be more favorable than that for acetic anhydride formation by about 8 eu (with reference to a standard state of 1 *M*). An entropy effect of this magnitude would change the equilibrium constant by only about 50 *M*.

If we have a reasonably good estimate of the equilibrium constant for addition of a straight-chain primary alcohol to a straight-chain aldehyde, we may estimate the ring effect in the formation of cyclic hemiacetals from hydroxyaldehydes of the type $HO(CH_2)_nCHO$. Since the equilibrium constants for hydration of straight-chain aldehydes[19] are near that for pyridine-4-carboxaldehyde,[20] the equilibrium constant for addition of ethanol to pyridine-4-carboxaldehyde (0.27 M^{-1})[20] may be taken as a fair approximation of the value for addition of a straight-chain alcohol to a straight-chain aldehyde. The equilibrium constants for cyclic hemiacetal formation from ω-hydroxy straight-chain aldehydes (Table 9-2) are larger by factors of 30 *M* and 60 *M* in the case of the formation of the five- and six-membered ring compounds, respectively. In the absence of additional data, however, we cannot tell to what extent these differences in equilibrium constants arise from enthalpy and to what extent they arise from entropy effects. The most abundant and reliable data on enthalpies and entropies of analogous cyclization and noncyclization reactions are probably those that may be calculated from the thermodynamic properties of the reactants and products in certain cases where we presently have no way of establishing the equilibrium experimentally. Consider the equilibrium constant K_c for the reaction in which two hydrogen atoms are split out between the two ends of *n*-hexane to give cyclohexane.

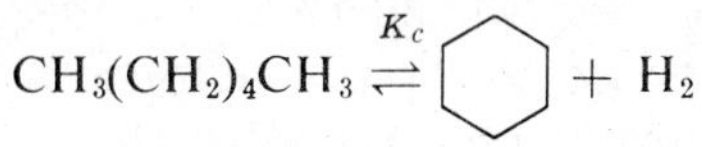

$$CH_3(CH_2)_4CH_3 \overset{K_c}{\rightleftharpoons} \quad + H_2$$

Let us compare this with the equilibrium constant K_a for splitting out two hydrogen atoms between the ends of two molecules of *n*-hexane to give *n*-dodecane.

$$2CH_3(CH_2)_4CH_3 \overset{K_a}{\rightleftharpoons} CH_3(CH_2)_{10}CH_3 + H_2$$

A compact way of making the comparison is to examine the value of the quotient K_c/K_a, which is simply the equilibrium constant for transformation of *n*-dodecane to cyclohexane and *n*-hexane.

$$CH_3(CH_2)_{10}CH_3 \overset{K_c/K_a}{\rightleftharpoons} \text{cyclohexane} + CH_3(CH_2)_4CH_3$$

According to the entropies and enthalpies of formation for the three compounds involved[12] in the gas phase at 25° C the enthalpy of reaction is 0.2 kcal/mole, which is within the experimental error of zero, and the entropy of reaction is 9 eu, on the basis of a standard state of 1 *M*. (The entropies listed in Ref. 12, which are based on a standard state of 1 atm pressure, like those in Tables 1-2 and 1-4, had to be corrected to the standard state of 1 *M*.) The entropy change in solution need not be much different from what it is in the gas phase. In fact, from the vapor pressures of the three hydrocarbons it may be shown that in a solvent in which all three form ideal solutions and which is 10 *M* in the pure state, $\Delta S°$ will also be 9 eu in a dilute solution. This means that there is an entropy factor that tends to make the equilibrium constant for the cyclization of *n*-hexane to cyclohexane larger than the equilibrium constant for the formation of *n*-dodecane from two *n*-hexane molecules by a factor of about 100 *M* at 25° C in either the gas phase or the ideal solvent referred to.

Similar comparison of the cyclization of *n*-pentane to give cyclopentane with the reaction of two *n*-pentane molecules to give *n*-decane shows that there is an entropy effect that, at 25° C in the absence of an enthalpy effect, would give the ratio K_c/K_a a value around 5×10^3 *M* in the gas phase or 2×10^3 *M* in an ideal solvent that is 10 *M* when pure. In this case there is an enthalpy effect that makes the ratios much smaller, but the fact that the entropy effect is larger than in the case of cyclohexane formation is to be expected. In the formation of cyclopentane we have the same advantage (of K_c over K_a) of forming two molecules from one, whereas the disadvantage owing to severely restricting rotation around five carbon-carbon bonds should be smaller than in the case of cyclohexane where rotation around six carbon-carbon bonds is severely restricted. On the basis of this argument the entropy effect should be even larger for the formation of smaller rings. An extreme example may be found by considering a double bond to be a two-membered ring. The value of $\Delta S°$ for loss of two hydrogen atoms from *n*-pentane to give 1-pentene at 25° C is about 30 eu larger in either the gas phase or an ideal solvent than for the loss of two hydrogen atoms from the ends of two *n*-pentane molecules to give *n*-decane (on the basis of a 1 M standard state). If the enthalpies of reaction were the same (actually decane formation is much more exothermic) this entropy difference would make the equilibrium constant for the formation of 1-pentene larger than that for the formation of decane by a factor of about 3×10^6 M.

Ratios of equilibrium constants of the type we have been discussing could be changed significantly by strong interactions between the solvent and the reactants and/or products. (In such a case the number of molecules directly involved in the reaction will not necessarily be the same as that shown in a stoichiometric equation.)

Entropy effects on cyclization reactions often may be estimated rather reliably from O'Neal and Benson's tabulation and discussion of the entropies and heat capacities of cyclic compounds.[21]

9-3. THE CHELATE EFFECT IN THE FORMATION OF METAL-ION COMPLEXES

Measurements on the coordination of organic bases with metal ions, largely made by scientists labeled inorganic chemists, provide the most abundant source of information on how equilibrium constants for cyclization reactions compare with those for analogous noncyclization reactions. The results have been discussed by a number of workers.[22–31] We shall give more attention here to changes in the base than to changes in the metal ion. Data on the coordination of Cu^{2+} ions with ammonia[33] and various polyamines[32–34] are listed in Table 9-5. Such data are not as greatly complicated by ionic strength and solvation effects as are those involving electrically charged bases. K_1 is the equilibrium constant, in M^{-1}, for the coordination of one molecule of base with the Cu^{2+} ion; K_{1n} is the cumulative constant, in M^{-n}, for the coordination of n molecules. Thus, K_{13} for ammonia is equal to the product $K_1K_2K_3$, or, in terms of concentrations (neglecting coordinated water),

$$K_{13} = \frac{[Cu(NH_3)_3{}^{2+}]}{[Cu^{2+}][NH_3]^3}$$

It may be argued that ammonia is a poor choice for the monodentate analogue of the various polydentate ligands listed in the table since it has no carbon attached to its nitrogen atom. On the other hand, methylamine and ethylamine, which appear to form weaker complexes with Cu^{2+} than ammonia does, are not as close to the bidentate ligand ethylenediamine in basicity per amino group as ammonia is. Furthermore, no very reliable data on the methylamine and ethylamine complexes appear to exist.

Table 9-5. Coordination of Cu^{2+} Ions with Ammonia and Polyamines in Aqueous Solution[a]

Base	Constant	log K	$-\Delta H°$ (kcal/mole)	$\Delta S°$ (eu)
Ammonia	K_1	4.3	5.9	0
	K_{12}	7.9	11.5	−3
	K_{13}	10.9	17.5	−9
	K_{14}	13.0	20.0	−8
$H_2NCH_2CH_2NH_2$[b]	K_1	10.6	12.6	6
	K_{12}	19.7	25.2	6
$H_2NCH_2CH_2CH_2NH_2$[c]	K_1	9.9	12.5	3
	K_{12}	17.0	24.6	−5
trans-1,2-Diaminocyclohexane	K_1	11.1	13.6	5
	K_{12}	20.7	25.9	8
$Me_2C(CH_2NH_2)_2$	K_1	10.1	12	7
	K_{12}	17.7	24	0
$O(CH_2CH_2NH_2)_2$	K_1	8.7	11.0	3
	K_{12}	13.0	14.7	11
$HN(CH_2CH_2NH_2)_2$	K_1	15.8	18.0	9
	K_{12}	21.0	26.2	2
$HN(CH_2CH_2CH_2NH_2)_2$	K_1	14.7	16.1	11
$H(NHCH_2CH_2)_3NH_2$[b]	K_1	20.1	21.6	20
$N(CH_2CH_2NH_2)_3$	K_1	18.8	20.4	18
$(CH_2SCH_2CH_2NH_2)_2$	K_1	10.6	15.5	−3
$H(NHCH_2CH_2)_4NH_2$	K_1	22.8	25.0	20
$CH_2(CH_2NHCH_2CH_2NH_2)_2$[b]	K_1	23.9	27.7	16
$(CH_2NHCH_2CH_2CH_2NH_2)_2$[b]	K_1	21.8	25.9	13
$CH_2(CH_2NHCH_2CH_2CH_2NH_2)_2$[b]	K_1	17.1	19.5	13

[a] Data at 25° C from Refs. 33 and 34 unless otherwise noted. The measurements on ammonia were made in the presence of 2 M NH_4NO_3. The other measurements were made at salt concentrations of 1 M or less. The constants denoted K_{1n} here are the ones denoted β_n in Refs. 33 and 34.
[b] From Ref. 32 or sources quoted therein.
[c] Average of three sets of values.[32–34]

Since ethylenediamine can form two copper-nitrogen bonds it is hardly surprising that its K_1 value is more than a million times as large as that for ammonia. A more reasonable comparison would be of K_1 for ethylenediamine with K_{12} for ammonia. The former is larger by the factor 500 M. Hence the equilibrium constant for the displacement of both molecules of ammonia from $Cu(NH_3)^{2+}$ by ethylenediamine (abbreviated en in Eq. 9-2) is 500 M.

$$Cu(NH_3)_2^{2+} + en \rightarrow Cuen^{2+} + 2NH_3 \tag{9-2}$$

Other entries in Table 9-5 show that $\Delta H°$ is -1.1 kcal/mole and $\Delta S°$ is 9 eu for this process. The value of $\Delta S°$ depends on the choice of 1 M as the standard state, of course. However, at the highest possible concentrations, using an equimolar mixture of ethylenediamine and ammonia as the solvent, the equilibrium in Eq. 9-2 would lie about 98% to the right if the equilibrium constant were the same as it is in aqueous solution.

Comparison of the equilibrium constants for trimethylenediamine with those for ethylenediamine shows that the five-membered ring compound is more stable than its six-membered analogue in the present case just as in the case of ketal formation from acetone (Table 9-3) but unlike the case of hemiacetal formation (Table 9-2). The lesser proton basicity of ethylenediamine suggests that its coordination with Cu^{2+} may be disfavored by a polar effect. However, the N—Cu—N bond angle of about 90° and the copper-nitrogen bond distances of about 2.0 Å may be more compatible with forming an unstrained ring with ethylenediamine than with trimethylenediamine. The small increases in equilibrium constants observed on going from ethylenediamine to *trans*-1,2-diaminocyclohexane and from trimethylenediamine to 1,3-diamino-2,2-dimethylpropane are attributed to destabilization of conformations that are unsuitable for ring formation.

Examination of the enthalpies of reaction shows that the formation of the average copper-nitrogen bond is accompanied by the release of roughly 6 kcal/mole. Thus the reaction of the first molecule of diethylenetriamine is exothermic by 18 kcal/mole. The reaction of two molecules gives off only 26 kcal/mole because copper forms strong bonds to only four ligand atoms. These are attached by four coplanar equatorial bonds; ligands attached to the two longer axial bonds are held more weakly.

Coordination of three or more atoms from a given base gives a polycyclic ring system. In view of the greater stability of a five-membered ring than a six-membered ring when monocyclic complexes are formed, it is hardly surprising that the complex with two five-membered rings formed from diethylenetriamine is more stable than the one with two six-membered rings formed from $NH(CH_2CH_2CH_2NH_2)_2$, or that the complex with three five-membered rings formed from triethylenetetramine is more stable than the one with three six-membered rings formed from $CH_2(CH_2NHCH_2CH_2CH_2NH_2)_2$. However, it may be surprising that the most stable tricyclic complex is the one formed from $CH_2(CH_2NHCH_2CH_2NH_2)_2$, which has a six-membered center ring and two five-membered outside rings (**2**). Examination of molecular models shows that the secondary nitrogen atoms can rather easily assume a conformation such that one bond to carbon is properly oriented to lead to a cis nitrogen atom via a

CH2
CH2 CH2
HOH
NH NH
CH2 CH2
Cu
CH2 CH2
NH2 NH2
HOH

2

two-carbon bridge and the other bond to carbon is properly oriented to lead to the other cis nitrogen atom via a three-carbon bridge.

The relative affinities of different bases for hexacoordinate metal ions that form six equivalent bonds to ligand atoms differ significantly from their relative affinities toward Cu^{2+}. This point is illustrated by the data in Table 9-6 on Ni^{2+} ions, which give measurable equilibrium constants for coordination with as many as six molecules of ammonia or three of ethylenediamine (even though the equilibrium constants for coordinating with the first molecules are smaller than the corresponding constants for Cu^{2+}). It is obvious why log K_{12} for diethylenetriamine exceeds log K_1 by a larger fraction (75%) in the case of Ni^{2+} than in the case of Cu^{2+} (33%). Triethylenetetramine gives a more stable complex than $N(CH_2CH_2NH_2)_3$ does with Cu^{2+} but a less stable one with Ni^{2+}. An important factor in producing this reversal in relative stabilities must be the ability of triethylenetetramine and the inability of $N(CH_2CH_2NH_2)_3$ to locate its four nitrogen atoms at the corners of a square that has the metal ion at its center. This is no handicap in coordinating with Ni^{2+}, which can form relatively strong

Table 9-6. Coordination of Ni^{2+} Ions with Ammonia and Polyamines in Aqueous Solutions[a]

Base	Constant	log K	Base	Constant	log K
Ammonia	K_1	2.4	$H_2NCH_2CH_2NH_2$	K_1	7.5
	K_{12}	4.3		K_{12}	13.9
	K_{13}	5.8		K_{13}	18.3
	K_{14}	7.0	$HN(CH_2CH_2NH_2)_2$	K_1	10.6
	K_{15}	7.9		K_{12}	18.6
	K_{16}	8.3	$H(NHCH_2CH_2)_3NH_2$	K_1	13.8
			$N(CH_2CH_2NH_2)_3$	K_1	14.6

[a] Data from Refs. 33 and 34 at 25° C and salt concentrations up to 1 M. The constants denoted K_{1n} here are the ones denoted β_n in Refs. 33 and 34.

Table 9-7. Coordination of Anions of the Type $X(CH_2CO_2^-)_2$ with Metal Cations[a]

Metal Cation	log K_1		
	$HN(CH_2CO_2^-)_2$	$O(CH_2CO_2^-)_2$	$S(CH_2CO_2^-)_2$
Ca^{2+}	2.7	3.4	1.4
Zn^{2+}	7.0	3.6	3.0
Co^{2+}	7.0	2.7	3.4
Ni^{2+}	8.2	2.8	4.1
Cu^{2+}	10.4	3.9	4.5

[a] Equilibrium constants in water with 0.1 *M* KCl at 30° C.[33]

bonds to four noncoplanar basic atoms if they are at four of the corners of an octahedron.

That hardness and softness (cf. Section 7-3) may also produce reversals in relative affinities of polydentate ligands for metal ions is suggested by the data in Table 9-7 on complexing by anions of the type $X(CH_2CO_2^-)_2$. The three X groups used probably rank O > NH > S in relative hardness. Toward Ca^{2+}, the hardest of the metal ions listed, the oxygen ligand has the greatest affinity. The nitrogen ligand has the greatest affinity for the other metals, probably more because of its strong general basicity than because of its hardness or softness.

PROBLEMS

1. Describe at least one factor that will tend to make the equilibrium constant for anhydride formation from tetramethylsuccinic acid smaller than for succinic acid itself. Discuss, in as quantitative a fashion as you can, other factors that must be working in the opposite direction to give the result shown in Table 9-1.

2. What fraction of the difference between log K_1 and log K_6 for the reaction of ammonia with Ni^{2+} ions (Table 9-6) is a symmetry effect? That is, how does the magnitude of log K_1 − log K_6 compare with that of log K_1^{chem} − log K_6^{chem}? [K_6 is the equilibrium constant for the formation of $Ni(NH_3)_6^{2+}$ from NH_3 and $Ni(NH_3)_5H_2O^{2+}$.]

3. Do the data in Table 9-5 support the hypothesis that none of the polyamine bases form more than four copper-nitrogen bonds of detectable strength or the alternative hypothesis that some of them do form a fifth and/or sixth bond whose strength is significant although less than that of the first four bonds? Explain.

REFERENCES

1. L. Eberson and H. Welinder, *J. Amer. Chem. Soc.*, **93,** 5821 (1971).
2. C. D. Hurd and W. H. Saunders, Jr., *J. Amer. Chem. Soc.*, **74,** 5324 (1952).
3. Cf. J. F. Stoddart, *Stereochemistry of Carbohydrates*, Wiley-Interscience, New York, 1971, Sec. 2.7.
4. M. Anteunis and Y. Rommelaere, *Bull. Soc. Chim. Belg.*, **79,** 523 (1970).
5. M. S. Newman and R. E. Dickson, *J. Amer. Chem. Soc.*, **92,** 6880 (1970).
6. P. v. R. Schleyer, *J. Amer. Chem. Soc.*, **83,** 1368 (1961).
7. R. M. Beesley, C. K. Ingold, and J. F. Thorpe, *J. Chem. Soc.*, **107,** 1080 (1915).
8. C. K. Ingold, *J. Chem. Soc.*, **119,** 305 (1921).
9. D. P. Stevenson and J. H. Morgan, *J. Amer. Chem. Soc.*, **70,** 2773 (1948).
10. P. v. R. Schleyer, J. E. Williams, and K. R. Blanchard, *J. Amer. Chem. Soc.*, **92,** 2377 (1970).
11. M. Davis and O. Hassel, *Acta Chem. Scand.*, **17,** 1181 (1963).
12. S. W. Benson, F. R. Cruickshank, D. M. Golden, G. R. Haugen, H. E. O'Neal, A. S. Rodgers, R. Shaw, and R. Walsh, *Chem. Rev.*, **69,** 279 (1969).
13. M. I. Page, *Chem. Soc. Rev.*, **2,** 295 (1973).
14. Cf. M. I. Page and W. P. Jencks, *Proc. Nat. Acad. Sci. U.S.*, **68,** 1678 (1971).
15. W. P. Jencks, F. Barley, R. Barnett, and M. Gilchrist, *J. Amer. Chem. Soc.*, **88,** 4464 (1966).
16. T. Higuchi, L. Eberson, and J. D. McRae, *J. Amer. Chem. Soc.*, **89,** 3001 (1967).
17. I. Wadsö, *Acta Chem. Scand.*, **16,** 471 (1962).
18. M. Ehrenberg, *Acta Cryst.*, **19,** 698 (1965).
19. R. P. Bell, *Advan. Phys. Org. Chem.*, **4,** 1 (1966).
20. E. G. Sander and W. P. Jencks, *J. Amer. Chem. Soc.*, **90,** 6154 (1968).
21. H. E. O'Neal and S. W. Benson, *J. Chem. Eng. Data*, **15,** 266 (1970).
22. J. Bjerrum and E. J. Nielsen, *Acta Chem. Scand.*, **2,** 297 (1948).
23. A. E. Martell and M. Calvin, *Chemistry of the Metal Chelate Compounds*, Prentice Hall, New York, 1952.
24. G. Schwarzenbach, *Helv. Chim. Acta*, **35,** 2344 (1952).
25. C. G. Spike and R. W. Parry, *J. Amer. Chem. Soc.*, **75,** 2726, 3770 (1953).
26. A. W. Adamson, *J. Amer. Chem. Soc.*, **76,** 1578 (1954).
27. H. Irving, R. J. P. Williams, D. J. Ferrett, and A. E. Williams, *J. Chem. Soc.*, 3494 (1954).
28. F. A. Cotton and F. E. Harris, *J. Phys. Chem.*, **59,** 1203 (1955); **60,** 1451 (1956).
29. F. H. Westheimer and L. L. Ingraham, *J. Phys. Chem.*, **60,** 1668 (1956).
30. F. J. C. Rossotti, in *Modern Coordination Chemistry*, J. Lewis and R. G. Wilkins, Eds., Interscience, New York, 1960, Chap. 1.
31. M. T. Beck, *Chemistry of Complex Equilibria*, Van Nostrand Reinhold, New York, 1970, Chap. 11.

32. R. Barbucci, L. Fabbrizzi, P. Paoletti, and A. Vacca, *JCS Dalton*, 1763 (1973).

33. L. G. Sillén and A. E. Martell, *Stability Constants of Metal-Ion Complexes*, Special Publication No. 17, The Chemical Society, London, 1964.

34. L. G. Sillén and A. E. Martell, *Stability Constants of Metal-Ion Complexes, Supplement No. 1*, Special Publication No. 25, The Chemical Society, London, 1971.

10

Equilibrium in the Formation of Free Radicals

10-1. FORMATION OF RELATIVELY STABLE CARBON RADICALS[1]

Gomberg's discovery that triphenylmethyl radicals may be generated in solution around room temperature at easily visible concentrations[2] initiated a flood of studies of this species and similar relatively stable free radicals. Determination of the equilibrium constants for the reactions of such radicals to give various nonradical species of known structure offers a way of learning much about structural effects on the stabilities of free radicals. However, such measurements present formidable experimental problems. In a typical determination of the equilibrium constant for dimerization of a triarylmethyl radical, for example, a solution of triarylmethyl chloride is treated with colloidal silver under nitrogen and the concentration of triarylmethyl radicals in the resulting solution is measured. The radical concentration may be measured spectrophotometrically, but the extinction coefficient is very hard to determine and absorption by other species present is difficult to correct for. The radical concentration may be measured by magnetic susceptibility determinations since radicals are paramagnetic whereas nonradicals are diamagnetic, but the problem of making the proper diamagnetic correction to the expected paramagnetic susceptibility does not appear to have been solved yet.[3,4] Apparent molecular weights may be determined but these often require the use of small differences between large numbers and the cryoscopic and ebulliometric methods are restricted to the solvent's melting point and boiling point, respectively. Electron-spin resonance (esr) measurements probably provide the most generally reliable method of measurement, but the concentrations obtained are often uncertain by as much as 50%. Even after the radical concentration has been measured it is necessary to know how much of the starting triarylmethyl chloride reacted and how much was transformed to by-products, such as triarylmethanol and triarylmethyl peroxide, in order to calculate the concentration of radical dimer that is present. After chemists had been making such measurements, generally by inferior methods and without all the desired precautions, on a variety of triarylmethyl radicals for more than 60 years, it was learned that the structure that had been assigned to the radical dimer was wrong in one case and probably wrong in most cases. Ultraviolet and nmr spectra support a 1-diphenylmethylene-4-triphenylmethyl-2,5-cyclohexadiene structure (**1**) for

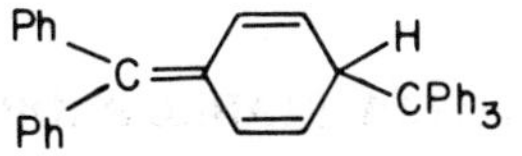

$$(Ph)_2C{=}C_6H_4(H)CPh_3 \qquad (10\text{-}1)$$

1

the dimer formed from unsubstituted triphenylmethyl radicals.[5] That is, the central carbon atom of one triphenylmethyl radical forms a bond to a para carbon atom of another. In addition, certain hindered diarymethyl radicals, such as bis(2,6-dimethylphenyl)methyl, which are stable enough to be detected directly in solution at room temperature, give analogous dimers. It does not follow, however, that the methylenecyclohexadiene type dimers are necessarily more stable than the polyarylethanes that had originally been thought to be the dimers. For example, tetrakis(2,6-dimethylphenyl)ethane has been prepared and found not to dissociate to bis(2,6-dimethylphenyl)methyl radicals or to be formed from these radicals at an appreciable rate at room temperature (although it does give the radicals at 230° C).[5] Hence the equilibrium constants that have been reported for the dissociation of hexaarylethanes to triarylmethyl radicals[1,6] may have nothing to do with hexaarylethanes. In fact, it is not clearly established that any hexaarylethanes are known compounds.

In view of the information about polar reactions and about the electronic effects of substituents that has been obtained by equilibrium measurements on polar reactions of meta- and para-substituted aromatic compounds, it seems particularly unfortunate that no very useful body of such information exists for the formation of free radicals, especially at carbon. The equilibrium measurements that have been made on triarylmethyl radicals have not covered a very wide range of meta and para substituents, and even for those substituents that have been studied the results are often complicated by uncertainty as to the structure of the dimer(s). The existing results do seem to make it clear that, unsurprisingly, the radicals can be stabilized relative to their dimers by steric inhibition of dimerization.[1,6] It is quite unclear what the electronic effects of the substituents are in the absence of such steric complications, however. Although most triarylmethyl radicals exist in the form of some kind of nonradical dimer in the solid, there are some in which the solid forms are radicals. It is sometimes stated that such species exist as radicals *even* in the solid, as if to imply that this fact guarantees a smaller tendency to dimerize than is the case with radicals that exist as such only in solution. Actually, the question of which species will separate as solid from a solution in which radical and dimer are present at equilibrium depends not only on the relative *amounts* of the two species present but also on their relative solubilities, that is, it also depends on crystal lattice energies.

10-2. FORMATION OF RELATIVELY STABLE RADICALS AT NITROGEN AND OXYGEN[1]

10-2a. Amino and Hydrazyl Radicals. Tetraphenylhydrazine is a nitrogen analogue of tetraphenylethane, which does not dissociate to diphenylmethyl radicals at an appreciable rate or to an appreciable extent at equilibrium at temperatures below 200° C. However, the fact that the nitrogen-nitrogen bond dissociation energy in hydrazine is about 71 kcal/mole, which is 17 kcal/mole smaller than the carbon-carbon bond dissociation energy of ethane, makes it plausible that tetraphenylhydrazine should dissociate more easily than tetraphenylethane. Apparently the equilibrium constants for such dissociation are increased by electron-donating substituents. Although there is kinetic evidence for the homolysis of tetraphenylhydrazine at temperatures below 100° C, at equilibrium neither this compound nor its 4,4,4′,4′-tetrafluoro or 4,4,4′,4′-tetrabromo derivatives gave enough diarylamino radicals to detect by esr measurements in xylene solution at temperatures up to 90° C; however, the 4,4,4′,4′-tetramethoxy and 4,4,4′,4′-tetrakis(dimethylamino) derivatives gave enough to detect at room temperature and the 4,4,4′,4′-tetramethyl derivative did at intermediate temperatures.[7] The marked effect of the *p*-dimethylamino substituent suggests that the bis(*p*-dimethylaminophenyl)amino radical is stabilized by significant contributions of resonance structure **3** as well as by structures such as **2**, which can be written for any diarylamino radical.

In permitting structure **3** to contribute to the total structure of the radical the *p*-phenylene group has acted merely as a conduit to allow the facile transfer of electrons from one nitrogen atom to the other. This transfer is also possible when an amino substituent is attached directly to

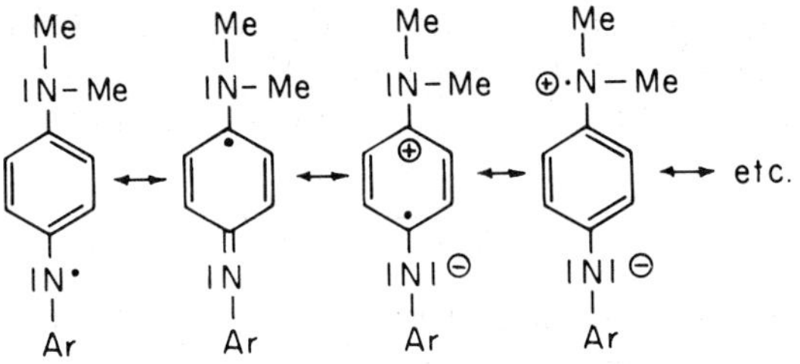

2 **3**

the nitrogen atom at which a radical is formed. Hence such resonance is also used to explain the relative stability of hydrazyl radicals. For example, hexaphenyltetrazane (**4**) is a white solid that gives blue solutions at room

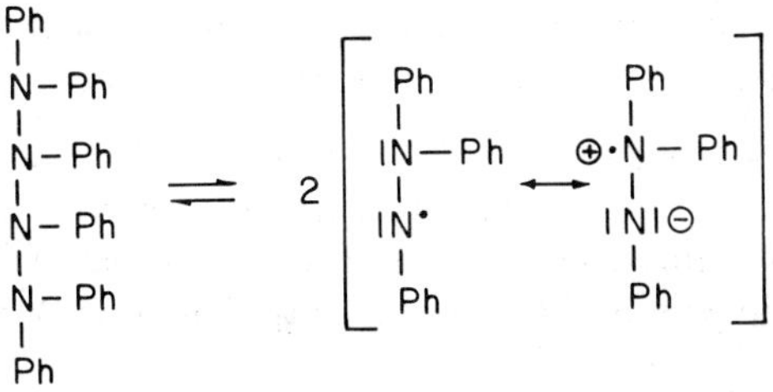

4 **5**

temperature because of dissociation to triphenylhydrazyl radicals.[8] It appears that because of the contribution of structures like **5** the diarylated nitrogen atom is made so positive and the monoarylated nitrogen atom so negative that 1,2,2-triarylhydrazyl radicals are stabilized by electron-withdrawing substituents on the 1-aryl ring and by electron-donating substituents on the 2-aryl rings. Such an electronic effect helps to explain why the 2,2-diphenyl-1-picrylhydrazyl radical is so stable that it does not dimerize appreciably in solution and is a radical in the solid state,[9,10] although steric effects must also be important. Even 1-acyl-2,2-diarylhydrazyl radicals are formed to measurable extents from the appropriate tetrazanes in solution at room temperature and below.[11,12]

Equilibrium constants for the dissociation of 1,1,4,4-tetraaryl-2,3-dibenzoyltetrazanes (Eq. 10-2) are increased by *p*-methyl and decreased

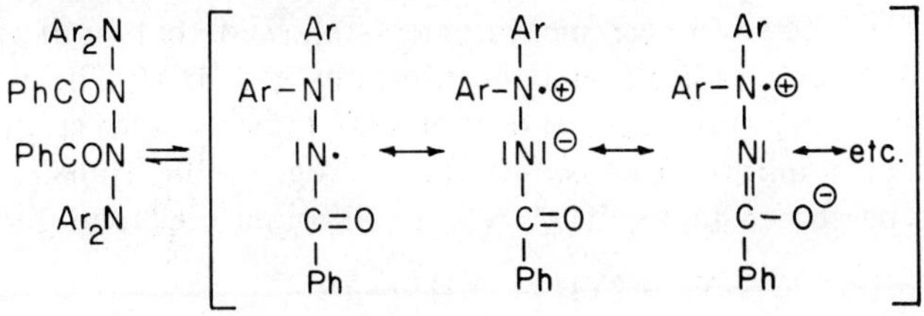

(10-2)

by *p*-bromo substituents in the aryl groups.[12] At $-30°$ C in acetone the values for the unsubstituted compound and four *p*-bromo or *p*-methyl derivatives obeyed Eq. 10-3, where $\sum\sigma$ is the sum of the σ values for the

$$\log \frac{K}{K_0} = -1.52 \sum \sigma \tag{10-3}$$

four substituents on the starting tetrazane. (If $\sum\sigma$ were defined as the sum of the σ values for the two substituents on the hydrazyl radical formed, ρ would be -3.04 rather than -1.52.) The good agreement with the Hammett equation that was observed weakens a common belief that radical formation from aromatic compounds should show large substituent effects that have only an accidental relationship to polar substituent effects. However, the present reaction is a rather special one, involving major contributing structures with much charge separation, and only *p*-bromo and *p*-methyl substituents were studied.

10-2b. Nitroxides. Nitroxides, the radicals formed by removal of the hydroxylic hydrogen atoms of hydroxylamine derivatives, are like hydrazyls in that they are stabilized by interaction of an unshared electron pair on nitrogen with the unpaired electron on an adjacent atom. When the nitrogen

$$R-\bar{N}(R)-\dot{\underline{O}}| \longleftrightarrow R-\overset{\oplus}{\dot{N}}(R)-\overset{\ominus}{\underline{\bar{O}}}|$$

atom is attached to tertiary or aromatic carbon atoms or to the bridgehead carbon atoms of fairly small bicyclic ring systems (so that certain disproportionation reactions are obviated) such radicals are often stable enough to be studied spectrally or to be isolated as pure species. The bicyclic nitroxides **6** and **7** differ markedly in the solid state in that **6** is a

6 **7**

yellow diamagnetic substance, whereas **7** is red and paramagnetic; presumably **6** exists in a dimeric nonradical form as a solid.[13] In isopentane solution at 25° C, however, esr measurements showed that both species are largely monomeric radicals at concentrations of 0.1 *M*. The relative stabilities of these two nitroxides and of similar radicals were studied by determination of equilibrium constants for hydrogen-atom transfer reactions of the type shown in Eq. 10-4. The equilibrium constants listed in

$$X\cdot + YH \rightleftharpoons XH + Y\cdot \tag{10-4}$$

Table 10-1 are for various Y's, all pitted against the same X, namely, $(t\text{-Bu})_2C{=}NO\cdot$.[14] The most stable radical, as measured by these equilibrium constants, is 2,2,6,6-tetramethylpiperidine *N*-oxyl (**8**), but all the

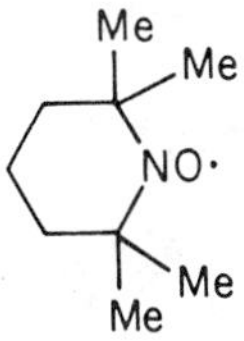

8

nitroxides are more stable than the iminoxy radical $(t\text{-Bu})_2C{=}NO\cdot$ with which they are compared. The lesser stability of iminoxy radicals may seem surprising in view of the extra contributing structure for the radical made possible by the carbon-nitrogen double bond of the reactant. It is ration-

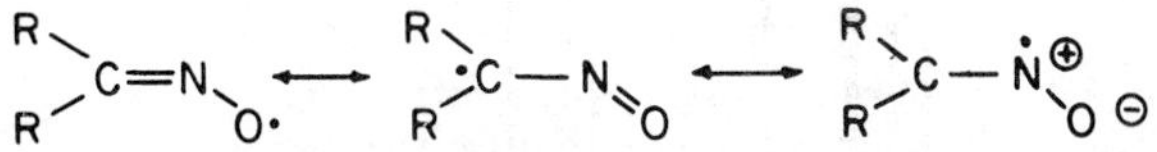

$$R_2C{=}N{-}O\cdot \longleftrightarrow R_2\dot{C}{-}N{=}O \longleftrightarrow R_2C{-}\dot{N}^{\oplus}{-}O^{\ominus}$$

alized in terms of the hypothesis that the structure with an unpaired electron on carbon is not a major contributor to the reasonance hybrid; the delocalization of the unpaired electron on oxygen comes largely from interaction with the unshared electron pair on nitrogen. This unshared pair in an iminoxy radical is in a nitrogen orbital that is roughly sp^2-hybridized; because of the greater effective electronegativity of such orbitals electrons in them are not as easily donated to other atoms as are electrons in orbitals with more *p* character. Structural studies on closely related nitroxides show that the nitrogen atom in nitroxide **8** is probably more nearly coplanar with the oxygen atom and two carbon atoms bonded to it than is the case with **6** and **7**. Hence the unshared electron pair on the nitrogen atom of **8** should be in an orbital with more *p* character, which would explain the greater stability of **8**.

10-2c. Phenoxy Radicals. Although there were enough stable phenoxy radicals to populate a 341-reference review of the subject in 1967,[15] very few equilibrium constants for their reactions have been determined. The most extensively studied example is probably the 2,4,6-tri-*t*-butylphenoxy radical, a deep blue substance that shows no tendency to dimerize in solution or in the solid state. This species, like any phenoxy radical, is stabilized by resonance, but much of its resistance to dimerization and to reaction by many other pathways must arise from steric hindrance. Equilibrium constants for the removal of a hydrogen atom by the 2,4,6-tri-*t*-butylphenoxy radical from 2,6-di-*t*-butyl-4-*t*-butoxyphenol,[16] 2,6-di-*t*-butyl-4-

Table 10-1. Equilibrium Constants for the Formation of Various Radicals from the Corresponding Hydrogen Compounds and $(t\text{-Bu})_2C{=}NO\cdot$ [a]

Radical	K
6	700
7	2900
8	$\geq 6 \times 10^{6}$ [b]
t-BuC(—i-Pr)=NO·	2.6×10^{-3}
2,2-Diphenyl-1-picryl-hydrazyl	~1 [c]

[a] In benzene at 25° C.[14]

[b] This is the quotient of the equilibrium constant for the reaction of $(t\text{-Bu})_2C{=}NO\cdot$ with **6**-H and the equilibrium constant for the reaction of **8** with **6**-H.

[c] This equilibrium was established too slowly to be measured reliably.

methoxyphenol,[16] 2,6-di-*t*-butyl-α-(3,5-di-*t*-butyl-4-oxo-2,5-cyclohexadien-1-ylidene)-*p*-cresol[17,18] (which will be referred to as H-G because the resulting radical, G·, is known trivially as galvinoxyl), 2,3′,5′,6-tetra-*t*-butylindophenol,[17,18] 1,1-diphenyl-2-picrylhydrazine,[18] and 1,2,3,4-tetrahydro-1-naphthyl hydroperoxide[19] are listed in Table 10-2. The first two equilibrium constants are larger than 1.0 because of the electron-donating character of *p*-alkoxy substituents. There is more extensive kinetic and

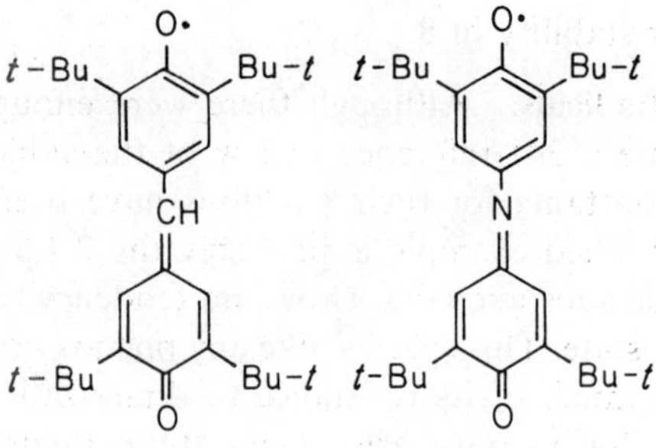

Galvinoxyl

2,3′,5′,6-Tetra-t-butyl-indophenoxy

Table 10-2. Equilibrium Constants for the Reaction R· + R′H ⇌ RH + R·′, where RH is 2,4,6-Tri-*t*-butylphenol[16–19]

R′H	K
2,6-Di-*t*-butyl-4-*t*-butoxyphenol	52
2,6-Di-*t*-butyl-4-methoxyphenol	~210
H-G	~ 80[b]
2,3′,5′,6-Tetra-*t*-butylindophenol	~400[b]
1,1-Diphenyl-2-picrylhydrazine	~ 10
1-Tetralyl hydroperoxide	$\sim 10^{-5}$[c]

[a] In benzene at 25° C unless otherwise noted.
[b] This equilibrium constant resulted from algebraic manipulation of the equilibrium constants involving 2,4,6-tri-*t*-butylphenol,[18] 1,1-diphenyl-2-picrylhydrazyl,[17,18] and 2,3′,5′,6-tetra-*t*-butylindophenol.[17]
[c] This equilibrium constant was determined somewhat indirectly, partly from rate measurements, in tetralin-chlorobenzene.[19]

other evidence that phenoxy radicals tend to be stabilized relative to the corresponding phenols by electron-donating substituents and destabilized by electron-withdrawing substituents.[15,20] This would be expected since the −O· group is electron withdrawing relative to the −OH group. [Note that in aqueous solution HX is about 10^5 times as strong an acid when −X is −O· as when it is −OH (cf. Section 6-1).] It is noteworthy that this electron donation is roughly as effective at stabilizing the phenoxy radical as is the much more extensive resonance delocalization of the unpaired electron that is possible in galvinoxyl and 2,3′,5′,6-tetra-*t*-butylindophenoxyl, whose two aromatic rings, however, must not quite be in the same plane.

10-3. STABILITIES OF HIGHLY REACTIVE RADICALS

10-3a. Thermodynamic Properties of Reactive Radicals. A large body of data is available on the stabilities of radicals that are ordinarily too reactive for their concentrations to be measured directly under equilibrium conditions. Most of these data come from kinetic studies and depend upon the validity of the mechanism in terms of which the raw data were

Table 10-3. Thermodynamic Functions of Gas Phase Radicals at 25° C

Radical	ΔH_f° [a] (kcal/mole)	S° [b] (eu)	Radical	ΔH_f° [a] (kcal/mole)	S° [b] (eu)
H	52.1	27.4	HC≡C	~122[b]	49.6
F	18.9	37.9	CH_2=CH	68.4	56.3
Cl	29.0	39.5	Ph	77.7	69.4
Br	26.7	41.8	$PhCH_2$	44.9	77.3
I	25.5	43.2	Allyl	41.4	63.4
HO	9.0	43.9	*c*-C_3H_5[c]	61.3	60.4
MeO	3.4	55	*c*-C_4H_7[c]	51.2	67.3
HCO_2	~−36[b]	57.3	Formyl	9.5	53.7
HS	34[b]	46.7	Ac	−5.8	63.5
MeS	~32.5[b]	~59	ClCO	−4.0[b]	63.5
MeNH	45.2	~57	CF_3	−112.5	63.8
Me	34.0	46.1	CCl_3	18.7	70.8
Et	25.7	59.8	CH_2Ac	~−5.5	~71.5
n-Pr	20.8	68.5			
i-Pr	17.8	66.7	CH_2OMe	−2.8	~64.7
t-Bu	7.5	74.6	CH_2NH_2	~36.5[b]	~57

[a] From Ref. 24 except where otherwise noted.
[b] From Ref. 22. The standard state for the entropy is the ideal gas at 1 atm pressure.
[c] *c* stands for cyclo.

analyzed and upon the validity of certain assumptions about activation energies.[21–23] For example, in studies of the reaction of RI with HI to give RH and I_2, Benson and co-workers described strong evidence that the reaction of R· with I_2 to give RI and I has an activation energy of 0 ± 1 kcal/mole for a wide range of R groups. Enthalpies of formation and entropies of a number of radicals, including some atoms, determined by kinetic studies and in other ways are listed in Table 10-3.[22,24] From these values and the thermodynamic functions for the appropriate compounds, a number of equilibrium constants for the formation of radicals may be calculated. Consider, for example, the homolysis of a carbon-hydrogen bond of ethane (Eq. 10-5). If ΔH_f° and S° for ethane are taken as twice

$$\mathrm{EtH} \rightleftharpoons \mathrm{Et\cdot} + \mathrm{H\cdot} \tag{10-5}$$

the (CH_3(X)] contributions from Table 1-4, combination with the values for Et· and H· gives ΔH° and ΔS° values of 98.0 kcal/mole and 26.4 eu, respectively, for the reaction shown in Eq. 10-5. These in turn give an equilibrium constant of 10^{-66} atm for the reaction in the gas phase at 25° C.

The values of $\Delta H_f°$ in Table 10-3 are measures of the stability of the various radicals relative to the elements in their standard states. The fact that trifluoromethyl is by far the most stable of those listed reflects mainly the great exothermicity of reactions of elemental fluorine in which bonds to carbon are formed. When organic chemists refer to the stability of a radical they more commonly are thinking of some related nonradical as the reference point. For example, the enthalpy of homolysis of RH to give R· and a hydrogen atom (the bond dissociation energy) may be taken as a definition of the stability of the radical R·.* In terms of this definition the fact that $\Delta H°$ for the homolysis of a carbon-hydrogen bond in dimethyl ether (Eq. 10-6) is 93.3 kcal/mole shows that the methoxymethyl radical is

$$CH_3OCH_3 \rightarrow CH_3OCH_2\cdot + H\cdot \qquad (10\text{-}6)$$

4.7 kcal/mole more stable than the ethyl radical. It should be noted that this definition with reference to RH does not necessarily give the same relative stabilities as if some other RX had been used as the reference compound. Consider, for example, the stability of R· with reference to that of ROMe. Equations 10-7 and 10-8 provide a comparison of the stabilities of ethyl and methoxymethyl radicals using this alternative definition of radical stabilities. According to this definition it is now the ethyl radical

	$\Delta H°$ (kcal/mole)	
$EtOMe \rightarrow Et\cdot + \cdot OMe$	80.8	(10-7)
$MeOCH_2OMe \rightarrow MeOCH_2\cdot + \cdot OMe$	83.9	(10-8)

that is more stable, by 3.1 kcal/mole. There has been a 7.8 kcal/mole reversal in the relative stabilities of the two radicals on changing from one definition to the other. Such a change is, of course, predictable on the basis of the effect of structure on thermodynamic properties discussed in Section 1-2. For example, none of the pairwise interactions as defined in Section 1-2 is lost on homolysis of a carbon-hydrogen bond. However, in the reaction shown in Eq. 10-7 a stabilizing (by 5.66 kcal/mole, cf. Table 1-3) interaction of carbon and oxygen across carbon is lost and in Eq. 10-8 a 13.09 kcal/mole stabilizing interaction of two oxygen atoms across carbon is lost. The 7.43 kcal/mole difference between these two interactions is well within the combined uncertainties of the 7.8 kcal/mole reversal in relative radical stabilities observed.

*A definition in terms of free energies could be given, of course, but differences in $T\Delta S°_{chem}$ for the homolysis of various RH's at moderate temperatures tend to be much smaller than the differences in $\Delta H°$. Hence there would not be much difference between the two definitions. Furthermore, $\Delta G°$ and $\Delta S°$ values for the appropriate reactions are less available than are the $\Delta H°$ values.

10-3b. Stabilities of Reactive Carbon Radicals. We shall discuss the relative stabilities of carbon radicals in terms of the dissociation energies of carbon-hydrogen bonds[22,24-26] listed in Table 10-4. The smallest value listed is for 1,4-pentadiene, which yields the pentadienyl radical (Eq. 10-9),

$$CH_2{=}CH{-}CH_2{-}CH{=}CH_2 \xrightarrow{-H\cdot} \left[CH_2{=}CH{-}\dot{C}H{-}CH{=}CH_2 \leftrightarrow \cdot CH_2{-}CH{=}CH{-}CH{=}CH_2 \leftrightarrow CH_2{=}CH{-}CH{=}CH{-}CH_2\cdot \right] \qquad (10\text{-}9)$$

for which three valence-bond structures may be written. Resonance also explains why the allyl and 1-methylallyl radicals, for which two valence-bond structures may be written, are formed so easily from propene and 1-butene, respectively. The benzyl radical, for which five major contributing structures may be written, is the most highly resonance-stabilized radical

Table 10-4. Enthalpies of the Gas Phase Transformation of RH to R· and H· at 25° C[a]

R	$\Delta H°$ (kcal/mole)	R	$\Delta H°$ (kcal/mole)	R	$\Delta H°$ (kcal/mole)
Me	104	Allyl	88.6	$PhCO_2CH_2$	100.2[c]
Et	98	$PhCH_2$	85	CH_2F	102.7
n-Pr	98	$CH_2{=}CHCHMe$	82.5	CHF_2	103.2
i-Pr	95	$HC{\equiv}CCH_2$	93.9	CF_3	105.9
i-Bu	98[b]	$(CH_2{=}CH)_2CH$	75.2	CH_2Cl	101.7
t-Bu	92	Formyl	87.5	$CHCl_2$	99.2
c-C_3H_5	100.7	Ac	86	CCl_3	96
c-C_4H_7	96.5	PhCO	86.9	CH_2Br	101.7
c-C_5H_9	94.8	CH_2Ac	~98.5	$CHBr_2$	~103
c-C_6H_{11}	95.5	CH_2OH	95.6	CBr_3	96
c-C_7H_{13}	92.5	CH_2OMe	93.3	CH_2I	102.7
$CH_2{=}CH$	108	CMe_2OH	90.1	CHI_2	102.7
Ph	110	CO_2H	90	CH_2NH_2	~94[b]
$HC{\equiv}C$	~120[b]	CO_2Me	95.3	CN	~129[d]

[a] From Ref. 24 unless otherwise noted.
[b] Calculated from a $\Delta H_f°$ value from Ref. 22.
[c] From Ref. 25.
[d] From Ref. 26.

in the table, but the reactant from which it is formed (toluene) is also significantly stabilized by resonance. The difference in resonance stabilization between benzyl radicals and toluene is enough to make the benzyl radical somewhat more easily formed than the allyl radical but not as easily formed as the pentadienyl radical.

Carbon-hydrogen bond dissociation energies for hydrocarbons are seen to increase steadily on going from sp^3-hybridized carbon (≤ 104 kcal/mole) to sp^2-hybridized carbon (108 and 110 kcal/mole for ethylene and benzene, respectively) to sp-hybridized carbon (~ 120 kcal/mole for acetylene). This increase in bond strength that accompanies increasing s character in the hybridization of the carbon atom probably arises, at least in part, from the same factors that cause the force constants for the carbon-hydrogen bonds to increase in the same sequence, namely, the increase in overlapping ability that occurs on going from sp^3 to sp^2 to sp orbitals.[27]

Although aldehydic hydrogen atoms are attached to sp^2-hybridized carbon they are removed more than 20 kcal/mole more easily than the hydrogen atoms of ethylene or benzene are. The resulting acyl radicals[28] are bent by about 120–130° at the carbon atom bearing the unpaired electron just as vinyl[29] and phenyl radicals are. Perhaps the relative stability of acyl radicals can be tied to some familiar fact by noting that the simplest acyl radical, formyl, is isoelectronic with protonated nitric oxide, and nitric oxide is a known stable radical that must have some ability to be protonated. Probably a significant part of the stabilization arises from the type of resonance used to explain the stability of hydrazyl radicals, which can be invoked any time an unpaired electron and an unshared electron pair are on adjacent atoms.

$$\mathrm{R{-}\dot{C}{=}\bar{O}|} \longleftrightarrow \mathrm{R{-}\overset{\ominus}{\bar{C}}{=}\overset{\oplus}{\dot{O}}|}$$

Such resonance must help make aminomethyl, hydroxymethyl, methoxymethyl, etc., radicals more stable than methyl radicals. However, such resonance should be more important for α-fluoro substituents than for any other α-halo substituents, and yet α-fluoro substituents (relative to α-hydrogen) are seen not to increase radical stability significantly, if at all. None of the α-halo substituents appear to be very good radical stabilizers but some seem better than fluorine. This situation apppears to arise not from a lack of interaction of α-fluorine with the radical center but from at least two types of interaction with opposed effects on radical stability. Ultraviolet,[30] electron spin resonance,[31] and infrared[32] spectral observations have shown that the methyl radical is planar or very nearly so. Arguments that, because of its great electronegativity, the α-fluoro substituent should tend to disfavor reactions in which sp^3-hybridized carbon becomes sp^2-hybridized have already been given (Section 6-5d). The ir and esr spectra of fluoromethyl,[33,34] difluoromethyl,[34,35] and trifluoromethyl[34,36] radicals give evidence for increasing deviations from planarity, with the bond angles for the trifluoromethyl radical probably being not far from those around an sp^3-hybridized carbon atom. Presumably replacement of the first hydrogen atom of the methyl radical by a fluorine atom markedly destabilizes the radical because of its sp^2-hybridized character and markedly stabilizes it by interactions between the orbital containing the unpaired electron and one containing an unshared electron pair on fluorine. Replacement of the third hydrogen atom by fluorine gives less destabilization because the difluoromethyl radical is already nonplanar, but it gives less stabilization because the rehybridization has made the interacting orbitals no longer parallel and has moved them further apart and because the unpaired electron in the difluoromethyl radical is in an orbital with more p character and therefore less effective electronegativity. The other α-halo substituents should have much less tendency to destabilize sp^2-hybridized carbon derivatives relative to sp^3-hybridized ones, but their unshared electron pairs, being in larger $3p$, $4p$, and $5p$ orbitals, should interact less with the unpaired electron in a $2p$ orbital of the carbon atom. In the absence of any quantitative measure of how these two factors, which operate in opposite directions, shrink on going down the Periodic Table from fluorine to iodine it is not possible to predict how the substituent effect of α-halogens on radical stability would vary.

The decrease in radical stability seen on going from cyclohexyl or cyclopentyl to cyclobutyl and then to cyclopropyl must be a reflection of the increase in s character in the carbon-hydrogen bond in the precursor (cf. the vinyl, phenyl, and ethynyl radicals).

The increase in radical stability found on going from methyl to primary to secondary to tertiary radicals would be expected in view of the tendency of α-alkyl substituents to stabilize sp^2-hybridized species relative to the corresponding sp^3-hybridized species (cf. Section 6-5d). However, resonance stabilization also appears to be important; esr spectra show that the unpaired electron density is distributed onto α-alkyl substituents as well as the central carbon atom.[37] It is not clear why the radical stability increases by 6 kcal/mole on going from methyl to ethyl and then by only 3 kcal/mole on going from ethyl to isopropyl or from isopropyl to *t*-butyl.

10-3c. Stabilities of Reactive Non-Carbon Radicals. The dissociation energies for bonds between hydrogen and various atoms other than

Table 10-5. Enthalpies of the Gas Phase Transformation of XH to X· and H· at 25° C[a]

X	$\Delta H°$ (kcal/mole)	X	$\Delta H°$ (kcal/mole)	X	$\Delta H°$ (kcal/mole)
H	104.2	EtO	104	NH_2	110
F	135.8	PhO	85	MeNH	103
Cl	103.1	AcO	110	Me_2N	95
Br	87.4	HCO_2	107[b]	PhNH	80[c]
I	71.3	HOO	90[b]	HS	91[b]
HO	119	ClO	98[b]	MeS	~90[b]
MeO	104	BrO	101[b]	SiH_3	94[d]

[a] From Ref. 24 unless otherwise noted.
[b] From Ref. 22.
[c] From Ref. 38.
[d] From Ref. 39.

carbon [22,24,38,39] listed in Table 10-5 show some of the same trends observed in Table 10-4. Relative to hydrogen, α-alkyl substituents stabilize radicals and α-phenyl substituents do so to a much greater extent. The effect of α-phenyl substituents is plausibly attributed largely to resonance, but that of alkyl substituents is harder to understand. The effect on the formation of nitrogen radicals is larger than for carbon radicals, and the effect on the formation of oxygen radicals is even larger. Yet, since the oxygen atom is attached to only one other atom in such a radical, the question of planarity or linearity, in terms of which some of the data on carbon and nitrogen radicals are rationalized, is meaningless. It is probably the electron-donating ability of alkyl groups relative to that of hydrogen that is important. Because of the differences in electronegativity hydrogen should donate electrons more to oxygen than to nitrogen or carbon. Therefore the electron-withdrawing power of an $-O\cdot$ group should exceed that of an $-OH$ group by a larger amount than the electron-withdrawing power of $-NH\cdot$ exceeds that of $-NH_2$ or the electron-withdrawing power of $-CH_2\cdot$ exceeds that of $-CH_3$. Hence an alkyl substituent should donate electrons more to $-O\cdot$ (relative to $-OH$) than to $-NH\cdot$ (relative to $-NH_2$) or to $-CH_2\cdot$ (relative to $-CH_3$), making the α-alkyl substituent a better radical stabilizer when attached to $-O\cdot$ than when attached to $-NH\cdot$ or $-CH_2\cdot$.

Table 10-5 also illustrates the strong tendency for elements further to the right or higher in the Periodic Table to form stronger bonds to hydrogen.

PROBLEMS

1. Suggest an explanation for why the *p-t*-butoxy substituent is not as effective at stabilizing a phenoxy radical as the *p*-methoxy substituent is (cf. Table 10-1).

2. Use data in the tables in this chapter to estimate the O–H bond energies in 2,4,6-tri-*t*-butylphenol and *t*-butyl isopropyl ketoxime. Discuss the assumptions and approximations made.

3. Give an explanation for why the methyl carbon-hydrogen bond dissociation energy listed in Table 10-4 for methyl benzoate is so much larger than that listed for dimethyl ether.

4. Give a qualitative molecular orbital description of the $R_2C{=}NO\cdot$ radical that explains why there should not be simultaneous major contributions of the three valence bond structures written for this radical in Section 10-2b.

REFERENCES

1. A. R. Forrester, J. M. Hay, and R. H. Thomson, *Organic Chemistry of Stable Free Radicals*, Academic Press, New York, 1968.
2. M. Gomberg, *Ber.*, **33,** 3150 (1900); *J. Amer. Chem. Soc.*, **22,** 757 (1900).
3. P. W. Selwood and R. M. Dobres, *J. Amer. Chem. Soc.*, **72,** 3860 (1950).
4. T.-L. Chu and S. I. Weissman, *J. Amer. Chem. Soc.*, **73,** 4462 (1951).
5. H. Lankamp, W. T. Nauta, and C. MacLean, *Tetrahedron Lett.*, 249 (1968).
6. Cf. G. W. Wheland, *Advanced Organic Chemistry*, 3rd ed., Wiley, New York, 1960, Section 15.10.
7. F. A. Neugebauer and P. H. H. Fischer, *Chem. Ber.*, **98,** 844 (1965).
8. S. Goldschmidt, *Ber.*, **53,** 44 (1920).
9. S. Goldschmidt and K. Renn, *Ber.*, **55,** 628 (1922).
10. J. Turkevich and P. W. Selwood, *J. Amer. Chem. Soc.*, **63,** 1077 (1941).
11. S. Goldschmidt, A. Wolf, E. Wolffhardt, I. Drimmer, and S. Nathan, *Justus Liebigs Ann. Chem.*, **437,** 194 (1924).
12. W. K. Wilmarth and N. Schwartz, *J. Amer. Chem. Soc.*, **77,** 4543 (1955).
13. G. D. Mendenhall and K. U. Ingold, *J. Amer. Chem. Soc.*, **95,** 6390 (1973).
14. L. R. Mahoney, G. D. Mendenhall, and K. U. Ingold, *J. Amer. Chem. Soc.*, **95,** 8610 (1973).
15. E. R. Altwicker, *Chem. Rev.*, **67,** 475 (1967).
16. C. D. Cook, C. B. Depatie, and E. S. English, *J. Org. Chem.*, **24,** 1356 (1959).
17. P. D. Bartlett and S. T. Purrington, *J. Amer. Chem. Soc.*, **88,** 3303 (1966).
18. P. B. Ayscough and K. E. Russell, *Can. J. Chem.*, **43,** 3039 (1965).
19. L. R. Mahoney and M. A. DaRooge, *J. Amer. Chem. Soc.*, **92,** 4063 (1970).

20. Cf. L. R. Mahoney and M. A. DaRooge, *J. Amer. Chem. Soc.*, **92,** 890 (1970).
21. S. W. Benson, *J. Chem. Educ.*, **42,** 502 (1965).
22. S. W. Benson, *Thermochemical Kinetics*, Wiley, New York, 1968.
23. D. M. Golden and S. W. Benson, *Chem. Rev.*, **69,** 125 (1969).
24. K. W. Egger and A. T. Cocks, *Helv. Chim. Acta*, **56,** 1516, 1537 (1973).
25. R. K. Solly and S. W. Benson, *Int. J. Chem. Kinet.*, **3,** 509 (1971).
26. H. T. Knight and J. P. Rink, *J. Chem. Phys.*, **35,** 199 (1961).
27. Cf. C. A. Coulson, *Valence*, Oxford Univ. Press, London, 1952, Section 8.5.
28. Cf. J. E. Bennett and B. Mile, *Trans. Faraday Soc.*, **67,** 1587 (1971).
29. Cf. P. H. Kasai, *J. Amer. Chem. Soc.*, **94,** 5950 (1972).
30. G. Herzberg, *Proc. Roy. Soc., Ser. A*, **262,** 291 (1961).
31. R. W. Fessenden, *J. Phys. Chem.*, **71,** 74 (1967).
32. L. Y. Tan, A. M. Winer, and G. C. Pimentel, *J. Chem. Phys.*, **57,** 4028 (1972).
33. J. I. Raymond and L. Andrews, *J. Phys. Chem.*, **75,** 3235 (1971).
34. R. W. Fessenden and R. H. Schuler, *J. Chem. Phys.*, **43,** 2704 (1965).
35. T. G. Carver and L. Andrews, *J. Chem. Phys.*, **50,** 5100 (1969).
36. G. A. Carlson and G. C. Pimentel, *J. Chem. Phys.*, **44,** 4053 (1966).
37. Cf. R. O. C. Norman and B. C. Gilbert, *Advan. Phys. Org. Chem.*, **5,** 53 (1967)
38. G. L. Esteban, J. A. Kerr, and A. F. Trotman-Dickenson, *J. Chem. Soc.*, 3879 (1963).
39. W. C. Steele, L. D. Nichols, and F. G. A. Stone, *J. Amer. Chem. Soc.*, **84,** 4441 (1962).

Name Index

Subject Index